Orthogonal Polynomials and Special Functions (Mathematics Essentials)

Orthogonal Polynomials and Special Functions (Mathematics Essentials)

Edited by
Alma Adams

www.willfordpress.com

Published by Willford Press,
118-35 Queens Blvd., Suite 400,
Forest Hills, NY 11375, USA

ISBN: 978-1-64728-529-6

Cataloging-in-Publication Data

Orthogonal polynomials and special functions (mathematics essentials) / edited by Alma Adams.
    p. cm.
Includes bibliographical references and index.
ISBN 978-1-64728-529-6
1. Orthogonal polynomials. 2. Functions, Special. 3. Mathematical analysis. I. Adams, Alma.
QA404.5 .O78 2023
515.55--dc23

For information on all Willford Press publications
visit our website at www.willfordpress.com

Contents

Preface

Orthogonal polynomials are a family of polynomials, wherein any two different polynomials in the sequence are orthogonal to each other under some inner product. Classical orthogonal polynomials, Hermite polynomials, Laguerre polynomials, Jacobi polynomials, and Gegenbauer polynomials are a few examples of orthogonal polynomials. These polynomials are used for least square approximations of a function, difference equations, and Fourier series. Another major application of orthogonal polynomials is error-correcting code and sphere packing. Orthogonal polynomials and special functions are useful mathematical functions, which have applications in various fields such as mathematical physics, statistics and probability, and engineering. These can be used to explain many physical and chemical phenomena. This book traces the recent studies in orthogonal polynomials and special functions. A number of latest researches have been included to keep the readers updated with the latest concepts in this area of study. With state-of-the-art inputs by acclaimed experts of mathematics, this book targets students and professionals.

This book is the end result of constructive efforts and intensive research done by experts in this field. The aim of this book is to enlighten the readers with recent information in this area of research. The information provided in this profound book would serve as a valuable reference to students and researchers in this field.

At the end, I would like to thank all the authors for devoting their precious time and providing their valuable contribution to this book. I would also like to express my gratitude to my fellow colleagues who encouraged me throughout the process.

Editor

New Stability Criteria for Discrete Linear Systems based on Orthogonal Polynomials

Luis E. Garza [1,*,†] ⓘ, **Noé Martínez** [2,†] **and Gerardo Romero** [2,†] ⓘ

[1]   Facultad de Ciencias, Universidad de Colima, Colima 28045, Mexico
[2]   Unidad Académica Multidisciplinaria Reynosa Rodhe, Universidad Autónoma de Tamaulipas,
      Reynosa 88779, Mexico; a2193728004@alumnos.uat.edu.mx (N.M.); gromero@docentes.uat.edu.mx (G.R.)
*   Correspondence: luis_garza1@ucol.mx
†   These authors contributed equally to this work.

Abstract: A new criterion for Schur stability is derived by using basic results of the theory of orthogonal polynomials. In particular, we use the relation between orthogonal polynomials on the real line and on the unit circle known as the Szegő transformation. Some examples are presented.

Keywords: orthogonal polynomials on the unit circle; Schur polynomials; Hurwitz polynomials; Szegő transformation

MSC: 42C05, 93C05

1. Introduction

1.1. Stability of Linear Systems

The stability problem of dynamical systems is of great interest because of its numerous applications, mainly in control systems. This property is very important because of both its relationship to the good performance of dynamical systems and the prevention of their physical damage. It is well known that for time-invariant linear dynamical systems, stability is determined by the roots of a characteristic equation that has a polynomial form. In the case of continuous systems, they are said to be stable if their roots have negative real part. This is known as Hurwitz stability. Now, for the discrete case, the system is stable if its roots are within a circle of radius equal to 1, this is known as Schur stability. Polynomials that satisfy the first condition are called Hurwitz polynomials, while those that satisfy the last one are called Schur polynomials. Therefore, the problem of verifying the stability property in dynamical systems is transformed into verifying the Hurwitz or Schur properties of polynomials. This contribution will address the problem of verifying the Schur stability property. The most common strategy to verify the Schur stability property is using a bilinear transformation (Möbius transformation, see [1,2]) and then applying Hurwitz stability tools to solve the original problem. However, [3] shows that their application is restricted by a pathological case and so the authors propose to use a biquadratic transformation to avoid this problem. Many papers that address this problem have been published, for example [4] presents conditions to verify the Schur stability property for the particular cases of real 2×2 matrices and real $n \times n$ tridiagonal matrices for which the concepts Schur D-stable and vertex stable are introduced and verified. In [5] some results are presented to determine regions defined by the gains of PID controllers that guarantee the stability property of discrete systems, those results use the bilinear transformation to exploit results of continuous systems and the linear programming technique to delimit the regions. On the other hand, there is another method that is also used to verify the Schur stability property, known as the Jury test (see [6,7]). A simplified proof for this test is presented in [8].

Another topic that has also been studied consists of verifying the Schur stability property by considering uncertainty in the polynomial coefficients, which is known as robust Schur stability. Some of the first results related to this topic were the extension of Kharitonov-like results using, in some of them, the Möbius transformation in order to transform the problem into a robust Hurwitz stability problem, see [9,10]. Other criteria used to verify the robust Schur stabiliy of a family of polynomials are Schur-Cohn and the Vandermonde matrix, see [11–14]. Also, there are other papers that deal with the verification of robust Schur stability of interval matrices by using the spectral radius to determine the stability conditions, see [15,16]. Other results use multivariate polynomials to verify the robust Schur stability property. For example [17] establishes conditions to use the Möbius transformation in multivariate polynomials, while in [18] the authors first transform the problem into verifying the positivity of particularly defined bivariate functions and then they use semi-defined programming. The main disadvantage of the latter method is that it presents only sufficient conditions of robust Schur stability. The Schur stability property of polynomials depends directly on their coefficients, thus it is possible to define regions in the coefficient space for which the polynomial meets the condition of Schur stability. These regions are generally represented by semi-algebraic sets, which in turn, makes it possible to use criteria such as the Jury test. The paper [19] introduces the concept of polynomial superlevel sets which are used as a tool to construct approximations of the semi-algebraic sets that may be applied in control problems, in particular, the Hurwitz and Schur stability properties.

Finally, in this paper we present a new criterion to verify the Schur stability property. This is derived by using the relation between orthogonal polynomials on the real line and on the unit circle known as the Szegő transformation. It is worth noting that this new approach to verify the Schur stability property can be generalized to define new semi-algebraic sets to be used in control theory. This approach is akin to the one presented in [19] but it has not been used before.

The structure of the manuscript is as follows. The remainder of this section contains some basic mathematical background on stable polynomials. Section 2 deals with basic results for orthogonal polynomials on the real line and on the unit circle that will be used in the sequel. The main contribution of our manuscript is contained in Section 3. There, we state a new criterion for Schur stability whose proof is based in orthogonality properties, and uses a well known mapping between the unit circle and the interval $[-2,2]$. Some illustrative examples are presented.

1.2. Stable Polynomials

A continuous linear system is stable if and only if its characteristic polynomial is a Hurwitz polynomial, i.e., the real part of all its zeros is strictly negative. As a consequence, Hurwitz polynomials are widely studied in the literature (see, for instance, [20,21]), and there are many criteria to determine if a given polynomial with real coefficients is Hurwitz without explicitly computing its zeros. Among many others, we have the Routh–Hurwitz criterion [20], the stability test [22], the continued fraction method [20], and the Hermite-Biehler theorem [23]. The latter determines the Hurwitz character of a polynomial $f(x)$ by verifying some properties of two polynomials associated with f. Indeed, if h and g are polynomials given by $f(x) = h(x^2) + xg(x^2)$, i.e., the even and odd parts of f, then the Hermite-Biehler theorem states that f is a Hurwitz polynomials if and only if h and g have real, negative and interlaced zeros. Moreover, if we consider the series expansions

$$\frac{g(x)}{h(x)} = \begin{cases} \frac{s_0}{x} - \frac{s_1}{x^2} + \frac{s_2}{x^3} - \cdots + \frac{s_{2n-2}}{x^{2n-1}} - \frac{s_{2n-1}}{x^{2n}} + \cdots, & \text{if} \quad \deg(f) \text{ is even,} \\ s_{-1} + \frac{s_0}{x} - \frac{s_1}{x^2} + \frac{s_2}{x^3} - \cdots + \frac{s_{2n-2}}{x^{2n-1}} - \frac{s_{2n-1}}{x^{2n}} + \cdots, & \text{if} \quad \deg(f) \text{ is odd,} \end{cases}$$

then the constants $\{s_0, s_1, \ldots, s_{2n-1}\}$ and $\{s_{-1}, s_0, s_1, \ldots, s_{2n-1}\}$ are known as Markov's parameters and it is known (see [20]) that f is a Hurwitz polynomial if and only if the matrices

$$H_{n-1} = \begin{pmatrix} s_0 & s_1 & \cdots & s_{n-1} \\ s_1 & s_2 & \cdots & s_n \\ \vdots & \vdots & \vdots & \vdots \\ s_{n-1} & s_n & \cdots & s_{2n-2} \end{pmatrix}, \quad H_{n-1}^{(1)} = \begin{pmatrix} s_1 & s_2 & \cdots & s_n \\ s_2 & s_3 & \cdots & s_{n+1} \\ \vdots & \vdots & \vdots & \vdots \\ s_n & s_{n+1} & \cdots & s_{2n-1} \end{pmatrix} \tag{1}$$

are both positive definite and $s_{-1} > 0$ in the odd case.

On the other hand, a discrete linear system is stable if and only if its characteristic polynomial is a Schur polynomial, i.e., all of its zeros are located in the open unit disc $\mathbb{D} = \{z \in \mathbb{C} : |z| < 1\}$. As we will see in a moment, the Hermite-Bielher theorem can be extended to this situation. More generally, an interlacing theorem can be stated for any stability region $A \subset \mathbb{C}$ such that the phase of any polynomial stable with respect to A varies monotonically through the boundary ∂A (see [22]).

In this section, we consider polynomials with real coefficients, unless otherwise stated. Without loss of generality, we will consider monic polynomials. Clearly, if

$$S_n(z) = z^n + p_{n-1}z^{n-1} + \ldots + p_1 z + p_0,$$

a necessary condition for $S_n(z)$ to be a Schur polynomial is that $|p_0| < 1$. On the other hand, a necessary and sufficient condition for $S_n(z)$ to be a Schur polynomial is that the graph of $S_n(e^{j\theta})$ circles n times around the origin or, equivalently, the graph of $e^{jn\theta}S_n(e^{-j\theta})$ does not circle around the origin when θ varies from 0 to 2π (see [22]). Notice that this result is valid even for polynomials with complex coefficients. For the case of real coefficients, consider the polynomials

$$R(\theta) = cos(n\theta) + p_{n-1}cos((n-1)\theta) + \ldots + cos(\theta)p_1 + p_0,$$

$$I(\theta) = sin(n\theta) + p_{n-1}sin((n-1)\theta) + \ldots + sin(\theta)p_1,$$

known as the real and imaginary parts of $S_n(z)$. The last criteria can be used to show that $S_n(z)$ (with $|p_0| < 1$) is a Schur polynomial if and only if (see [22])

(i) $R(\theta)$ has exactly n zeros in $[0, \pi]$,
(ii) $I(\theta)$ has exactly $n+1$ zeros in $[0, \pi]$,
(iii) The zeros of $R(\theta)$ and $I(\theta)$ interlace.

Another interlacing theorem is the following. Let $S_n(z) = S_n^s(z) + S_n^a(z)$, where

$$S_n^s(z) = \frac{1}{2}[S_n(z) + z^n S_n(\frac{1}{z})], \quad S_n^a(z) = \frac{1}{2}[S_n(z) - z^n S_n(\frac{1}{z})]. \tag{2}$$

Then, $S_n(z)$ is a Schur polynomial if and only if

(i) $S_n^s(z)$ and $S_n^a(z)$ are polynomials of degree n with coefficients of the same sign.
(ii) $S_n^s(z)$ and $S_n^a(z)$ have simple and interlaced zeros on the unit circle $\mathbb{T} = \{z \in \mathbb{C} : |z| = 1\}$.

Notice that, since the coefficients are real, the zeros of $S_n^s(z)$ and $S_n^a(z)$ appear in conjugate pairs. The polynomial $z^n S_n(\frac{1}{z})$ is usually called the *reciprocal (or reversed) polynomial* of $S_n(z)$ and it is denoted by $S_n^*(z)$. We say that a polynomial $A(z)$ is symmetric if $A(z) = A^*(z)$ and anti-symmetric if $A(z) = -A^*(z)$. This explains the notation in (2) since $S_n^s(z)$ and $S_n^a(z)$ are clearly symmetric and anti-symmetric polynomials, respectively. The previous criterion can be generalized as follows.

Theorem 1 ([24]). *Let $S_n(z)$ be a real Schur polynomial and $k \geq 0$. Then, the polynomials $\mathcal{A}$ and $\mathcal{B}$ defined by*

$$\mathcal{A}(z) = S_n(z) + z^{-k}S_n^*(z), \tag{3}$$

$$\mathcal{B}(z) = S_n(z) - z^{-k}S_n^*(z), \tag{4}$$

have interlaced zeros in $\mathbb{T}$ and $\mathcal{A}$ has a conjugate pair of zeros closer to $z = 1$. Conversely, if A and B are any two real polynomials of the same degree, such that one is symmetric and the other is anti-symmetric, with interlaced zeros on $\mathbb{T}$, then $A(z) + B(z)$ is a Schur polynomial.

A different criterion to determine whether or not a given polynomial is a Schur polynomial uses a recursive algorithm, as follows. Notice that it is equivalent to the Jury criteria, and its proof is based in the so-called Boundary Crossing Theorem (see [6,7], and also [22]). It constitutes a discrete analogue of the Routh–Hurwitz stability criterion.

Observe that each application of step 5 results in a 1 degree reduction in the computed polynomial, with respect to the previous one. Thus, after $n - 1$ cycles the resulting polynomial has degree 1.

Finally, it is well known that the Möbius (also called bilinear) transformation $z = \frac{x+1}{x-1}$ maps the open unit disc into the open left half plane. Thus, it can be used to transform a polynomial $S_n(z)$ into a polynomial $f_n(x)$ by using

$$(x - 1)^n S_n \left(\frac{x+1}{x-1} \right) = f_n(x). \tag{5}$$

Then, if the leading coefficient of f_n is not zero, the transformation preserves the degree and in this situation $S(z)$ is a Schur polynomials if and only if $f(x)$ is a Hurwitz polynomial [22]. As a consequence, it is possible to determine if a given polynomial is Schur by using Hurwitz criteria.

2. Orthogonal Polynomials and the Szegő Transformation

2.1. Orthogonal Polynomials on the Real Line

Let μ be a positive measure supported in some subset $E \subset \mathbb{R}$ with infinite points and such that

$$\left| \int_E x^k d\mu(x) \right| < \infty, \quad k \geq 0.$$

If $\mu(x)$ is an absolutely continuous measure, then we can write $d\mu(x) = \omega(x)dx$, where $\omega(x)$ is an integrable, non-negative function such that $\int_E \omega(x)dx > 0$. It is well known that there exists a sequence of real polynomials $\{p_n\}_{n\geq 0}$ such that

$$\int_E p_n(x)p_m(x)d\mu(x) = \gamma_n \delta_{n,m}, \quad \gamma_n > 0, \quad n, m \geq 0, \tag{6}$$

where $\delta_{m,n}$ is the Kronecker's delta. $\{p_n\}_{n\geq 0}$ is called the sequence of polynomials orthogonal with respect to μ (or with respect to the weight function ω) and it is unique up to constant multiplications (see [25]). If $\gamma_n = 1$ for every n, then $\{p_n\}_{n\geq 0}$ is called an orthonormal sequence. We can also consider a monic sequence $\{P_n\}_{n\geq 0}$ by dividing each $p_n(x)$ by its leading coefficient. This is the normalization that will be used in the manuscript.

Orthogonal polynomials on the real line have many applications and many useful properties. In particular, it is well known that each $P_n(x)$ has real and simple zeros, which are located in the interior of the convex hull of E. Moreover, if we denote by $x_{n,k}$ the zeros of $P_n(x)$, then they satisfy the following interlacing property (see [26])

$$x_{n,k} < x_{n-1,k} < x_{n,k+1}, \quad 1 \leq k \leq n - 1. \tag{7}$$

Furthermore, for any polynomial P_n of degree n with real and simple zeros, it is possible to find a positive measure μ such that P_n belongs to a sequence of polynomials orthogonal with respect to μ, although $P_{n-1}, P_{n-2}, \ldots, P_1, P_0$ are not uniquely determined. However, if P_n and P_{n-1} are arbitrary polynomials with degree n and $n - 1$, respectively, whose zeros are real, simple and interlaced, then there exists a measure μ such that they are orthogonal with respect to μ, and $P_n, P_{n-1}, P_{n-2}, \ldots, P_0$ are uniquely determined. This is known as the Geronimus-Wendroff Theorem (see [26,27]).

On the other hand, the location of the zeros of orthogonal polynomials has also been studied when the orthogonality measure is perturbed. In the literature, the case when the measure is multiplied by a polynomial is called the Christoffel transformation. Let $\alpha(x)$ be a polynomial which is non-negative in E of the form

$$\alpha(x) = \prod_{k=1}^{m} (x - x_k).$$

If we have $x_k \neq x_j$ for $k \neq j$, then the (monic) polynomials $P_n^{[m]}(x)$ defined by

$$A_{n,m}\alpha(x)P_n^{[m]}(x) = \begin{vmatrix} P_n(x_1) & P_{n+1}(x_1) & \cdots & P_{n+m}(x_1) \\ P_n(x_2) & P_{n+1}(x_2) & \cdots & P_{n+m}(x_2) \\ \vdots & \vdots & \vdots & \vdots \\ P_n(x_m) & P_{n+1}(x_m) & \cdots & P_{n+m}(x_m) \\ P_n(x) & P_{n+1}(x) & \cdots & P_{n+m}(x) \end{vmatrix}, \tag{8}$$

with

$$A_{n,m} = \begin{vmatrix} P_n(x_1) & P_{n+1}(x_1) & \cdots & P_{n+m-1}(x_1) \\ P_n(x_2) & P_{n+1}(x_2) & \cdots & P_{n+m-1}(x_2) \\ \vdots & \vdots & \vdots & \vdots \\ P_n(x_m) & P_{n+1}(x_m) & \cdots & P_{n+m-1}(x_m) \end{vmatrix}, \tag{9}$$

are orthogonal with respect to the perturbed measure $\alpha(x)d\mu(x)$ in E, and $P_n^{[m]}(x)$ has degree n. In particular, if $\{x_{n,k}^{[1]}\}_{k=0}^{n}$ denote the zeros of the polynomial $P_n^{[1]}(x)$ orthogonal with respect to $(x - c)d\mu(x)$ on $E = [a,b]$, then the zeros of the polynomials $P_n^{[1]}(x)$, $P_n(x)$ and $P_{n+1}(x)$ satisfy the following interlacing property (see [26], where the polynomials $\{P_n^{[1]}\}_{n\geq 0}$ are called *kernel polynomials* [28,29]):

- If $c \leq a$, then
$$x_{n+1,1} < x_{n,1} < x_{n,1}^{[1]} < x_{n+1,2} < \cdots < x_{n,n} < x_{n,n}^{[1]} < x_{n+1,n+1}, \tag{10}$$

- If $c \geq b$, then
$$x_{n+1,1} < x_{n,1}^{[1]} < x_{n,1} < \cdots < x_{n+1,n} < x_{n,n}^{[1]} < x_{n,n} < x_{n+1,n+1}. \tag{11}$$

Finally, we point out that there is a close relation between Hurwitz polynomials and orthogonal polynomials on the real line. Indeed, any Hurwitz polynomial can be expressed in terms of an orthogonal polynomial and its associated polynomial (see [26]). Conversely, it is possible to construct a Hurwitz polynomial by using orthogonal polynomials. For more details, we refer the reader to [20,30–33] and, more recently, [34]. In the latter, the authors construct sequences of Hurwitz polynomials from a sequence of orthogonal polynomials, and show several algebraic properties of the constructed family. Also, classical orthogonal polynomials are used in [35] to construct families of Hurwitz polynomials that are robustly stable.

2.2. Orthogonal Polynomials on the Unit Circle

Let σ be a positive, non-trivial measure supported on the unit circle $\mathbb{T}$. Generally, σ is assumed to be a probability measure, i.e., $\int_{\mathbb{T}} d\sigma(z) = 1$. Then, there exists a sequence of complex polynomials $\{\phi_n\}_{n\geq 0}$ with $deg(\phi_n) = n$ such that

$$\int_{\mathbb{T}} \phi_n(z) \overline{\phi_m(z)} \, d\sigma(z) = k_n \delta_{n,m}, \qquad \forall\, n, m \geq 0, \tag{12}$$

where $k_n > 0$ for every $n \geq 0$. We will assume that they are monic. $\{\phi_n\}_{n\geq 0}$ is called the monic sequence of polynomials orthogonal with respect to σ, and satisfies the following recurrence relations [25,27]

(i) *Forward recurrence:*

$$\phi_{n+1}(z) = z\phi_n(z) + \phi_{n+1}(0)\phi_n^*(z), \tag{13}$$

(ii) *Backward recurrence:*

$$\phi_{n+1}(z) = (1 - |\phi_{n+1}(0)|^2)z\phi_n(z) + \phi_{n+1}(0)\phi_{n+1}^*(z), \tag{14}$$

where $\phi_n^*(z) = z^n\overline{\phi_n}(z^{-1})$.

Notice that $\phi_n^*(z)$ is the reciprocal polynomial as defined in the previous section, except that here the coefficients are in general complex and the conjugate has to be taken. The complex numbers $\{\phi_n(0)\}_{n\geq 0}$ are called *Verblunsky* (Schur, reflection) coefficients, and satisfy $|\phi_n(0)| < 1$ for every $n \geq 1$.

It is well known that every positive, nontrivial measure σ in $\mathbb{T}$ determines a unique sequence of Verblunsky coefficients $\{\phi_n(0)\}_{n\geq 1}$. Conversely, given an arbitrary sequence $\{a_n\}_{n\geq 1}$ of complex numbers satisfying $|a_n| < 1$ for $n \geq 1$, there exists a unique measure σ supported on $\mathbb{T}$ such that its corresponding monic orthogonal sequence $\{\phi_n(z)\}_{n\geq 1}$ satisfies $\phi_n(0) = a_n$ for $n \geq 1$. In other words, any complex sequence $\{a_n\}_{n>1}$ in $\mathbb{D}$ is the sequence of Verblunsky coefficients for some measure σ. This is known as Verblunsky's theorem (see [27]).

Moreover, the zeros of each $\phi_n(z)$ are located in $\mathbb{D}$. Conversely, given any complex polynomial $\phi_n(z)$ with zeros in $\mathbb{D}$, it is possible to find a measure σ supported in $\mathbb{T}$ such that $\phi_n(z)$ is orthogonal with respect to σ. Furthermore, the previous polynomials in the sequence, $\phi_{n-1}, \ldots, \phi_0$, are completely determined. This is a consequence of the backward recurrence relation. Indeed, this is the unit circle analogue of the Wendroff-Geronimus theorem discussed above (see [27]). Notice that this implies that any Schur polynomial is an orthogonal polynomial, and vice versa. This is an important difference with respect to the relation between Hurwitz and orthogonal polynomials on the real line.

2.3. The Szegő Transformation

The Szegő transformation establishes a relation between orthogonal polynomials on the real line and on the unit circle by defining a correspondence between measures supported on the interval $[-2, 2]$ and measures supported on $\mathbb{T}$ (see [25,27,36]).

The mappings $z = e^{i\theta} \mapsto 2\cos(\theta)$, with $\theta \in [0, 2\pi)$ and $x \mapsto \arccos(x/2)$ define a two-one correspondence between $\mathbb{T}$ and $[-2, 2]$, that can be used to define a mapping between probability measures supported in $\mathbb{T}$ and probability measures supported in $[-2, 2]$. If we restrict this mapping to measures in $\mathbb{T}$ that are even, i.e., $d\sigma(\theta) = d\sigma(-\theta)$, then the correspondence is one to one. Such measures are called symmetric. This mapping is commonly referred to as the Szegő transformation and it is denoted by Sz. More precisely, we say $d\mu = Sz(d\sigma)$ if and only if $d\sigma(\theta) = d\sigma(-\theta)$ and

$$\int_0^{2\pi} f(\theta)d\sigma(\theta) = \int_{-2}^2 f(arcos(x/2))d\mu(x),$$

for any function f such that $f(\theta) = f(-\theta)$. Notice that because of the symmetry we have [36]

$$d\sigma \text{ is even} \Leftrightarrow \overline{\phi_n(z)} = \phi_n(\bar{z}) \Leftrightarrow \phi_n(0) \in \mathbb{R}$$

for all $n \geq 1$. That is, the sequence $\{\phi_n\}_{n\geq 0}$ orthogonal with respect to σ has real coefficients and, in particular, real Verblunsky coefficients. The next theorem establishes a relation between the sequences of orthogonal polynomials associated with μ and σ, when they are related through the Szegő transformation. The proof can be found in [25,36].

Theorem 2. *Let $d\mu = Sz(d\sigma)$, and denote by $\{P_n\}_{n \geq 0}$ and $\{\phi_n\}_{n \geq 0}$ the monic orthogonal sequences associated with μ and σ, respectively. Then,*

$$P_n(z + \frac{1}{z}) = [1 + \phi_{2n}(0)]^{-1}z^{-n}[\phi_{2n}(z) + \phi_{2n}^*(z)], \quad n \geq 0, \tag{15}$$

$$\|P_n\|_{L^2(d\mu)}^2 = 2[1 + \phi_{2n}(0)]^{-1}\|\phi_{2n}\|_{L^2(d\sigma)}^2, \quad n \geq 0, \tag{16}$$

$$P_n(z + \frac{1}{z}) = z^{-n}[z\phi_{2n-1}(z) + \phi_{2n-1}^*(z)], \quad n \geq 1, \tag{17}$$

$$\|P_n\|_{L^2(d\mu)}^2 = 2[1 - \phi_{2n}(0)]\|\phi_{2n-1}\|_{L^2(d\sigma)}^2, \quad n \geq 1, \tag{18}$$

where $L^2(d\mu)$ is the space of measurable functions such that $\int_{[-2,2]} |f(x)|^2 d\mu(x) < \infty$, and $L^2(d\sigma)$ is defined in a similar way.

There is a second family related to $\{\phi_n\}_{n \geq 0}$, orthogonal with respect to the measure

$$d\mu_1(x) = \frac{1}{4}(4 - x^2)d\mu(x).$$

If we denote by $\{Q_n(x)\}_{n \geq 0}$ the corresponding monic orthogonal polynomials, we have for $n \geq 1$

$$\begin{aligned} Q_{n-1}(z + \frac{1}{z}) &= (1 - \phi_{2n}(0))^{-1}z^{-n}\frac{\phi_{2n}(z) - \phi_{2n}^*(z)}{z - z^{-1}}, \\ &= z^{-n}\frac{z\phi_{2n-1}(z) - \phi_{2n-1}^*(z)}{z - z^{-1}}, \end{aligned} \tag{19}$$

and, from Equations (15), (17) and (19) we have

$$\begin{aligned} \phi_{2n}(z) &= \frac{z^n}{2}[(1 + \phi_{2n}(0))P_n(z + \frac{1}{z}) + (1 - \phi_{2n}(0))(z - z^{-1})Q_{n-1}(z + \frac{1}{z})], \\ \phi_{2n-1}(z) &= \frac{z^{n-1}}{2}[P_n(z + \frac{1}{z}) + (z - z^{-1})Q_{n-1}(z + \frac{1}{z})]. \end{aligned} \tag{20}$$

3. Stability Criteria via Orthogonality

In this section, we use the basic results of the previous section to study stability criteria for Schur polynomials. First, notice that the expression in step 5 of Algorithm 1 is equivalent to the Szegő backward relation Equation (14). Thus, by applying Algorithm 1 to a polynomial of degree n with complex coefficients we are in fact computing a sequence $\{S_k\}_{k=0}^n$ of polynomials that satisfy the Szegő recursion. The criterion asserts that the initial polynomial is Schur if and only if we have $|S_k(0)| < 1$ for $k = 1, \ldots, n$. Thus, Verblunsky's theorem and the analogue of the Geronimus-Wendroff theorem for the unit circle directly lead to another proof for the validity of the algorithm. This was already discussed in [27].

Algorithm 1: Algorithm to determine if a given polynomial is Schur

    Input: Any monic polynomial $S_n(z) = z^n + p_{n-1}z^{n-1} + \ldots + p_0$ (coefficients can be complex).

    Output: Determination of the Schur character of $S_n(z)$.

1 initialization;

2 **for** $i = 0, 2, \ldots n-1$ **do**

3      $S_n^{(i=0)} = S_n(z)$,

4   Verify $|p_0^{(i)}| < 1$,

5   Compute $S_n^{(i+1)}(z) = \frac{1}{z}\left[\frac{S_n^{(i)}(z) - S_n^{(i)}(0)z^{n-i}\overline{S_n^{(i)}(\frac{1}{z})}}{(1-|S_n^{(i)}(0)|^2)}\right]$,

6   Return to step 4 until the condition is not satisfied. In such a case, $S_n(z)$ is not a Schur polynomial. If the condition in step 4 holds for $p_0^{(0)}, p_0^{(1)}, \ldots, p_0^{(n-1)}$, then $S(z)$ is a Schur polynomial.

7   **return**

Before stating our main result, we need the following lemma.

Lemma 1. *Let $\{P_n(x)\}_{n\geq 0}$ be the sequence of monic orthogonal polynomials with respect to a measure $d\mu$ supported on the interval $E = [-c, c]$, for some $c > 0$, and denote by $\{P_n^{[2]}(x)\}_{n\geq 0}$ the monic orthogonal sequence with respect to $d\mu^{[2]} = (x^2 - c^2)d\mu$. Then, if we denote by $\{x_{n,k}^{[2]}\}_{k=1}^n$ the zeros of $P_n^{[2]}(x)$ and by $\{x_{n+1,k}\}_{k=1}^{n+1}$ the zeros of $P_{n+1}(x)$, we have the interlacing property*

$$x_{n+1,k} < x_{n,k}^{[2]} < x_{n+1,k+1}, \quad 1 \leq k \leq n.$$

Proof. Denote by $\{x_{n,k}^{[1]}\}_{k=0}^n$ the zeros of $P_n^{[1]}(x)$, orthogonal with respect to the measure $d\mu^{[1]} = (x - c)d\mu$. Then, they satisfy the interlacing property Equation (11). Since $\{P_n^{[2]}(x)\}_{n\geq 0}$ is orthogonal with respect to $d\mu^{[2]} = (x + c)d\mu^{[1]}$, then from Equation (10) we see that the zeros $\{x_{n,k}^{[2]}\}_{k=0}^n$ of $P_n^{[2]}(x)$ satisfy

$$x_{n+1,1}^{[1]} < x_{n,1}^{[1]} < x_{n,1}^{[2]} < x_{n+1,2}^{[1]} < \ldots < x_{n,n}^{[1]} < x_{n,n}^{[2]} < x_{n+1,n+1}^{[1]}.$$

As a consequence, we have

$$x_{n+1,1} < x_{n,1}^{[2]} < x_{n+1,2} < \ldots < x_{n+1,n} < x_{n,n}^{[2]} < x_{n+1,n+1}.$$

$\square$

As mentioned before, a useful tool to determine the Schur character of a polynomial is the Möbius transformation from $\mathbb{D}$ to the left half-plane of the complex plane. Then, the problem of determining the Schur stability becomes a problem of determining Hurwitz stability. Our next result establishes a novel criterion to determine Schur stability by using the Szegő transformation defined in the previous section.

Theorem 3. *Let $S_m(z)$ be a m-th degree monic polynomial with real coefficients satisfying $|S_m(0)| < 1$. If $m = 2n$, define*

$$P_n(z + \frac{1}{z}) = [1 + S_{2n}(0)]^{-1}z^{-n}[S_{2n}(z) + S_{2n}^*(z)], \tag{21}$$

$$Q_{n-1}(z + \frac{1}{z}) = [(1 - S_{2n}(0))(z - z^{-1})]^{-1}z^{-n}[S_{2n}(z) - S_{2n}^*(z)]. \tag{22}$$

If $m = 2n - 1$, define

$$P_n(z + \frac{1}{z}) = z^{-n}[zS_{2n-1}(z) + S_{2n-1}^*(z)],$$ (23)

$$Q_{n-1}(z + \frac{1}{z}) = (z - z^{-1})^{-1}z^{-n}[zS_{2n-1}(z) - S_{2n-1}^*(z)].$$ (24)

Then, $S_m(z)$ is a Schur polynomial if and only if $P_n(x)$ and $Q_{n-1}(x)$ are real polynomials with real, simple, and interlaced roots in $(-2, 2)$.

Proof. Notice that Equations (21)–(24) are Equations (15), (17) and (19). If $S_m(z)$ is a Schur polynomial with real coefficients, then it is orthogonal with respect to some even function σ supported on $\mathbb{T}$. Then, by applying the Szegő transformation, we obtain a measure μ supported on $[-2, 2]$, with an associated monic orthogonal sequence $\{P_n\}_{n \geq 0}$. Notice that the zeros of each P_n are real, simple, and located in $(-2, 2)$. Moreover, the sequence $\{Q_n\}_{n \geq 0}$ is orthogonal with respect to $\frac{1}{4}(4 - x^2)d\mu$, and their zeros are also simple and lie in $(-2, 2)$. By the previous lemma, $P_n(x)$ and $Q_{n-1}(x)$ have interlaced roots.

Conversely, assume that $P_n(x)$ and $Q_{n-1}(x)$ are real polynomials with real, simple and interlaced roots in $(-2, 2)$. Then, $P_n(z + \frac{1}{z})$ and $(z - z^{-1})Q_{n-1}(z + \frac{1}{z})$ clearly have simple and interlacing zeros in $\mathbb{T}$. If $m = 2n - 1$ (resp. $m = 2n$), then it follows from Equations (17) and (19) (resp. from Equations (15) and (19)) that $[zS_m(z) + S_m^*(z)]$ and $[zS_m(z) - S_m^*(z)]$ (resp. $[S_m(z) + S_m^*(z)]$ and $[S_m(z) - S_m^*(z)]$) have interlaced zeros in $\mathbb{T}$. Notice that $[zS_m(z) + S_m^*(z)]$ is a symmetric polynomial and $[zS_m(z) - S_m^*(z)]$ is an anti-symmetric polynomial (resp. $[S_m(z) + S_m^*(z)]$ is symmetric and $[S_m(z) - S_m^*(z)]$ is anti-symmetric) and both polynomials have degree $m + 1$ (resp. m). Since $zS(z) = \frac{1}{2}[zS_m(z) + S_m^*(z) + zS_m(z) - S_m^*(z)]$ (resp. $S(z) = \frac{1}{2}[S_m(z) + S_m^*(z) + S_m(z) - S_m^*(z)]$) it follows from Theorem 1 that $S_m(z)$ is a Schur polynomial. $\square$

The following result follows at once. It establishes a necessary condition for the Schur character of a real polynomial.

Corollary 1. *Let $S_m(z)$ be a m-th degree monic Schur polynomial with real coefficients. Define $P_n(x)$ as in the previous theorem. Then, $P_n(x + 2)$ is a Hurwitz polynomial with zeros in $(-4, 0)$.*

In other words, given any real polynomial $S_m(z)$ of degree m, we can compute $P_n(x)$ as in the previous theorem and then check $P_n(x + 2)$ for Hurwitz stability. If it is not Hurwitz, then $S_m(z)$ is not Schur. As another straightforward consequence, we can state the following Wendroff-Geronimus type theorem.

Corollary 2. *Let $P_n(x)$ and $Q_{n-1}(x)$ be two monic, real polynomials with degree n and $n - 1$, respectively. If P_n and Q_{n-1} have real, simple and interlaced zeros in $(-2, 2)$, then there exists a positive measure μ supported in $[-2, 2]$ such that P_n is orthogonal with respect to μ and Q_{n-1} is orthogonal with respect to the Christoffel transformation $(x^2 - 4)d\mu$. Moreover, the orthogonal sequences $P_{n-1}, \ldots, P_0$ and $Q_{n-2}, \ldots, Q_0$ are uniquely determined.*

Proof. If $P_n(x)$ and $Q_{n-1}(x)$ have simple and interlaced roots in $(-2, 2)$, then the polynomial $\phi_{2n-1}(z)$ obtained from Equation (20) is a Schur (and thus orthogonal) monic polynomial with real coefficients, from Theorem 3. By using Equations (23) and (24) for $\phi_{2n-1}(z)$ we obtain again $P_n(x)$ and $Q_{n-1}(x)$, and therefore $P_n(x)$ is orthogonal with respect to some μ supported on $[-2, 2]$ and $Q_{n-1}(x)$ is orthogonal with respect to $(x^2 - 4)d\mu$. The polynomials $P_{n-1}, \ldots, P_0$ and $Q_{n-2}, \ldots, Q_0$ are uniquely determined by the polynomials $\phi_{2n-3}, \phi_{2n-5}, \ldots, \phi_1$ obtained by applying the backward recurrence relation to $\phi_{2n-1}(z)$. $\square$

The following lemma shows how to compute the coefficients of the polynomials P_n and Q_{n-1} in Theorem 3. Notice that instead of computing the coefficients of $Q_{n-1}(x)$, we will find the coefficients of the polynomial $(x^2 - 4)Q_{n-1}(x)$, which has two additional zeros at $x = \pm 2$.

Lemma 2. *Let $S_m(z) = z^m + p_{m-1}z^{m-1} + \ldots + p_1 z + p_0$ and $P_n(x) = x^n + \gamma_{n-1}x^{n-1} + \ldots + \gamma_1 x + \gamma_0$ be the polynomials in Theorem 3. Then, the coefficients of $P_n(x)$ are determined by the linear system*

$$
\begin{pmatrix}
\binom{0}{0} & 0 & \binom{2}{1} & 0 & \binom{4}{2} & \cdots & & \cdots \\
0 & \binom{1}{0} & 0 & \binom{3}{1} & 0 & \cdots & & \cdots \\
0 & 0 & \binom{2}{0} & 0 & \binom{4}{1} & \cdots & & \cdots \\
\vdots & \vdots & \vdots & \ddots & \vdots & \vdots & \vdots & \vdots \\
0 & 0 & 0 & 0 & \ddots & 0 & \binom{n-1}{1} & 0 \\
0 & 0 & 0 & 0 & \cdots & \binom{n-2}{1} & 0 & \binom{n}{1} \\
0 & 0 & 0 & 0 & \cdots & 0 & \binom{n-1}{0} & 0 \\
0 & 0 & 0 & 0 & \cdots & 0 & 0 & \binom{n}{0}
\end{pmatrix}
\begin{pmatrix}
\gamma_0 \\ \gamma_1 \\ \gamma_2 \\ \vdots \\ \\ \\ \gamma_n
\end{pmatrix}
=
\begin{pmatrix}
c_0 \\ c_1 \\ c_2 \\ \vdots \\ \\ \\ c_n
\end{pmatrix},
\tag{25}
$$

where $c_i = \frac{1}{1+p_0}(p_{n+i} + p_{n-i})$ for $0 \le i \le n$ if $m = 2n$, and $c_i = p_{n+i-1} + p_{n-i-1}$ for $0 \le i \le n$ if $m = 2n - 1$, and defining $\gamma_n = p_m = 1$ and $p_{-1} = 0$. On the other hand, the coefficients of $(x^2 - 4)Q_{n-1}(x) = x^{n+1} + \eta_n x^n + \ldots + \eta_1 x + \eta_0$ are given by the linear system

$$
\begin{pmatrix}
\binom{0}{0} & 0 & \binom{2}{1} & 0 & \binom{4}{2} & \cdots & & \cdots \\
0 & \binom{1}{0} & 0 & \binom{3}{1} & 0 & \cdots & & \cdots \\
0 & 0 & \binom{2}{0} & 0 & \binom{4}{1} & \cdots & & \cdots \\
\vdots & \vdots & \vdots & \ddots & \vdots & \vdots & \vdots & \vdots \\
0 & 0 & 0 & 0 & \ddots & 0 & \binom{n}{1} & 0 \\
0 & 0 & 0 & 0 & \cdots & \binom{n-1}{1} & 0 & \binom{n+1}{1} \\
0 & 0 & 0 & 0 & \cdots & 0 & \binom{n}{0} & 0 \\
0 & 0 & 0 & 0 & \cdots & 0 & 0 & \binom{n+1}{0}
\end{pmatrix}
\begin{pmatrix}
\eta_0 \\ \eta_1 \\ \eta_2 \\ \vdots \\ \\ \\ \eta_{n+1}
\end{pmatrix}
=
\begin{pmatrix}
-2l_1 \\ -l_2 \\ -l_3 + l_1 \\ -l_4 + l_2 \\ \vdots \\ -l_{n-1} + l_{n-3} \\ -l_n + l_{n-2} \\ l_{n-1} \\ l_n
\end{pmatrix},
\tag{26}
$$

where $l_i = \frac{1}{1-p_0}(p_{n+i} - p_{n-i})$ for $1 \le i \le n$ if $m = 2n$, and $l_i = p_{n+i-1} - p_{n-i-1}$ for $1 \le i \le n$ if $m = 2n - 1$, with $\eta_{n+1} = 1$.

Proof. Assume $S_m(z)$ has even degree $m = 2n$. Then, $P_n(x)$ has degree n and we have

$$
P_n(z + 1/z) = \sum_{k=0}^{n} \gamma_k (z + 1/z)^k = \sum_{k=0}^{n} \gamma_k \sum_{j=0}^{k} \binom{k}{j} z^{k-2j}.
\tag{27}
$$

On the other hand, from Equation (21) we get

$$
P_n(z + 1/z) = \frac{1}{p_0 + 1}\left[(p_0 + p_{2n})(z^n + z^{-n}) + (p_1 + p_{2n-1})(z^{n-1} + z^{-(n-1)}) \right.
$$
$$
\left. + (p_2 + p_{2n-2})(z^{n-2} + z^{-(n-2)}) + \cdots + 2p_n\right].
\tag{28}
$$

Thus, by comparing the coefficients of the positive powers in Equations (27) and (28) we easily get Equation (25). Now, applying the change of variable $x = z + 1/z$ in $(x^2 - 4)Q_{n-1}(x)$, we get $(z - z^{-1})^2 Q_{n-1}(z + 1/z)$. Then, from Equation (22) we obtain

$$
\begin{aligned}
(z - z^{-1})^2 Q_{n-1}(z + 1/z) = \frac{1}{1 - p_0} & [(1 - p_0)(z^{n+1} + z^{-(n+1)}) + (p_{2n-1} - p_1)(z^n + z^{-n}) \\
& + (p_{2n-2} - p_2 - (1 - p_0))(z^{n-1} + z^{-(n-1)}) \\
& + (p_{2n-3} - p_3 - (p_{2n-1} - p_1))(z^{n-2} + z^{-(n-2)}) \\
& + \cdots + (p_{n+1} - p_{n-1} - (p_{n+3} - p_{n-3}))(z^2 + z^{-2}) \\
& + (p_{n-2} - p_{n+2})(z + z^{-1}) + 2(p_{n-1} - p_{n+1})].
\end{aligned}
\tag{29}
$$

Thus, comparing coefficients in Equations (27) and (29) we obtain Equation (26). Notice that the matrix in the linear system has size $(n + 2) \times (n + 2)$ since the polynomial $(x^2 - 4)Q_{n-1}(x)$ has degree $n + 1$. The odd case $m = 2n - 1$ follows in a similar way. $\square$

Finally, we illustrate the criterion in Theorem 3 with the following examples. Notice that the matrices of the linear systems in the previous lemma are lower triangular and therefore their solutions can be computed efficiently.

Example 1. *Consider the polynomial* $S_4(z) = z^4 - z^3 + \frac{3}{4}z^2 + z + \frac{1}{2}$. *The corresponding linear system is*

$$
\begin{pmatrix} \binom{0}{0} & 0 & \binom{2}{1} \\ 0 & \binom{1}{0} & 0 \\ 0 & 0 & \binom{2}{0} \end{pmatrix} \begin{pmatrix} \gamma_0 \\ \gamma_1 \\ \gamma_2 \end{pmatrix} = \begin{pmatrix} 1 \\ 0 \\ 1 \end{pmatrix},
$$

and has solution $(-1, 0, 1)^t$. *On the other hand, the linear system*

$$
\begin{pmatrix} \binom{0}{0} & 0 & \binom{2}{1} & 0 \\ 0 & \binom{1}{0} & 0 & \binom{3}{1} \\ 0 & 0 & \binom{2}{0} & 0 \\ 0 & 0 & 0 & \binom{3}{0} \end{pmatrix} \begin{pmatrix} \eta_0 \\ \eta_1 \\ \eta_2 \\ \eta_3 \end{pmatrix} = \begin{pmatrix} 8 \\ -1 \\ -4 \\ 1 \end{pmatrix},
$$

has solution $(16, -4, -4, 1)^t$. *As a consequence, the polynomials* $P_2(x)$ *and* $Q_1(x)$ *of Theorem 3 are* $P_2(x) = x^2 - 1$ *and* $(x^2 - 4)Q_1(x) = x^3 - 4x^2 - 4x + 16$. *Notice that* $P_2(x)$ *has zeros at* $x = \pm 1$, *and* $(x^2 - 4)Q_1(x)$ *has zeros at* $2, -2$ *and* 4. *Thus,* $S_4(z)$ *is not a Schur polynomial. Indeed,* $S(z)$ *has zeros* $z_1 \approx -0.391713 - 0.335138i, z_2 \approx -0.391713 + 0.335138i, z_3 \approx 0.891713 - 1.04224i, z_4 \approx 0.891713 + 1.04224i$.

Notice that the previous example shows that $P_n(x)$ having zeros in $(-2, 2)$ is not sufficient to guarantee that $S_{2n}(z)$ is a Schur polynomial.

Example 2. *Consider* $S_{11}(z) = z^{11} - \frac{9}{4}z^{10} + \frac{157}{32}z^9 - \frac{55}{8}z^8 + \frac{4637}{512}z^7 - \frac{9485}{1024}z^6 + \frac{8909}{1024}z^5 - \frac{6717}{1024}z^4 + \frac{2261}{512}z^3 - \frac{37}{16}z^2 + \frac{31}{32}z - \frac{1}{4}$. *Then, the linear system*

$$
\begin{pmatrix} \binom{0}{0} & 0 & \binom{2}{1} & 0 & \binom{4}{2} & 0 & \binom{6}{3} \\ 0 & \binom{1}{0} & 0 & \binom{3}{1} & 0 & \binom{5}{2} & 0 \\ 0 & 0 & \binom{2}{0} & 0 & \binom{4}{1} & 0 & \binom{6}{2} \\ 0 & 0 & 0 & \binom{3}{0} & 0 & \binom{5}{1} & 0 \\ 0 & 0 & 0 & 0 & \binom{4}{0} & 0 & \binom{6}{1} \\ 0 & 0 & 0 & 0 & 0 & \binom{5}{0} & 0 \\ 0 & 0 & 0 & 0 & 0 & 0 & \binom{6}{0} \end{pmatrix} \begin{pmatrix} \gamma_0 \\ \gamma_1 \\ \gamma_2 \\ \gamma_3 \\ \gamma_4 \\ \gamma_5 \\ \gamma_6 \end{pmatrix} = \begin{pmatrix} 2\frac{8909}{1024} \\ -\frac{9485}{1024} - \frac{6717}{1024} \\ \frac{4637}{512} + \frac{2261}{512} \\ -\frac{55}{8} - \frac{37}{16} \\ \frac{157}{32} + \frac{31}{32} \\ -\frac{9}{4} - \frac{1}{4} \\ 1 \end{pmatrix},
$$

has solution $(\frac{105}{512}, -\frac{389}{512}, -\frac{263}{256}, \frac{53}{16}, -\frac{1}{8}, -\frac{5}{2}, 1)^t$, *and the linear system*

$$
\begin{pmatrix}
\binom{0}{0} & 0 & \binom{2}{1} & 0 & \binom{4}{2} & 0 & \binom{6}{3} & 0 \\
0 & \binom{1}{0} & 0 & \binom{3}{1} & 0 & \binom{5}{2} & 0 & \binom{7}{3} \\
0 & 0 & \binom{2}{0} & 0 & \binom{4}{1} & 0 & \binom{6}{2} & 0 \\
0 & 0 & 0 & \binom{3}{0} & 0 & \binom{5}{1} & 0 & \binom{7}{2} \\
0 & 0 & 0 & 0 & \binom{4}{0} & 0 & \binom{6}{1} & 0 \\
0 & 0 & 0 & 0 & 0 & \binom{5}{0} & 0 & \binom{7}{1} \\
0 & 0 & 0 & 0 & 0 & 0 & \binom{6}{0} & 0 \\
0 & 0 & 0 & 0 & 0 & 0 & 0 & \binom{7}{0}
\end{pmatrix}
\begin{pmatrix}
\eta_0 \\ \eta_1 \\ \eta_2 \\ \eta_3 \\ \eta_4 \\ \eta_5 \\ \eta_6 \\ \eta_7
\end{pmatrix}
=
\begin{pmatrix}
-2(-\frac{9485}{1024} + \frac{6717}{1024}) \\
-(\frac{4637}{512} - \frac{2261}{512}) \\
-(-\frac{55}{8} + \frac{37}{16}) - \frac{9485}{1024} + \frac{6717}{1024} \\
-(\frac{157}{32} - \frac{31}{32}) + \frac{4637}{512} - \frac{2261}{512} \\
-(-\frac{9}{4} + \frac{1}{4}) - \frac{55}{8} + \frac{37}{16} \\
-1 + \frac{157}{32} - \frac{31}{32} \\
-\frac{9}{4} + \frac{1}{4} \\
1
\end{pmatrix},
$$

has solution $(\frac{9}{16}, \frac{15}{16}, -\frac{377}{64}, \frac{1}{64}, \frac{151}{16}, -\frac{65}{16}, -2, 1)^t$. *Therefore, the polynomials* $P_6(x)$ *and* $Q_5(x)$ *of Theorem 3 are*

$$
P_6(x) = x^6 - \frac{5}{2} - \frac{x^4}{8} + \frac{53}{16}x^3 - \frac{263}{256}x^2 - \frac{389}{512}x + \frac{105}{512},
$$

and

$$
(x^2 - 4)Q_5(x) = x^7 - 2x^6 - \frac{65}{16}x^5 + \frac{151}{16}x^4 + \frac{x^3}{64} - \frac{377}{64}x^2 + \frac{15}{16}x + \frac{9}{16}.
$$

It is not difficult to show that P_6 and Q_5 have real, simple and interlaced roots in $(-2, 2)$ and therefore S_{11} is a Schur polynomial. The zeros of the polynomials involved are plotted in the following Figure 1.

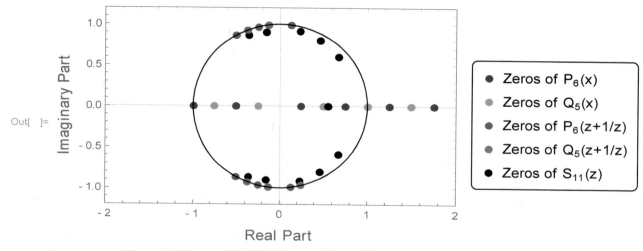

Out[ ]=

Figure 1. Zeros of $P_6(x), P_6(z + 1/z), Q_5(x), Q_5(z + 1/z)$ and $S_{11}(z)$.

On the other hand, if T denotes the transformation defined by Equation (5), define

$$
\hat{S}(x) := TS(z) = (x - 1)^n S\left(\frac{x+1}{x-1}\right).
$$

Then we have the following straightforward Schur analogue of Markov's parameters criterion for Hurwitz polynomials.

Proposition 1. *Let* $S_m(z)$ *be a polynomial of degree* m *with real coefficients such that* $S_m(1) \neq 0$. *Then,* $S_m(z)$ *is a Schur polynomial if and only if the constants* s_i *of the expansion*

$$
\frac{\hat{S}_m^a(x^{1/2})}{x^{1/2}\hat{S}_m^s(x^{1/2})} =
\begin{cases}
\frac{s_0}{x} - \frac{s_1}{x^2} + \frac{s_2}{x^3} - \ldots + \frac{s_{2n-2}}{x^{2n-1}} - \frac{s_{2n-1}}{x^{2n}} + \ldots, & \text{if } m = 2n, \\
s_{-1} + \frac{s_0}{x} - \frac{s_1}{x^2} + \frac{s_2}{x^3} - \ldots + \frac{s_{2n-2}}{x^{2n-1}} - \frac{s_{2n-1}}{x^{2n}} + \ldots, & \text{if } m = 2n+1.
\end{cases}
$$

are such that the matrices

$$H_{n-1} = \begin{pmatrix} s_0 & s_1 & \cdots & s_{n-1} \\ s_1 & s_2 & \cdots & s_n \\ \vdots & \vdots & \vdots & \vdots \\ s_{n-1} & s_n & \cdots & s_{2n-2} \end{pmatrix}, \quad H_{n-1}^{(1)} = \begin{pmatrix} s_1 & s_2 & \cdots & s_n \\ s_2 & s_3 & \cdots & s_{n+1} \\ \vdots & \vdots & \vdots & \vdots \\ s_n & s_{n+1} & \cdots & s_{2n-1} \end{pmatrix}, \quad (30)$$

are positive definite. If $m = 2n + 1$ it is also required that $s_{-1} > 0$.

Proof. $S_m(z)$ is a Schur polynomial if and only if

$$\hat{S}_m^s(x^{1/2}) = h(x), \quad x^{-1/2}\hat{S}_m^a(x^{1/2}) = g(x),$$

where $f(x) = h(x^2) + xg(x^2)$ is a Hurwitz polynomial (see [22]). The result follows from Markov's parameters criterion. $\square$

Notice that the above means that under the action of the Möbius transformation Equation (5), the symmetric (resp. asymmetric) part of a Schur polynomial is related to the even (resp. odd) part of a Hurwitz polynomial.

Example 3. *Consider the polynomial $S_4(z) = z^4 - 2z^3 + \frac{67}{36}z^2 - \frac{31}{36}z + \frac{13}{72}$. Then,*

$$S_4^s = \frac{85}{144}z^4 - \frac{103}{72}z^3 + \frac{67}{36}z^2 - \frac{103}{72}z + \frac{85}{144}, \quad and \quad S_4^a = \frac{59}{144}z^4 - \frac{41}{72}z^3 + \frac{41}{72}z - \frac{59}{144},$$

and by applying Equation (5) we get

$$\hat{S}_4^s = \frac{1}{72}(425 + 242z^2 + 13z^4), \quad \hat{S}_4^a = \frac{50}{9}z + z^3.$$

Thus,

$$\frac{\hat{S}_4^a(x^{1/2})}{x^{1/2}\hat{S}_4^s(x^{1/2})} = \frac{400 + 72x}{425 + 242x + 13x^2} = \frac{72}{13x} - \frac{12{,}224}{169x^2} + \frac{2{,}560{,}408}{2197x^3} - \frac{552{,}081{,}136}{28{,}561x^4} + \cdots,$$

and the matrices

$$H_1 = \begin{pmatrix} \frac{72}{13} & \frac{12{,}224}{169} \\ \frac{12{,}224}{169} & \frac{2{,}560{,}408}{2197} \end{pmatrix}, \quad H_1^{(1)} = \begin{pmatrix} \frac{12{,}224}{169} & \frac{2{,}560{,}408}{2197} \\ \frac{2{,}560{,}408}{2197} & \frac{552{,}081{,}136}{28{,}561} \end{pmatrix} \quad (31)$$

are positive definite. As a consequence, $S_4(z)$ is a Schur polynomial.

4. Conclusions and Further Remarks

We have obtained a criterion in Theorem 3 to determine whether or not a given polynomial $S_m(z)$ is Schur. To the best of our knowledge, this method is not known in the literature. Notice that it is a Hermite-Biehler type criterion in the sense that it involves verifying the interlacing of a pair of polynomials that are obtained from S_m. Our result is an improvement of Theorem 1, since there it is required to verify the interlacing of two polynomials of degree n, whereas in our method the interlacing is verified for polynomials of degree $m/2$ (resp. $(m + 1)/2$) when n is even (resp. odd). As an immediate consequence, we also obtain a simple necessary condition for Schur stability in Corollary 1. A criterion related to the positivity of Hankel matrices is obtained in Proposition 1.

Author Contributions: Formal analysis, L.E.G., N.M. and G.R.; Funding acquisition, G.R.; Software, N.M.; Writing—original draft, L.E.G. and N.M.; Writing—review and editing, L.E.G. and G.R. All authors have read and agreed to the published version of the manuscript.

References

1. Åström, K.J.; Wittenmark, B. *Computer-Controlled Systems*; Prentice Hall: Upper Saddle River, NJ, USA, 1990.
2. Kuo, B.C. *Digital Control Systems*; Oxford University Press: Oxford, UK, 1992.
3. Jalili-Kharaajoo, M.; Araabi, B.N. The Schur stability via the Hurwitz stability analysis using a biquadratic transformation. *Automatica* **2005**, *41*, 173–176. [CrossRef]
4. Fleming, R.; Grossman, G.; Lenker, T.; Narayan, S. On Schur D-Stable matrices. *Linear Algebra Appl.* **1988**, *279*, 39–50. [CrossRef]
5. Xu, H.; Datta, A.; Bhattacharyya, S.P. Computation of all stabilizing PID gains for digital control systems. *IEEE Trans. Autom. Control* **2001**, *46*, 647–652. [CrossRef]
6. Jury, E.I.; Blanchard, J. A stability test for linear discrete system in table form. *Proc. IRE* **1961**, *49*, 1947–1948.
7. Jury, E.I.; Blanchard, J. A stability test for linear discrete system using a simple division. *Proc. IRE* **1961**, *49*, 1948–1949.
8. Choo, Y. An elementary proof of the Jury test for real polynomials. *Automatica* **2011**, *47*, 249–252. [CrossRef]
9. Pérez, F.; Abdallah, C.; Docampo, D. Extreme-point stability test for discrete-time polynomials. In Proceedings of the 31st Conference on Decision and Control, Tucson, AZ, USA, 16–18 December 1992; pp. 1552–1553.
10. Shiomi, K.; Otsuka, N.; Inaba, H.; Ishii, R. The property of bilinear transformation matrix and Schur stability for a linear combination of polynomials. *J. Frankl. Inst.* **1999**, *336*, 533–541. [CrossRef]
11. Ackermann, J.E.; Barmish, B.R. Robust Schur stability of a polytope of polynomials. *IEEE Trans. Autom. Control* **1988**, *33*, 984–986. [CrossRef]
12. Bose, N.K.; Jury, E.I.; Zeheb, E. On robust hurwitz and Schur polynomials. *IEEE Trans. Autom. Control* **1988**, *33*, 1166–1168. [CrossRef]
13. Greiner, R. Necessary conditions for Schur-stability of interval polynomials. *IEEE Trans. Autom. Control* **2004**, *49*, 740–744. [CrossRef]
14. Kraus, F.; Mansour, M.; Jury, E.I. Robust Schur stability of interval polynomials. *IEEE Trans. Autom. Control* **1992**, *37*, 141–143. [CrossRef]
15. Shih, M.H.; Pang, C.T. Simultaneous Schur stability of interval matrices. *Automatica* **2008**, *44*, 2621–2627. [CrossRef]
16. Pastravanu, O.; Matcovschi, M.H. Sufficient conditions for Schur and Hurwitz diagonal stability of complex interval matrices. *Linear Algebra Appl.* **2015**, *467*, 149–173. [CrossRef]
17. Torres-Muñoz, J.A.; Rodríguez-Ángeles, E.; Kharitonov, V.L. On Schur stable multivariate polynomials. *IEEE Trans. Circuits Syst. I Regul. Pap.* **2006**, *53*, 1166–1173. [CrossRef]
18. Dumitrescu, B.; Chang, B.C. Robust Schur stability with polynomial parameters. *IEEE Trans. Circuits Syst. II Exp. Briefs* **2006**, *53*, 535–537. [CrossRef]
19. Dabbene, F.; Henrion, D.; Lagoa, C.M. Simple approximations of semialgebraic sets and their applications to control. *Automatica* **2017**, *78*, 110–118. [CrossRef]
20. Gantmacher, F.R. *The Theory of Matrices*; Chelsea Publishing Co.: New York, NY, USA, 1959; Volumes 1–2.
21. Lancaster, P.; Tismenetsky, M. *The Theory of Matrices and Applications*; Academic Press: Cambridge, MA, USA, 1985.
22. Bhattacharyya, S.P.; Chapellat, H.; Keel, L.H. *Robust Control: The Parametric Approach*; Prentice-Hall: Upper Saddle River, NJ, USA, 1995.
23. Hermite, C. Sur le nombre des racines dune equation algebrique comprise entre des limites donnes. *J. Reine Angew. Math.* **1856**, *52*, 39–51.
24. Bäckström, T.; Magi, C. Properties of line spectrum pair polynomials—A review. *Signal Process.* **2006**, *86*, 3286–3298. [CrossRef]
25. Szegő, G. *Orthogonal Polynomials*, 4th ed.; American Mathematical Society Colloquium Publications: Providence, RI, USA, 1975; Volume 23.
26. Chihara, T.S. *An Introduction to Orthogonal Polynomials*; Mathematics and Its Applications Series; Gordon and Breach: New York, NY, USA, 1978.

27.   Simon, B. *Orthogonal Polynomials on the Unit Circle*; American Mathematical Society Colloquium Publications: Providence, RI, USA, 2005; Volume 54.

28.   Ismail, M.E.H. *Classical and Quantum Orthogonal Polynomials in One Variable*; Encyclopedia of Mathematics and Its Applications; Cambridge University Press: Cambridge, UK, 2005; Volume 98.

29.   Huertas, E.J.; Marcellán, F.; Rafaeli, F.R. Zeros of orthogonal polynomials generated by canonical perturbations of measures. *Appl. Math. Comput.* **2012**, *218*, 7109–7127. [CrossRef]

30.   Brezinski, C. *Padé-Type Approximation and General Orthogonal Polynomials*; International Series of Numerical Mathematics, SO; Birkhäuser Verlag: Basel, Switzerland, 1980.

31.   Genin, I.V. Euclid algorithm, orthogonal polynomials, and generalized routh-hurwitz algorithm. *Linear Algebra Appl.* **1996**, *246*, 131–158. [CrossRef]

32.   Holtz, O. Hermite-Biehler, Routh–Hurwitz, and total positivity. *Linear Algebra Appl.* **2003**, *372*, 105–110. [CrossRef]

33.   Lange, L.J. Continued fraction applications to zero location. In *Analytic Theory of Continued Fractions II, Lecture Notes in Math., Vol. 1199, Proceedings of a Seminar-Worship, Pitlochry and Aviemore, Scotland, UK, 13–29 June 1985*; Thron, W.J., Ed.; Springer: Berlin/Heidelberg, Germany, 1986; pp. 220–262.

34.   Martínez, N.; Garza, L.E.; Aguirre-Hernández, B. On sequences of Hurwitz polynomials related to orthogonal polynomials. *Linear Multilinear A* **2019**, *67*, 2191–2208. [CrossRef]

35.   Arceo, A.; Garza, L.E.; Romero, G. Robust stability of hurwitz polynomials associated with modified classical weights. *Mathematics* **2019**, *7*, 818. [CrossRef]

36.   Simon, B. *Szegő's Theorem and Its Descendants: Spectral Theory for L2 Perturbations of Orthogonal Polynomials*; M. B. Porter Lectures; Princeton University Press: Princeton, NJ, USA, 2011.

2

Multiple Meixner Polynomials on a Non-Uniform Lattice

Jorge Arvesú *⬥ and Andys M. Ramírez-Aberasturis

Department of Mathematics, Universidad Carlos III de Madrid, Avda. de la Universidad, 30,
28911 Leganés, Madrid, Spain; andysramirezaberasturis@gmail.com
* Correspondence: jarvesu@math.uc3m.es

Abstract: We consider two families of type II multiple orthogonal polynomials. Each family has orthogonality conditions with respect to a discrete vector measure. The r components of each vector measure are q-analogues of Meixner measures of the first and second kind, respectively. These polynomials have lowering and raising operators, which lead to the Rodrigues formula, difference equation of order $r + 1$, and explicit expressions for the coefficients of recurrence relation of order $r + 1$. Some limit relations are obtained.

Keywords: Hermite–Padé approximation; multiple orthogonal polynomials; discrete orthogonality; recurrence relations

MSC: 42C05; 33C47; 33E99

1. Introduction

Hermite's proof [1] of the transcendence of the number e uses the notion of simultaneous approximation, which was subsequently studied in approximation theory and number theory [2–8]. Multiple orthogonal polynomials are polynomials that satisfy orthogonality conditions shared with respect to a set of measures [9–17]. They are related to the simultaneous rational approximation of a system of r analytic functions [18,19] and play an important role both in pure and applied mathematics (see for instance [20–22] as well as [23–27]). In this context, some families of continuous and discrete multiple orthogonal polynomials have been studied [3,28–30] as well as some multiple q-orthogonal polynomials [31–33]. The goal of the present paper is to study some multiple Meixner polynomials on a non-uniform lattice $x(s) = q^s - 1/q - 1, s = 0, 1, \ldots$

The paper is structured as follows. Section 2 is devoted to introduce the necessary background material. In Section 3, we consider two families of multiple q-orthogonal polynomials, namely, multiple q-Meixner polynomials of the first and second kind, respectively. They are analogous to the discrete multiple Meixner polynomials studied in [28]. We obtain the raising and lowering q-difference operators as well as the Rodrigues-type formula, which lead to an explicit expression for the multiple q-Meixner polynomials. Then, the recurrence relations as well as the q-difference equations with respect to the independent variable $x(s)$ are obtained. In Section 4, some limit relations as the parameter q approaches 1 are studied. An appendix to the Section 3 is considered in Section 5, in which the AT-property of the involved system of q-discrete measures is addressed. We make concluding remarks in Section 6.

2. Background Material

Let $\vec{\mu} = (\mu_1, \ldots, \mu_r)$ be a vector of r positive Borel measures supported on $\mathbb{R}$ with finite moments. By Ω_i we denote the smallest interval that contains supp (μ_i). Define a multi-index $\vec{n} = (n_1, \ldots, n_r) \in \mathbb{N}^r$, where $\mathbb{N}$ stands for the set of nonnegative integers. For the multi-index $\vec{n}$, a type II multiple orthogonal polynomial $P_{\vec{n}}$ is a polynomial of degree $\leq |\vec{n}| = n_1 + \cdots + n_r$, which satisfies the orthogonality conditions [34]

$$\int_{\Omega_i} P_{\vec{n}}(x) x^k d\mu_i(x) = 0, \qquad k = 0, \ldots, n_i - 1, \qquad i = 1, \ldots, r. \tag{1}$$

Special attention is paid to a unique solution of (1) (up to a multiplicative factor) with $\deg P_{\vec{n}}(x) = |\vec{n}|$ for every $\vec{n}$. In this situation the index is said to be normal [34]. In particular, if the above system of measures forms an AT system [34], then every multi-index is normal.

The polynomial $P_{\vec{n}}(z)$ is the common denominator of the simultaneous rational approximants $\frac{Q_{\vec{n},i}(z)}{P_{\vec{n}}(z)}$, to Cauchy transforms

$$\hat{\mu}_i(z) = \int_{\Omega_i} \frac{d\mu_i(x)}{z - x}, \quad z \notin \Omega_i \quad i = 1, \ldots, r, \tag{2}$$

of the vector components of $\vec{\mu} = (\mu_1, \ldots, \mu_r)$, i.e., for function (2) we have the following simultaneous rational approximation with prescribed order near infinity [34]

$$P_{\vec{n}}(z)\hat{\mu}_i(z) - Q_{\vec{n},i}(z) = \frac{\zeta_i}{z^{n_i+1}} + \cdots = \mathcal{O}(z^{-n_i-1}), \quad i = 1, \ldots, r.$$

If the measures in (1) are discrete

$$\mu_i = \sum_{k=0}^{N_i} \omega_{i,k} \delta_{x_{i,k}}, \qquad \omega_{i,k} > 0, \qquad x_{i,k} \in \mathbb{R}, \qquad N_i \in \mathbb{N} \cup \{+\infty\}, \qquad i = 1, 2, \ldots, r, \tag{3}$$

where $\delta_{x_{i,k}}$ denotes the Dirac delta function and $x_{i_1,k} \neq x_{i_2,k}$, $k = 0, \ldots, N_i$, whenever $i_1 \neq i_2$, the corresponding polynomial solution $P_{\vec{n}}(x)$ of the linear system of Equation (1) is called discrete multiple orthogonal polynomial (see [28] and the examples therein). In particular, the paper [28] considers discrete multiple orthogonal polynomial on the linear lattice $x(k) = k, k = 1, \ldots, N, N \in \mathbb{N} \cup \{+\infty\}$.

We will deal only with systems of discrete measures, for which $\Omega_i = \Omega \subset \mathbb{R}^+$ (the set of nonnegative reals) for each $i = 1, 2, \ldots, r$. Recall that the system of positive discrete measures $\mu_1, \mu_2, \ldots, \mu_r$, given in (3), forms an AT system if there exist r continuous functions $v_1, \ldots, v_r$ on Ω with $v_i(x_k) = \omega_{i,k}, k = 0, \ldots, N_i$, $i = 1, 2, \ldots, r$, such that the $|\vec{n}|$ functions

$$v_1(x), xv_1(x), \ldots, x^{n_1-1} v_1(x), \ldots, v_r(x), xv_r(x), \ldots, x^{n_r-1} v_r(x),$$

form a Chebyshev system on Ω for each multi-index $\vec{n}$ with $|\vec{n}| < N + 1$, i.e., every linear combination $\sum_{i=1}^{r} Q_{n_i-1}(x) v_i(x)$, where $Q_{n_i-1} \in \mathbb{P}_{n_i-1} \setminus \{0\}$, has at most $|\vec{n}| - 1$ zeros on Ω. Here $\mathbb{P}_m \subset \mathbb{P}$ denotes the linear subspace (of the space $\mathbb{P}$) of polynomials of degree at most $m \in \mathbb{Z}^+$.

In the sequel we will consider discrete multiple orthogonal polynomials on a non-uniform lattice $x(s) = q^s - 1/q - 1$ (see [35,36]).

Definition 1. *A polynomial $P_{\vec{n}}(x(s))$ on the lattice $x(s) = c_1 q^s + c_3, q \in \mathbb{R}^+ \setminus \{1\}, c_1, c_3 \in \mathbb{R}$, is said to be a multiple q-orthogonal polynomial of a multi-index $\vec{n} \in \mathbb{N}^r$ with respect to positive discrete measures $\mu_1, \mu_2, \ldots, \mu_r$ (with finite moments) such that $\mathrm{supp}\,(\mu_i) \subset \Omega_i \subset \mathbb{R}, i = 1, 2, \ldots, r$, if the following conditions hold:*

$$\deg P_{\vec{n}}(x(s)) \leq |\vec{n}| = n_1 + n_2 + \cdots + n_r,$$

$$\sum_{s=0}^{N_i} P_{\vec{n}}(x(s)) x(s)^k d\mu_i = 0, \qquad k = 0, \ldots, n_i - 1, \qquad N_i \in \mathbb{N} \cup \{+\infty\}. \tag{4}$$

In Section 3 we will deal with particular measures involving the q-Gamma function, which is defined as follows

$$\Gamma_q(s) = \begin{cases} f(s;q) = (1-q)^{1-s} \dfrac{\prod\limits_{k\geq 0}(1-q^{k+1})}{\prod\limits_{k\geq 0}(1-q^{s+k})}, & 0 < q < 1, \\[2em] q^{\frac{(s-1)(s-2)}{2}} f(s;q^{-1}), & q > 1. \end{cases} \tag{5}$$

See also [37,38] for the definition of the q-Gamma function. In addition, we use the q-analogue of the Stirling polynomials denoted by $[s]_q^{(k)}$, which is a polynomial of degree k in the variable $x(s) = (q^s - 1)/(q-1)$, i.e.,

$$[s]_q^{(k)} = \prod_{j=0}^{k-1} \frac{q^{s-j}-1}{q-1} = x(s)x(s-1)\cdots x(s-k+1) \quad \text{for} \quad k > 0, \quad \text{and} \quad [s]_q^{(0)} = 1. \tag{6}$$

Hereafter, confusion should be avoided between (6) and the notation for the q-analogue of a complex number $z \in \mathbb{C}$,

$$[z] = \frac{q^z - q^{-z}}{q - q^{-1}}. \tag{7}$$

The relation between (6) and (7) is as follows: $[z] = q^{1-z}[2z]_q^{(1)}/(q+1)$. The term q-analogue means that the expression $[z]$ tends to z, as q approaches 1. In general, we say that the function $f_q(s)$ is a q-analogue to the function $f(s)$ if for any sequence $(q_n)_{n\geq 0}$ approaching to 1, the corresponding sequence $\left(f_{q_n}(s)\right)_{n\geq 0}$ tends to $f(s)$ (see Section 4).

The following difference operators are used throughout this paper

$$\Delta \stackrel{\text{def}}{=} \frac{\triangle}{\triangle x(s-1/2)}, \qquad \nabla \stackrel{\text{def}}{=} \frac{\nabla}{\nabla x(s+1/2)}, \tag{8}$$

$$\nabla^{n_j} = \underbrace{\nabla \cdots \nabla}_{n_j \text{ times}}, \quad n_j \in \mathbb{N}, \tag{9}$$

where $\nabla f(x) = f(x) - f(x-1)$ and $\triangle f(x) = \nabla f(x+1)$ denote the backward and forward difference operators, respectively. When convenient, a less common notation taken from [38] will also be used: $\nabla x_1(s) \stackrel{\text{def}}{=} \nabla x(s+1/2) = \triangle x(s-1/2) = q^{s-1/2}$.

Observe that

$$\nabla^m (f(s)g(s)) = \sum_{k=0}^{m} \binom{m}{k} \left(\nabla^k f(s)\right)\left(\nabla^{m-k}g(s-k)\right), \quad m \in \mathbb{N}, \tag{10}$$

is a discrete analogue of the well-known Leibniz formula (product rule for derivatives). In particular,

$$\nabla^m f(s) = \sum_{k=0}^{m}(-1)^k \binom{m}{k} f(s-k). \tag{11}$$

Finally, we will make use of the following notations for multi-indices: The multi-index $\vec{e}_i$ denotes the standard r-dimensional unit vector with the i-th entry equals 1 and 0 otherwise, the multi-index $\vec{e}$ with all its r-entries equal 1. In addition, for any vector $\vec{\alpha} \in \mathbb{C}^r$ and number $p \in \mathbb{C}$,

$$\vec{\alpha}_{i,p} \stackrel{\text{def}}{=} \vec{\alpha} - \alpha_i(1-p)\vec{e}_i = (\alpha_1, \ldots, p\alpha_i, \ldots, \alpha_r). \tag{12}$$

Multiple Meixner Polynomials of the First and Second Kind

In [28], for multiple Meixner polynomials, it was considered two vector measures $\vec{\mu} = (\mu_1, \ldots, \mu_r)$ and $\vec{\nu} = (\nu_1, \ldots, \nu_r)$, where in both cases each component is a Pascal distribution (negative binomial distribution) with different parameters

$$\mu_i = \sum_{x=0}^{\infty} v^{\alpha_i, \beta}(x)\delta_x, \quad v^{\alpha_i, \beta}(x) = \begin{cases} \dfrac{\Gamma(\beta + x)}{\Gamma(\beta)} \dfrac{\alpha_i^x}{\Gamma(x+1)}, & x \in \mathbb{R} \setminus (\mathbb{Z}^- \cup \{-\beta, -\beta - 1, \beta - 2, \ldots\}), \\ 0, & \text{otherwise,} \end{cases}$$

$$\nu_i = \sum_{x=0}^{\infty} v^{\alpha, \beta_i}(x)\delta_x, \quad i = 1, \ldots, r.$$

Notice that $v^{\alpha, \beta_i}(x)$ is a C^∞-function on $\mathbb{R} \setminus \{-\beta_i, -\beta_i - 1, -\beta_i - 2, \ldots\}$ with simple poles at the points in $\{-\beta_i, -\beta_i - 1, -\beta_i - 2, \ldots\}$. For the above measures $0 < \alpha, \alpha_i < 1$, with all the α_i different, and $\beta, \beta_i > 0$ ($\beta_i - \beta_j \notin \mathbb{Z}$ for all $i \neq j$). Under these conditions for both $\vec{\mu}$ and $\vec{\nu}$ the multi-index $\vec{n} \in \mathbb{N}^r$ is normal.

For the monic multiple Meixner polynomial of the first kind [28] corresponding to the multi-index $\vec{n} \in \mathbb{N}^r$ and the vector measure $\vec{\mu}$, define the monic polynomial $M_{\vec{n}}^{\vec{\alpha}, \beta}(x)$ of degree $|\vec{n}|$ and different positive parameters $\alpha_1, \ldots, \alpha_r$ (indexed by $\vec{\alpha} = (\alpha_1, \ldots, \alpha_r)$) and the same $\beta > 0$ which satisfies the orthogonality conditions

$$\sum_{x=0}^{\infty} M_{\vec{n}}^{\vec{\alpha}, \beta}(x)(-x)_j v^{\alpha_i, \beta}(x) = 0, \qquad j = 0, \ldots, n_i - 1, \qquad i = 1, \ldots, r,$$

where $(x)_j = (x)(x+1)\cdots(x+j-1)$, $(x)_0 = 1$, $j \geq 1$, denotes the Pochhammer symbol. This polynomial of degree j is used to deal more conveniently with the orthogonality conditions (1)–(3) on the linear lattice $\{x = 0, 1, \ldots\}$.

For the monic *multiple Meixner polynomial of the second kind* [28] corresponding to the multi-index $\vec{n} \in \mathbb{N}^r$ and the vector measure $\vec{\nu}$, define the monic polynomial $M_{\vec{n}}^{\alpha, \vec{\beta}}(x)$ of degree $|\vec{n}|$ and $\vec{\beta} = (\beta_1, \ldots, \beta_r)$, with different components, which satisfies the orthogonality conditions

$$\sum_{x=0}^{\infty} M_{\vec{n}}^{\alpha, \vec{\beta}}(x)(-x)_j v^{\alpha, \beta_i}(x) = 0, \qquad j = 0, \ldots, n_i - 1, \qquad i = 1, \ldots, r.$$

For both families of multiple orthogonal polynomials the following r raising operators were found

$$\mathcal{L}^{\alpha_i, \beta}\left(M_{\vec{n}}^{\vec{\alpha}, \beta}(x)\right) = -M_{\vec{n} + \vec{e}_i}^{\vec{\alpha}, \beta - 1}(x), \tag{13}$$

$$\mathcal{L}^{\alpha, \beta_i}\left(M_{\vec{n}}^{\alpha, \vec{\beta}}(x)\right) = -M_{\vec{n} + \vec{e}_i}^{\alpha, \vec{\beta} - \vec{e}_i}(x), \tag{14}$$

where

$$\mathcal{L}^{\sigma, \tau} \stackrel{\text{def}}{=} \frac{\sigma(\tau - 1)}{(1 - \sigma) v^{\sigma, \tau - 1}(x)} \nabla v^{\sigma, \tau}(x), \quad (\sigma, \tau) \in \{(\alpha_i, \beta)\} \cup \{(\alpha, \beta_i)\}, \quad i = 1, \ldots, r.$$

As a consequence of (13) and (14), there holds the Rodrigues-type formulas

$$M_{\vec{n}}^{\vec{\alpha}, \beta}(x) = (\beta)_{|\vec{n}|} \left(\prod_{i=1}^{r} \left(\frac{\alpha_i}{\alpha_i - 1}\right)^{n_i}\right) \frac{\Gamma(\beta)\Gamma(x+1)}{\Gamma(\beta + x)} M_{\vec{n}}^{\vec{\alpha}}\left(\frac{\Gamma(\beta + |\vec{n}| + x)}{\Gamma(\beta + |\vec{n}|)\Gamma(x+1)}\right), \tag{15}$$

$$M_{\vec{n}}^{\alpha, \vec{\beta}}(x) = \left(\frac{\alpha}{\alpha - 1}\right)^{|\vec{n}|} \left(\prod_{i=1}^{r} (\beta_i)_{n_i}\right) \frac{\Gamma(x+1)}{\alpha^x} N_{\vec{n}}^{\vec{\beta}}\left(\frac{\alpha^x}{\Gamma(x+1)}\right), \tag{16}$$

where $\mathcal{M}_{\vec{n}}^{\vec{\alpha}} = \prod\limits_{i=1}^{r} \left(\alpha_i^{-x} \nabla^{n_i} \alpha_i^x \right)$ and $\mathcal{N}_{\vec{n}}^{\vec{\beta}} = \prod_{i=1}^{r} \frac{\Gamma(\beta_i)}{\Gamma(\beta_i + x)} \nabla^{n_i} \frac{\Gamma(\beta_i + n_i + x)}{\Gamma(\beta_i + n_i)}$. Then, from (10) and (11) the above

Formulas (15) and (16) provide an explicit expressions for the above polynomials $M_{\vec{n}}^{\vec{\alpha},\beta}(x)$ and $M_{\vec{n}}^{\alpha,\vec{\beta}}(x)$.

Two important algebraic properties are known for multiple Meixner polynomials [28], namely the $(r+1)$-order linear difference equations [39]

$$\prod_{i=1}^{r} \mathcal{L}^{\alpha_i, \beta+i+1-r} \left(\triangle M_{\vec{n}}^{\vec{\alpha},\beta}(x) \right) = -\sum_{i=1}^{r} n_i \prod_{\substack{j=1 \\ j \neq i}}^{r} \mathcal{L}^{\alpha_j, \beta+j+1-r} \left(M_{\vec{n}}^{\vec{\alpha},\beta}(x) \right), \tag{17}$$

$$\prod_{i=1}^{r} \mathcal{L}^{\alpha, \beta_i+1} \left(\triangle M_{\vec{n}}^{\alpha,\vec{\beta}}(x) \right) = -\sum_{i=1}^{r} \frac{d_i \prod\limits_{l=1}^{r} (n_l + \beta_l - \beta_i)}{\prod\limits_{k=1, k \neq i}^{r-1} (\beta_i - \beta_k) \prod\limits_{l=i+1}^{r} (\beta_l - \beta_i)} \prod_{\substack{j=1 \\ j \neq i}}^{r} \mathcal{L}^{\alpha, \beta_j+1} \left(M_{\vec{n}}^{\alpha,\vec{\beta}}(x) \right), \tag{18}$$

where

$$d_i = \sum_{j=1}^{r} \frac{(-1)^{i+j} \prod\limits_{k=1}^{r} (n_j + \beta_j - \beta_k)}{(n_j + \beta_j - \beta_i) \prod\limits_{k=1, k \neq j}^{r-1} (n_k - n_j + \beta_k - \beta_j) \prod\limits_{l=j+1}^{r} (n_j - n_l + \beta_j - \beta_l)},$$

and the recurrence relations [28]

$$x M_{\vec{n}}^{\vec{\alpha},\beta}(x) = M_{\vec{n}+\vec{e}_k}^{\vec{\alpha},\beta}(x) + \left((\beta + |\vec{n}|) \left(\frac{\alpha_k}{1 - \alpha_k} \right) + \sum_{i=1}^{r} \frac{n_i}{1 - \alpha_i} \right) M_{\vec{n}}^{\vec{\alpha},\beta}(x)$$

$$+ \sum_{i=1}^{r} \frac{\alpha_i n_i (\beta + |\vec{n}| - 1)}{(\alpha_i - 1)^2} M_{\vec{n}-\vec{e}_i}^{\vec{\alpha},\beta}(x), \tag{19}$$

$$x M_{\vec{n}}^{\alpha,\vec{\beta}}(x) = M_{\vec{n}+\vec{e}_k}^{\alpha,\vec{\beta}}(x) + \left((n_k + \beta_k) \left(\frac{\alpha}{1 - \alpha} \right) + \frac{|\vec{n}|}{1 - \alpha} \right) M_{\vec{n}}^{\alpha,\vec{\beta}}(x)$$

$$+ \alpha \sum_{i=1}^{r} \frac{n_i (\beta_i + n_i - 1)}{(1 - \alpha)^2} \prod_{j \neq i}^{r} \frac{n_i + \beta_i - \beta_j}{n_i - n_j + \beta_i - \beta_j} M_{\vec{n}-\vec{e}_i}^{\alpha,\vec{\beta}}(x). \tag{20}$$

Note that each relation (19) and (20) involve r relations of nearest-neighbor polynomials. Moreover, each family of multiple Meixner polynomials $M_{\vec{n}}^{\vec{\alpha},\beta}(x)$ and $M_{\vec{n}}^{\alpha,\vec{\beta}}(x)$ forms common eigenfunctions of the above two linear difference operators of order $(r+1)$, namely (17)–(20), respectively.

3. Multiple Meixner Polynomials on a Non-Uniform Lattice

Some algebraic properties will be studied in this section: The Rodrigues-type formula, some recurrence relations and the difference equations with respect to the independent discrete variable $x(s)$. For the q-difference equation (of order $r + 1$) we will proceed as follows. First, we define an r-dimensional subspace $\mathbb{V}$ of polynomials of degree at most $|\vec{n}| - 1$ in the variable $x(s)$ by using some interpolation conditions. Then, we find the lowering operator and express its action on the polynomials as a linear combination of the basis vectors of $\mathbb{V}$. This operator depends on the specific family of multiple orthogonal polynomials, therefore some 'ad hoc' computations are needed. Finally, we combine the lowering and the raising operators to derive the q-difference equation. A similar procedure is given in [31,32,36,39–41]. Finally, the recurrence relations will be derived from some specific difference operators used in Theorems 2 and 4.

3.1. On Some q-Analogues of Multiple Meixner Polynomials of the First Kind

Consider the following vector measure $\vec{\mu}_q$ with positive q-discrete components on $\mathbb{R}^+$,

$$\mu_i = \sum_{s=0}^{\infty} \omega_i(k)\delta(k-s), \qquad \omega_i > 0, \qquad i = 1,2,\ldots,r. \tag{21}$$

Here $\omega_i(s) = v_q^{\alpha_i,\beta}(s) \triangle x(s - 1/2)$, and

$$v_q^{\alpha_i,\beta}(s) = \begin{cases} \dfrac{\alpha_i^s \Gamma_q(\beta+s)}{\Gamma_q(s+1)}, & \text{if } s \in \mathbb{R}^+ \cup \{0\}, \\ 0, & \text{otherwise,} \end{cases} \tag{22}$$

where $0 < \alpha_i < 1$, $\beta > 0$, $i = 1,2,\ldots,r$, and with all the α_i different.

The system of measures $\mu_1, \mu_2, \ldots, \mu_r$ given in (21) forms an AT system on $\mathbb{R}^+$ (see Lemma 9).

Definition 2. *A polynomial $M_{q,\vec{n}}^{\vec{\alpha},\beta}(s)$, with multi-index $\vec{n} \in \mathbb{N}^r$ and degree $|\vec{n}|$, that verifies the orthogonality conditions*

$$\sum_{s=0}^{\infty} M_{q,\vec{n}}^{\vec{\alpha},\beta}(s)[s]_q^{(k)} v_q^{\alpha_i,\beta}(s) \triangle x(s-1/2) = 0, \qquad 0 \le k \le n_i - 1, \qquad i = 1,\ldots,r, \tag{23}$$

is said to be the q-Meixner multiple orthogonal polynomial of the first kind. See also (4) with respect to measure (21).

Notice that for $r = 1$ we recover the scalar q-Meixner polynomials given in [35] and that the orthogonality conditions (4) have been written more conveniently as (23), in which the monomials $x(s)^k$ were replaced by $[s]_q^{(k)}$. In addition, because we have an AT-system of positive discrete measures the q-Meixner multiple orthogonal polynomial of the first kind $M_{q,\vec{n}}^{\vec{\alpha},\beta}(s)$ has exactly $|\vec{n}|$ different zeros on $\mathbb{R}^+$ (see [28], theorem 2.1, pp. 26–27). Finally, in Section 4 we will recover the multiple Meixner polynomials of the first kind given in [28] as a limiting case of $M_{q,\vec{n}}^{\vec{\alpha},\beta}(s)$.

Let us replace $[s]_q^{(k)}$ in (23) by

$$[s]_q^{(k)} = \frac{q^{k-1/2}}{[k+1]_q^{(1)}} \nabla [s+1]_q^{(k+1)}, \tag{24}$$

then, we have

$$\sum_{s=0}^{\infty} M_{q,\vec{n}}^{\vec{\alpha},\beta}(s)\nabla [s+1]_q^{(k+1)} v_q^{\alpha_i,\beta}(s) \triangle x(s-1/2) = 0, \qquad 0 \le k \le n_i - 1, \qquad i = 1,\ldots,r.$$

Using summation by parts and condition $v_q^{\alpha_i}(-1) = v_q^{\alpha_i}(\infty) = 0$, we have that for any two polynomials ϕ and ψ in the variable $x(s)$,

$$\sum_{s=0}^{\infty} \triangle\phi(s)\psi(s)v_q^{\alpha_i,\beta}(s) \nabla x_1(s) = -\sum_{s=0}^{\infty} \phi(s)\nabla\left(\psi(s)v_q^{\alpha_i,\beta}(s)\right) \triangle x(s-1/2). \tag{25}$$

Thus, the following relation

$$\sum_{s=0}^{\infty} \nabla \left(M_{q,\vec{n}}^{\vec{\alpha},\beta}(s) v_q^{\alpha_i,\beta}(s) \right) [s]_q^{(k+1)} \triangle x(s-1/2) = -\sum_{s=0}^{\infty} M_{q,\vec{n}}^{\vec{\alpha},\beta}(s) v_q^{\alpha_i,\beta}(s) \Delta [s]_q^{(k+1)} \triangle x(s-1/2)$$

$$= -\sum_{s=0}^{\infty} M_{q,\vec{n}}^{\vec{\alpha},\beta}(s) v_q^{\alpha_i,\beta}(s) \nabla [s+1]_q^{(k+1)} \triangle x(s-1/2),$$

holds. Equivalently,

$$\sum_{s=0}^{\infty} \nabla \left(M_{q,\vec{n}}^{\vec{\alpha},\beta}(s) v_q^{\alpha_i,\beta}(s) \right) [s]_q^{(k+1)} \triangle x(s-1/2) = 0, \qquad 0 \le k \le n_i - 1, \qquad i = 1, \dots, r.$$

Observe that

$$\nabla \left(M_{q,\vec{n}}^{\vec{\alpha},\beta}(s) v_q^{\alpha_i,\beta}(s) \right) = q^{-|\vec{n}|+1/2} \frac{c_{q,\vec{n}}^{\alpha_i,\beta-1}}{\alpha_i x(\beta-1)} v_q^{\alpha_i/q,\beta-1}(s) \mathcal{Q}_{q,\vec{n}+\vec{e}_i}(s),$$

where

$$c_{q,\vec{n}}^{\alpha_i,\beta} = \left(\alpha_i q^{|\vec{n}|+\beta} - 1 \right). \tag{26}$$

This coefficient will be extensively used throughout the paper and $\mathcal{Q}_{q,\vec{n}+\vec{e}_i}(s)$ represents a monic polynomial $x^{|\vec{n}|+1}$ + lower degree terms. Consequently,

$$\sum_{s=0}^{\infty} \mathcal{Q}_{q,\vec{n}+\vec{e}_i}(s) v_q^{\alpha_i/q,\beta-1}(s) [s]_q^{(k+1)} \triangle x(s-1/2) = \sum_{s=0}^{\infty} \nabla \left(M_{q,\vec{n}}^{\vec{\alpha},\beta}(s) v_q^{\alpha_i,\beta}(s) \right) [s]_q^{(k+1)} \triangle x(s-1/2) = 0. \tag{27}$$

From the next Lemma 1 we will conclude that $\mathcal{Q}_{q,\vec{n}+\vec{e}_i}(s) = M_{q,\vec{n}+\vec{e}_i}^{\alpha_1,\dots,\alpha_i/q,\dots,\alpha_r,\beta-1}(s)$.

Lemma 1. *Let the vector subspace $\mathbb{W} \subset \mathbb{P}$ of polynomials $W(s)$ of degree at most $|\vec{n}| + 1$ in the variable $x(s)$ be defined by conditions*

$$\sum_{s=0}^{\infty} W(s) [s]_q^{(k)} v_q^{\alpha_j/q,\beta-1}(s) \triangledown x_1(s) = 0, \qquad 0 \le k \le n_j, \qquad j = 1, \dots, r,$$

$$W(-1) \ne 0.$$

Then, the spanning set of the system $\left\{ M_{q,\vec{n}+\vec{e}_j}^{\vec{\alpha}_{j,1/q},\beta-1}(s) \right\}_{j=1}^{r}$ coincides with $\mathbb{W}$ (see notation (12) for the index $\vec{\alpha}_{i,1/q}$).

Proof. The polynomials $M_{q,\vec{n}+\vec{e}_j}^{\vec{\alpha}_{j,1/q},\beta-1}(-1) \ne 0$, $j = 1, \dots, r$, because they have exactly $|\vec{n}| + 1$ different zeros on $\mathbb{R}^+$. Moreover, from orthogonality relations

$$\sum_{s=0}^{\infty} M_{q,\vec{n}+\vec{e}_j}^{\vec{\alpha}_{j,1/q},\beta-1}(s) [s]_q^{(k)} v_q^{\alpha_j/q,\beta-1}(s) \triangledown x_1(s) = 0, \qquad 0 \le k \le n_j, \qquad j = 1, \dots, r,$$

we have that the system of polynomials $M_{q,\vec{n}+\vec{e}_j}^{\vec{\alpha}_{j,1/q},\beta-1}(s)$, $j = 1, \dots, r$, belongs to $\mathbb{W}$.

Assume that there exist numbers λ_j, $j = 1, \dots, r$, such that

$$\sum_{j=1}^{r} \lambda_j M_{q,\vec{n}+\vec{e}_j}^{\vec{\alpha}_{j,1/q},\beta-1}(s) = 0, \qquad \text{where} \qquad \sum_{j=1}^{r} |\lambda_j| > 0. \tag{28}$$

Multiplying the previous equation by $[s]_q^{(n_k-1)} v_q^{\alpha_k,\beta-1}(s) \bigtriangledown x_1(s)$ and then summing from $s = 0$ to ∞, one gets

$$\sum_{j=1}^{r} \lambda_j \sum_{s=0}^{\infty} M_{q,\vec{n}+\vec{e}_j}^{\vec{\alpha}_{j,1/q},\beta-1}(s)[s]_q^{(n_k-1)} v_q^{\alpha_k,\beta-1}(s) \bigtriangledown x_1(s) = 0.$$

Thus, from relations

$$\sum_{s=0}^{\infty} M_{q,\vec{n}+\vec{e}_j}^{\vec{\alpha}_{j,1/q},\beta-1}(s)[s]_q^{(n_k-1)} v_q^{\alpha_k,\beta-1}(s) \bigtriangledown x_1(s) = c\delta_{j,k}, \qquad c \in \mathbb{R} \setminus \{0\}, \tag{29}$$

one concludes that $\lambda_k = 0$ for $k = 1,\ldots,r$. Here $\delta_{j,k}$ denotes the Kronecker delta symbol. Thus, the assumption (28) is false, so the system $\left\{ M_{q,\vec{n}+\vec{e}_j}^{\vec{\alpha}_{j,1/q},\beta-1}(s) \right\}_{j=1}^{r}$ is linearly independent in $\mathbb{W}$. Moreover, we know that any polynomial from vector subspace $\mathbb{W}$ is determined by its $|\vec{n}| + 2$ coefficients while $(|\vec{n}| + 2 + r)$ conditions are imposed on $\mathbb{W}$. Consequently the dimension of $\mathbb{W}$ is at most r. Therefore, span $\left\{ M_{q,\vec{n}+\vec{e}_i}^{\vec{\alpha}_{i,1/q},\beta-1}(s) \right\}_{i=1}^{r} = \mathbb{W}$. $\square$

From Equation (27) and Lemma 1 we have

$$\bigtriangledown \left(M_{q,\vec{n}}^{\vec{\alpha},\beta}(s) v_q^{\alpha_i,\beta}(s) \right) = q^{-|\vec{n}|+1/2} \frac{c_{q,\vec{n}}^{\alpha_i,\beta-1}}{\alpha_i x(\beta-1)} v_q^{\alpha_i/q,\beta-1}(s) M_{q,\vec{n}+\vec{e}_i}^{\alpha_1,\ldots,\alpha_i/q,\ldots,\alpha_r,\beta-1}(s).$$

Then, for monic q-Meixner multiple orthogonal polynomials of the first kind we have r raising operators

$$\mathcal{D}_q^{\alpha_i,\beta} M_{q,\vec{n}}^{\vec{\alpha},\beta}(s) = -q^{1/2} M_{q,\vec{n}+\vec{e}_i}^{\vec{\alpha}_{i,1/q},\beta-1}(s), \qquad i = 1,\ldots,r, \tag{30}$$

where

$$\mathcal{D}_q^{\alpha_i,\beta} \stackrel{\text{def}}{=} -\frac{\alpha_i x (\beta-1)}{q^{-|\vec{n}|} c_{q,\vec{n}}^{\alpha_i,\beta-1}} \left(\frac{1}{v_q^{\alpha_i/q,\beta-1}(s)} \bigtriangledown v_q^{\alpha_i,\beta}(s) \right).$$

Furthermore,

$$\mathcal{D}_q^{\alpha_i,\beta} f(s) = \frac{q^{|\vec{n}|+1/2}}{c_{q,\vec{n}}^{\alpha_i,\beta-1}} \left(\left(\alpha_i q^{\beta-1} \left(x(1-\beta) - x(s) \right) + x(s) \right) \mathcal{I} - x(s) \bigtriangledown \right) f(s),$$

for any function $f(s)$ defined on the discrete variable s. Here $\mathcal{I}$ denotes the identity operator. We call $\mathcal{D}_q^{\alpha_i,\beta}$ a raising operator since the i-th component of the multi-index $\vec{n}$ in (30) is increased by 1.

In the sequel we will only consider monic q-Meixner multiple orthogonal polynomials of the first kind.

Proposition 1. *The following q-analogue of Rodrigues-type formula holds:*

$$M_{q,\vec{n}}^{\vec{\alpha},\beta}(s) = \mathcal{G}_q^{\vec{n},\vec{\alpha},\beta} \frac{\Gamma_q(\beta)\Gamma_q(s+1)}{\Gamma_q(\beta+s)} \mathcal{M}_{q,\vec{n}}^{\vec{\alpha}} \left(\frac{\Gamma_q(\beta+|\vec{n}|+s)}{\Gamma_q(\beta+|\vec{n}|)\Gamma_q(s+1)} \right), \tag{31}$$

where

$$\mathcal{M}_{q,\vec{n}}^{\vec{\alpha}} = \prod_{i=1}^{r} \mathcal{M}_{q,n_i}^{\alpha_i}, \quad \mathcal{M}_{q,n_i}^{\alpha_i} = (\alpha_i)^{-s} \nabla^{n_i} (\alpha_i q^{n_i})^s, \tag{32}$$

and

$$\mathcal{G}_q^{\vec{n},\vec{\alpha},\beta} = (-1)^{|\vec{n}|} [-\beta]_q^{(|\vec{n}|)} q^{-\frac{|\vec{n}|}{2}} \left(\prod_{i=1}^{r} \frac{\alpha_i^{n_i} \prod_{j=1}^{n_i} q^{|\vec{n}|_i + \beta + j - 1}}{\prod_{j=1}^{n_i} (\alpha_i q^{|\vec{n}| + \beta + j - 1} - 1)} \right) \left(\prod_{i=1}^{r} q^{n_i \sum_{j=i}^{r} n_j} \right), \tag{33}$$

with $|\vec{n}|_i = n_1 + \cdots + n_{i-1}$, $|\vec{n}|_1 = 0$.

Proof. For $i = 1, \ldots, r$, applying k_i-times the raising operators (30) in a recursive way one obtains

$$\prod_{i=1}^{r} \left(\frac{\alpha_i}{q^{k_i}} \right)^{-s} \nabla^{k_i} \alpha_i^s \frac{\Gamma_q(\beta + s)}{\Gamma_q(\beta)\Gamma_q(s+1)} M_{q,\vec{n}}^{\vec{\alpha},\beta}(s) = [\beta - 1]_q^{(|\vec{k}|)} q^{|\vec{k}|/2} \left(\prod_{i=1}^{r} \frac{\prod_{j=1}^{k_i} \left(\alpha_i q^{|\vec{n}| + \beta - j} - 1 \right)}{\alpha_i^{k_i}} \right)$$

$$\times \prod_{i=1}^{r} q^{-n_i \sum_{j=i}^{r} k_j} \prod_{i=1}^{r-1} q^{-k_i \sum_{j=i+1}^{r} k_j} M_{q,\vec{n}+\vec{k}}^{\alpha_1/q^{k_1},\ldots,\alpha_r/q^{k_r}, \beta - |\vec{k}|}(s) \frac{\Gamma_q(\beta - |\vec{k}| + s)}{\Gamma_q(\beta - |\vec{k}|)\Gamma_q(s+1)}.$$

Taking $n_1 = n_2 = \cdots = n_r = 0$ and replacing β by $\beta + |\vec{k}|$, α_i by $\alpha_i q^{k_i}$, and k_i by n_i, for $i = 1, \ldots, r$, yields the Formula (31). $\square$

3.2. q-Difference Equation for the q-Analogue of Multiple Meixner Polynomials of the First Kind

We will find a lowering operator for the q-Meixner multiple orthogonal polynomials of the first kind. We will follow a similar strategy used in [32].

Lemma 2. *Let* $\mathbb{V}$ *be the linear subspace of polynomials* $Q(s)$ *on the lattice* $x(s)$ *of degree at most* $|\vec{n}| - 1$ *defined by the following conditions*

$$\sum_{s=0}^{\infty} Q(s)[s]_q^{(k)} v_q^{q\alpha_j, \beta+1}(s) \triangledown x_1(s) = 0, \qquad 0 \le k \le n_j - 2 \qquad and \qquad j = 1, \ldots, r.$$

Then, the system $\{ M_{q,\vec{n}-\vec{e}_i}^{\vec{\alpha}_{i,q}, \beta+1}(s) \}_{i=1}^{r}$, *where* $\vec{\alpha}_{i,q} = (\alpha_1, \ldots, q\alpha_i, \ldots, \alpha_r)$, *is a basis for* $\mathbb{V}$.

Proof. From orthogonality relations

$$\sum_{s=0}^{\infty} M_{q,\vec{n}-\vec{e}_j}^{\vec{\alpha}_{j,q}, \beta+1}(s)[s]_q^{(k)} v_q^{q\alpha_j, \beta+1}(s) \triangledown x_1(s) = 0, \qquad 0 \le k \le n_j - 2, \qquad j = 1, \ldots, r,$$

we have that polynomials $M_{q,\vec{n}-\vec{e}_i}^{\vec{\alpha}_{i,q}, \beta+1}(s)$, $i = 1, \ldots, r$, belong to $\mathbb{V}$.

Now, aimed to get a contradiction, let us assume that there exist constants λ_i, $i = 1, \ldots, r$, such that

$$\sum_{i=1}^{r} \lambda_i M_{q,\vec{n}-\vec{e}_i}^{\vec{\alpha}_{i,q}, \beta+1}(s) = 0, \qquad where \qquad \sum_{i=1}^{r} |\lambda_i| > 0.$$

Then, multiplying the previous equation by $[s]_q^{(n_k-1)} v_q^{\alpha_k,\beta}(s) \bigtriangledown x_1(s)$ and then taking summation on s from 0 to ∞, one gets

$$\sum_{i=1}^{r} \lambda_i \sum_{s=0}^{\infty} M_{q,\vec{n}-\vec{e}_i}^{\vec{\alpha}_{i,q},\beta+1}(s)[s]_q^{(n_k-1)} v_q^{\alpha_k,\beta}(s) \bigtriangledown x_1(s) = 0.$$

Thus, from relations

$$\sum_{s=0}^{\infty} M_{q,\vec{n}-\vec{e}_i}^{\vec{\alpha}_{i,q},\beta+1}(s)[s]_q^{(n_k-1)} v_q^{\alpha_k,\beta}(s) \bigtriangledown x_1(s) = c\delta_{i,k}, \qquad c \in \mathbb{R} \setminus \{0\}, \tag{34}$$

we deduce that $\lambda_k = 0$ for $k = 1,\ldots,r$. Here $\delta_{i,k}$ represents the Kronecker delta symbol. Therefore, the vectors $\left\{ M_{q,\vec{n}-\vec{e}_i}^{\vec{\alpha}_{i,q},\beta+1}(s) \right\}_{i=1}^{r}$ are linearly independent in $\mathbb{V}$. Furthermore, we know that any polynomial of $\mathbb{V}$ can be determined with $|\vec{n}|$ coefficients while $(|\vec{n}| - r)$ linear conditions are imposed on $\mathbb{V}$. Consequently the dimension of $\mathbb{V}$ is at most r. Hence, the system $\left\{ M_{q,\vec{n}-\vec{e}_i}^{\vec{\alpha}_{i,q},\beta+1}(s) \right\}_{i=1}^{r}$ spans $\mathbb{V}$, which completes the proof. $\square$

Now we will prove that the operator (8) is indeed a lowering operator for the sequence of q-Meixner multiple orthogonal polynomials of the first kind $M_{q,\vec{n}}^{\vec{\alpha},\beta}(s)$.

Lemma 3. *The following relation holds:*

$$\Delta M_{q,\vec{n}}^{\vec{\alpha},\beta}(s) = \sum_{i=1}^{r} q^{|\vec{n}|-n_i+1/2} \frac{1-\alpha_i q^{n_i+\beta}}{1-\alpha_i q^{|\vec{n}|+\beta}} [n_i]_q^{(1)} M_{q,\vec{n}-\vec{e}_i}^{\vec{\alpha}_{i,q},\beta+1}(s). \tag{35}$$

Proof. Using summation by parts we have

$$\sum_{s=0}^{\infty} \Delta M_{q,\vec{n}}^{\vec{\alpha},\beta}(s)[s]_q^{(k)} v_q^{q\alpha_j,\beta+1}(s) \bigtriangledown x_1(s) = -\sum_{s=0}^{\infty} M_{q,\vec{n}}^{\vec{\alpha},\beta}(s) \nabla ([s]_q^{(k)} v_q^{q\alpha_j,\beta+1}(s)) \bigtriangledown x_1(s)$$

$$= -\sum_{s=0}^{\infty} M_{q,\vec{n}}^{\vec{\alpha},\beta}(s) \varphi_{j,k}(s) v_q^{\alpha_j,\beta}(s) \bigtriangledown x_1(s), \tag{36}$$

where

$$\varphi_{j,k}(s) = q^{1/2} \left(\frac{q^\beta x(s)}{x(\beta)} + 1 \right) [s]_q^{(k)} - q^{-1/2} \frac{x(s)}{\alpha_j x(\beta)} [s-1]_q^{(k)},$$

is a polynomial of degree $\leq k+1$ in the variable $x(s)$. Consequently, from the orthogonality conditions (23) we get

$$\sum_{s=0}^{\infty} \Delta M_{q,\vec{n}}^{\vec{\alpha},\beta}(s)[s]_q^{(k)} v_q^{q\alpha_j,\beta+1}(s) \bigtriangledown x_1(s) = 0, \qquad 0 \leq k \leq n_j-2, \qquad j=1,\ldots,r.$$

Hence, from Lemma 2, $\Delta M_{q,\vec{n}}^{\vec{\alpha},\beta}(s) \in \mathbb{V}$. Moreover, $\Delta M_{q,\vec{n}}^{\vec{\alpha},\beta}(s)$ can be expressed as a linear combination of polynomials $\{M_{q,\vec{n}-\vec{e}_i}^{\vec{\alpha}_{i,q},\beta+1}(s)\}_{i=1}^{r}$, i.e.,

$$\Delta M_{q,\vec{n}}^{\vec{\alpha},\beta}(s) = \sum_{i=1}^{r} \xi_i M_{q,\vec{n}-\vec{e}_i}^{\vec{\alpha}_{i,q},\beta+1}(s), \qquad \sum_{i=1}^{r} |\xi_i| > 0. \tag{37}$$

Multiplying both sides of the Equation (37) by $[s]_q^{(n_k-1)} v_q^{q\alpha_k,\beta+1}(s) \, \triangledown \, x_1(s)$ and using relations (34) one has

$$\sum_{s=0}^{\infty} \Delta M_{q,\vec{n}}^{\vec{\alpha},\beta}(s)[s]_q^{(n_k-1)} v_q^{q\alpha_k,\beta+1}(s) \, \triangledown \, x_1(s) = \sum_{i=1}^{r} \xi_i \sum_{s=0}^{\infty} M_{q,\vec{n}-\vec{e}_i}^{\vec{\alpha}_{i,q},\beta+1}(s)[s]_q^{(n_k-1)} v_q^{q\alpha_k,\beta+1}(s) \, \triangledown \, x_1(s)$$

$$= \xi_k \sum_{s=0}^{\infty} M_{q,\vec{n}-\vec{e}_k}^{\vec{\alpha}_{k,q},\beta+1}(s)[s]_q^{(n_k-1)} v_q^{q\alpha_k,\beta+1}(s) \, \triangledown \, x_1(s). \quad (38)$$

If we replace $[s]_q^{(k)}$ by $[s]_q^{(n_k-1)}$ in the left-hand side of Equation (36), then Equation (38) transforms into

$$\sum_{s=0}^{\infty} \Delta M_{q,\vec{n}}^{\vec{\alpha},\beta}(s)[s]_q^{(n_k-1)} v_q^{q\alpha_k,\beta+1}(s) \, \triangledown \, x_1(s) = -\sum_{s=0}^{\infty} M_{q,\vec{n}}^{\vec{\alpha},\beta}(s)\varphi_{k,n_k-1}(s) v_q^{\alpha_k,\beta}(s) \, \triangledown \, x_1(s)$$

$$= \frac{q^{-1/2}(1-\alpha_k q^{n_k+\beta})}{\alpha_k x(\beta)} \sum_{s=0}^{\infty} M_{q,\vec{n}}^{\vec{\alpha},\beta}(s)[s]_q^{(n_k)} v_q^{\alpha_k,\beta}(s) \, \triangledown \, x_1(s). \quad (39)$$

For this transformation we have used that $x(s)[s-1]_q^{(n_k-1)} = [s]_q^{(n_k)}$ to get

$$\varphi_{k,n_k-1}(s) = -\frac{q^{-1/2}(1-\alpha_k q^{n_k+\beta})}{\alpha_k x(\beta)}[s]_q^{(n_k)} + \text{lower degree terms}.$$

On the other hand, from (30) one has that

$$\frac{q^{-1/2}(1-\alpha_k q^{|\vec{n}|+\beta})}{\alpha_k x(\beta)} v_q^{\alpha_k,\beta}(s) M_{q,\vec{n}}^{\vec{\alpha},\beta}(s) = -q^{|\vec{n}|-1/2}\triangledown\left(v_q^{q\alpha_k,\beta+1}(s) M_{q,\vec{n}-\vec{e}_k}^{\vec{\alpha}_{k,q},\beta+1}(s)\right). \quad (40)$$

Considering (40) and using once more summation by parts on the right-hand side of Equation (39) we obtain

$$\sum_{s=0}^{\infty} \Delta M_{q,\vec{n}}^{\vec{\alpha},\beta}(s)[s]_q^{(n_k-1)} v_q^{q\alpha_k,\beta+1}(s) \, \triangledown \, x_1(s)$$

$$= -q^{|\vec{n}|-1}\frac{1-\alpha_k q^{n_k+\beta}}{1-\alpha_k q^{|\vec{n}|+\beta}} \sum_{s=0}^{\infty}[s]_q^{(n_k)}\triangledown\left(v_q^{q\alpha_k,\beta+1}(s) M_{q,\vec{n}-\vec{e}_k}^{\vec{\alpha}_{k,q},\beta+1}(s)\right) \, \triangledown \, x_1(s)$$

$$= q^{|\vec{n}|-1}\frac{1-\alpha_k q^{n_k+\beta}}{1-\alpha_k q^{|\vec{n}|+\beta}} \sum_{s=0}^{\infty} M_{q,\vec{n}-\vec{e}_k}^{\vec{\alpha}_{k,q},\beta+1}(s)\left(\Delta[s]_q^{(n_k)}\right) v_q^{q\alpha_k,\beta+1}(s) \, \triangledown \, x_1(s).$$

Since $\Delta[s]_q^{(n_k)} = q^{3/2-n_k}[n_k]_q^{(1)}[s]_q^{(n_k-1)}$, we have

$$\sum_{s=0}^{\infty} \Delta M_{q,\vec{n}}^{\vec{\alpha},\beta}(s)[s]_q^{(n_k-1)} v_q^{q\alpha_k,\beta+1}(s) \, \triangledown \, x_1(s)$$

$$= q^{|\vec{n}|-n_k+1/2}\frac{1-\alpha_k q^{n_k+\beta}}{1-\alpha_k q^{|\vec{n}|+\beta}}[n_k]_q^{(1)} \sum_{s=0}^{\infty} M_{q,\vec{n}-\vec{e}_k}^{\vec{\alpha}_{k,q},\beta+1}(s)[s]_q^{(n_k-1)} v_q^{q\alpha_k,\beta+1}(s) \, \triangledown \, x_1(s).$$

Comparing this equation with (38), we obtain the coefficients in the expansion (37), i.e.,

$$\xi_k = q^{|\vec{n}|-n_k+1/2}\frac{1-\alpha_k q^{n_k+\beta}}{1-\alpha_k q^{|\vec{n}|+\beta}}[n_k]_q^{(1)}.$$

Therefore, relation (35) holds. □

Theorem 1. *The q-Meixner multiple orthogonal polynomial of the first kind $M_{q,\vec{n}}^{\vec{\alpha},\beta}(s)$ satisfies the following $(r+1)$-order q-difference equation*

$$\prod_{i=1}^{r}\mathcal{D}_q^{q\alpha_i,\beta+1}\Delta M_{q,\vec{n}}^{\vec{\alpha},\beta}(s) = -\sum_{i=1}^{r}q^{|\vec{n}|-n_i+1}\frac{1-\alpha_i q^{n_i+\beta}}{1-\alpha_i q^{|\vec{n}|+\beta}}[n_i]_q^{(1)}\prod_{\substack{j=1\\j\neq i}}^{r}\mathcal{D}_q^{q\alpha_j,\beta+1}M_{q,\vec{n}}^{\vec{\alpha},\beta}(s). \tag{41}$$

Proof. Since the operators (30) commute, we write

$$\prod_{i=1}^{r}\mathcal{D}_q^{q\alpha_i,\beta+1} = \left(\prod_{\substack{j=1\\j\neq i}}^{r}\mathcal{D}_q^{q\alpha_j,\beta+1}\right)\mathcal{D}_q^{q\alpha_i,\beta+1}. \tag{42}$$

Using (30) when acting on Equation (35) with the product of operators (42), we obtain (41), i.e.,

$$\prod_{i=1}^{r}\mathcal{D}_q^{q\alpha_i,\beta+1}\Delta M_{q,\vec{n}}^{\vec{\alpha},\beta}(s) = \sum_{i=1}^{r}q^{|\vec{n}|-n_i+1/2}\frac{1-\alpha_i q^{n_i+\beta}}{1-\alpha_i q^{|\vec{n}|+\beta}}[n_i]_q^{(1)}\prod_{\substack{j=1\\j\neq i}}^{r}\mathcal{D}_q^{q\alpha_j,\beta+1}\left(\mathcal{D}_q^{q\alpha_i,\beta+1}M_{q,\vec{n}-\vec{e}_i}^{\vec{\alpha}_{i,q},\beta+1}(s)\right)$$

$$= -\sum_{i=1}^{r}q^{|\vec{n}|-n_i+1}\frac{1-\alpha_i q^{n_i+\beta}}{1-\alpha_i q^{|\vec{n}|+\beta}}[n_i]_q^{(1)}\prod_{\substack{j=1\\j\neq i}}^{r}\mathcal{D}_q^{q\alpha_j,\beta+1}M_{q,\vec{n}}^{\vec{\alpha},\beta}(s).$$

This completes the proof of the theorem. □

3.3. Recurrence Relation for q-Meixner Multiple Orthogonal Polynomials of the First Kind

In this section we will study the nearest neighbor recurrence relation for any multi-index $\vec{n}$. The approach presented here differs from those used in [28,42]. We begin by defining the following linear difference operator

$$\mathcal{F}_{q,n_i} := g_{q,i}^{-1}(s)\nabla^{n_i}g_{q,k}(s), \tag{43}$$

where n_i is the i-th entry of the vector index $\vec{n}$ and $g_{q,k}$ is defined in the variable s and depends on the i-th component of the vector orthogonality measure $\vec{\mu}$. In the case that $g_{q,k}$ depends also on the i-th component of $\vec{n}$, then the index $k = n_i$; otherwise $k = i$.

Lemma 4. *Let n_i be a positive integer and let $f(s)$ be a function defined on the discrete variable s. The following relation is valid*

$$\mathcal{F}_{q,n_i}x(s)f_q(s) = q^{-n_i+1/2}x(n_i)g_{q,i}^{-1}(s)\nabla^{n_i-1}g_{q,k}(s)f_q(s) + q^{-n_i}(x(s)-x(n_i))\mathcal{F}_{q,n_i}f_q(s). \tag{44}$$

Proof. Let us act n_i-times with backward difference operators (9) on the product of functions $x(s)f(s)$. Assume that $n_i \geq N > 1$,

$$\nabla^{n_i}x(s)f(s) = \nabla^{n_i-1}(\nabla x(s)f(s)) = \nabla^{n_i-1}(q^{-1/2}f(s)+x(s-1)\nabla f(s))$$

$$= q^{-1/2}\nabla^{n_i-1}f(s) + \nabla^{n_i-1}(x(s-1)\nabla f(s))$$

$$= q^{-1/2}\nabla^{n_i-1}f(s) + \nabla^{n_i-2}(\nabla x(s-1)\nabla f(s)). \tag{45}$$

Repeating this process, but on the second term of the right-hand side of Equation (45)

$$\nabla^{n_i}x(s)f(s) = (q^{1/2-n_i}+\cdots+q^{-5/2}+q^{-3/2}+q^{-1/2})\nabla^{n_i-1}f(s) + x(s-n_i)\nabla^{n_i}f(s)$$

$$= q^{1/2-n_i}x(n_i)\nabla^{n_i-1}f(s) + x(s-n_i)\nabla^{n_i}f(s).$$

Thus,

$$\nabla^{n_i} x(s) f(s) = q^{-n_i+1/2} x(n_i) \nabla^{n_i-1} f(s) + q^{-n_i} \left(x(s) - x(n_i) \right) \nabla^{n_i} f(s), \qquad n_i \geq 1. \tag{46}$$

Now, to involve the difference operator $\mathcal{F}_{q,n_i}$ in the above equation, we multiply the Equation (46) from the left by $g_{q,i}(s)^{-1}$ and replace $f(s)$ by $g_{q,k}(s)f(s)$. Therefore, the Equation (46) transforms into (44). $\square$

Theorem 2. *The q-Meixner multiple orthogonal polynomials of the first kind satisfy the following $(r+2)$-term recurrence relation*

$$x(s) M_{q,\vec{n}}^{\vec{\alpha},\beta}(s) = M_{q,\vec{n}+\vec{e}_k}^{\vec{\alpha},\beta}(s) + b_{\vec{n},k} M_{q,\vec{n}}^{\vec{\alpha},\beta}(s) + \sum_{i=1}^{r} \frac{x(n_i) \alpha_i q^{|\vec{n}|+n_i-1} x(\beta+|\vec{n}|-1)}{c_{q,\vec{n}+n_i\vec{e}_i}^{\alpha_i,\beta-1} c_{q,\vec{n}+n_i\vec{e}_i}^{\alpha_i,\beta-2}} B_{\vec{n},i} M_{q,\vec{n}-\vec{e}_i}^{\vec{\alpha},\beta}(s), \tag{47}$$

where

$$b_{\vec{n},k} = -\alpha_k q^{|\vec{n}|+n_k+1} \frac{x(\beta+|\vec{n}|)}{c_{q,\vec{n}+n_k\vec{e}_k}^{\alpha_k,\beta+1}} + (q-1) \prod_{i=1}^{r} \frac{x(n_i)}{c_{q,\vec{n}+n_i\vec{e}_i}^{\alpha_i,\beta}} \left(q^{|\vec{n}|+\beta} \prod_{i=1}^{r} \alpha_i q^{n_i} - 1 \right)$$

$$+ \sum_{i=1}^{r} \frac{x(n_i)}{q^{-|\vec{n}|}} \left(\frac{\alpha_i q^{n_i} - 1}{c_{q,\vec{n}}^{\alpha_i,\beta}} \prod_{i=1}^{r} \frac{c_{q,\vec{n}}^{\alpha_i,\beta}}{c_{q,\vec{n}+n_i\vec{e}_i}^{\alpha_i,\beta}} - \frac{\alpha_i q^{|\vec{n}|+\beta+n_i-1}}{c_{q,\vec{n}+n_i\vec{e}_i}^{\alpha_i,\beta-1}} \frac{\alpha_i q^{n_i} - 1}{c_{q,\vec{n}+n_i\vec{e}_i}^{\alpha_i,\beta}} \prod_{j\neq i}^{r} \frac{\alpha_i q^{|\vec{n}|} - \alpha_j q^{n_j}}{\alpha_i q^{n_i} - \alpha_j q^{n_j}} \right)$$

and

$$B_{\vec{n},i} = \frac{\alpha_i q^{n_i} - 1}{c_{q,\vec{n}+n_i\vec{e}_i}^{\alpha_i,\beta}} \prod_{j\neq i}^{r} \frac{\alpha_i q^{|\vec{n}|} - \alpha_j q^{n_j}}{\alpha_i q^{n_i} - \alpha_j q^{n_j}} \prod_{i=1}^{r} \frac{c_{q,\vec{n}}^{\alpha_i,\beta-1}}{c_{q,\vec{n}+n_i\vec{e}_i}^{\alpha_i,\beta}}.$$

Proof. Let

$$f_{\mathbf{n}}(s;\beta) = \frac{\Gamma_q(\beta+\mathbf{n}+s)}{\Gamma_q(\beta+\mathbf{n})\Gamma_q(s+1)}, \quad \text{where} \quad \mathbf{n} = |\vec{n}|.$$

We will use Lemma 4 involving this function $f_{\mathbf{n}}(s;\beta)$ as well as difference operator (32). Consider equation

$$(\alpha_k)^{-s} \nabla^{n_k+1} (\alpha_k q^{n_k+1})^s f_{\mathbf{n}+1}(s;\beta) = (\alpha_k)^{-s} \nabla^{n_k} \left(q^{-s+1/2} \nabla \left((\alpha_k q^{n_k+1})^s f_{\mathbf{n}+1}(s;\beta) \right) \right)$$

$$= q^{1/2}(\alpha_k)^{-s} \nabla^{n_k} \left((\alpha_k q^{n_k})^s \left(1 + \frac{c_{q,\vec{n}+n_k\vec{e}_k}^{\alpha_k,\beta+1}}{(\alpha_k q^{n_k+1}) x(\beta+|\vec{n}|)} x(s) \right) f_{\mathbf{n}}(s;\beta) \right),$$

which can be rewritten in terms of difference operators (32) as follows

$$q^{-1/2} \mathcal{M}_{q,n_k+1}^{\alpha_k} f_{\mathbf{n}+1}(s;\beta) = \mathcal{M}_{q,n_k}^{\alpha_k} f_{\mathbf{n}}(s;\beta) + \frac{c_{q,\vec{n}+n_k\vec{e}_k}^{\alpha_k,\beta+1}}{(\alpha_k q^{n_k+1}) x(\beta+|\vec{n}|)} \mathcal{M}_{q,n_k}^{\alpha_k} x(s) f_{\mathbf{n}}(s;\beta). \tag{48}$$

Since operators (32) commute, the multiplication of Equation (48) from the left-hand side by the product $\prod_{\substack{i=1 \\ i\neq k}}^{r} \mathcal{M}_{q,n_i}^{\alpha_i}$ yields the following relation

$$\mathcal{M}_{q,\vec{n}}^{\vec{\alpha}} x(s) f_{\mathbf{n}}(s;\beta) = \frac{(\alpha_k q^{n_k+1}) x(\beta+|\vec{n}|)}{c_{q,\vec{n}+n_k\vec{e}_k}^{\alpha_k,\beta+1}} \left(q^{-1/2} \mathcal{M}_{q,\vec{n}+\vec{e}_k}^{\vec{\alpha}} f_{\mathbf{n}+1}(s;\beta) - \mathcal{M}_{q,\vec{n}}^{\vec{\alpha}} f_{\mathbf{n}}(s;\beta) \right). \tag{49}$$

Let us recursively use Lemma 4 involving the product of r difference operators acting on the function $f_{\mathbf{n}}(s; \beta)$, which in this case is the operator $\mathcal{M}^{\vec{\alpha}}_{q,\vec{n}}$ (see expression (32)). Thus,

$$
\left(q^{|\vec{n}|} \mathcal{M}^{\vec{\alpha}}_{q,\vec{n}} x(s) - q^{1/2} \sum_{i=1}^{r} \prod_{j \neq i}^{r} \frac{\alpha_i q^{|\vec{n}|} - \alpha_j q^{n_j}}{\alpha_i q^{n_i} - \alpha_j q^{n_j}} \frac{x(n_i) c^{\alpha_j,\beta}_{q,\vec{n}+n_j \vec{e}_j}}{(\alpha_i q^{n_i} - 1)^{-1} \prod_{\nu=1}^{r} c^{\alpha_\nu,\beta}_{q,\vec{n}}} \prod_{l=1}^{r} \mathcal{M}^{\alpha_l}_{q,n_l - \delta_{l,i}} \right) f_{\mathbf{n}}(s; \beta)
$$

$$
= \left(x(s) \prod_{i=1}^{r} \frac{c^{\alpha_i,\beta}_{q,\vec{n}+n_i \vec{e}_i}}{c^{\alpha_i,\beta}_{q,\vec{n}}} - \sum_{i=1}^{r} \frac{q^{|\vec{n}|} x(n_i)}{c^{\alpha_i,\beta}_{q,\vec{n}} (\alpha_i q^{n_j} - 1)^{-1}} + \frac{1 - q^{2|\vec{n}|+\beta} \prod_{i=1}^{r} \alpha_i}{(q-1)^{-1}} \prod_{i=1}^{r} \frac{x(n_i)}{c^{\alpha_i,\beta}_{q,\vec{n}}} \right) \mathcal{M}^{\vec{\alpha}}_{q,\vec{n}} f_{\mathbf{n}}(s; \beta). \tag{50}
$$

Using the expressions (49) and (50) one gets

$$
x(s) \mathcal{M}^{\vec{\alpha}}_{q,\vec{n}} f_{\mathbf{n}}(s; \beta) = q^{|\vec{n}|-1/2} \prod_{i=1}^{r} \frac{c^{\alpha_i,\beta}_{q,\vec{n}}}{c^{\alpha_i,\beta}_{q,\vec{n}+n_i \vec{e}_i}} \frac{(\alpha_k q^{n_k+1}) x(\beta + |\vec{n}|)}{c^{\alpha_k,\beta+1}_{q,\vec{n}+n_k \vec{e}_k}} \mathcal{M}^{\vec{\alpha}}_{q,\vec{n}+\vec{e}_k} f_{\mathbf{n}+1}(s; \beta)
$$

$$
+ \prod_{i=1}^{r} \frac{c^{\alpha_i,\beta}_{q,\vec{n}}}{c^{\alpha_i,\beta}_{q,\vec{n}+n_i \vec{e}_i}} \left(\sum_{i=1}^{r} \frac{q^{|\vec{n}|} x(n_i)}{(\alpha_i q^{n_j} - 1)^{-1} c^{\alpha_i,\beta}_{q,\vec{n}}} - \frac{1 - q^{2|\vec{n}|+\beta} \prod_{i=1}^{r} \alpha_i}{(q-1)^{-1}} \prod_{i=1}^{r} \frac{x(n_i)}{c^{\alpha_i,\beta}_{q,\vec{n}}} - \frac{q^{|\vec{n}|} \alpha_k x(\beta + |\vec{n}|)}{q^{-n_k-1} c^{\alpha_k,\beta+1}_{q,\vec{n}+n_k \vec{e}_k}} \right) \mathcal{M}^{\vec{\alpha}}_{q,\vec{n}} f_{\mathbf{n}}(s; \beta)
$$

$$
- q^{1/2} \sum_{i=1}^{r} \prod_{j \neq i}^{r} \frac{\alpha_i q^{|\vec{n}|} - \alpha_j q^{n_j}}{\alpha_i q^{n_i} - \alpha_j q^{n_j}} \frac{x(n_i)(\alpha_i q^{n_i} - 1)}{c^{\alpha_i,\beta}_{q,\vec{n}+n_i \vec{e}_i}} \prod_{l=1}^{r} \mathcal{M}^{\alpha_l}_{q,n_l - \delta_{l,i}} f_{\mathbf{n}}(s; \beta).
$$

Observe that when $l = i$ in the above expression we have

$$
\mathcal{M}^{\alpha_i}_{q,n_i - 1} f_{\mathbf{n}}(s; \beta) = q^{-1/2} \frac{\alpha_i q^{\beta+|\vec{n}|+n_i-1}}{c^{\alpha_i,\beta-1}_{q,\vec{n}+n_i \vec{e}_i}} \mathcal{M}^{\alpha_i}_{q,n_i} f_{\mathbf{n}}(s; \beta) - \frac{1}{c^{\alpha_i,\beta-1}_{q,\vec{n}+n_i \vec{e}_i}} \mathcal{M}^{\alpha_i}_{q,n_i - 1} f_{\mathbf{n}-1}(s; \beta).
$$

Therefore,

$$
x(s) \mathcal{M}^{\vec{\alpha}}_{q,\vec{n}} f_{\mathbf{n}}(s; \beta)
$$

$$
= q^{|\vec{n}|-1/2} \prod_{i=1}^{r} \frac{c^{\alpha_i,\beta}_{q,\vec{n}}}{c^{\alpha_i,\beta}_{q,\vec{n}+n_i \vec{e}_i}} \frac{(\alpha_k q^{n_k+1}) x(\beta + |\vec{n}|)}{c^{\alpha_k,\beta+1}_{q,\vec{n}+n_k \vec{e}_k}} \mathcal{M}^{\vec{\alpha}}_{q,\vec{n}+\vec{e}_k} f_{\mathbf{n}+1}(s; \beta) + b_{\vec{n},k} \mathcal{M}^{\vec{\alpha}}_{q,\vec{n}} f_{\mathbf{n}}(s; \beta)
$$

$$
- q^{1/2} \sum_{i=1}^{r} \prod_{j \neq i}^{r} \frac{\alpha_i q^{|\vec{n}|} - \alpha_j q^{n_j}}{\alpha_i q^{n_i} - \alpha_j q^{n_j}} \frac{x(n_i)(\alpha_i q^{n_i} - 1)}{c^{\alpha_i,\beta}_{q,\vec{n}+n_i \vec{e}_i}} \frac{1}{c^{\alpha_i,\beta-1}_{q,\vec{n}+n_i \vec{e}_i}} \prod_{l=1}^{r} \mathcal{M}^{\alpha_l}_{q,n_l - \delta_{l,i}} f_{\mathbf{n}-1}(s; \beta).
$$

Finally, multiplying from the left both sides of the previous expression by $G^{\vec{n},\vec{\alpha},\beta}_q \frac{\Gamma_q(\beta) \Gamma_q(s+1)}{\Gamma_q(\beta+s)}$ and using Rodrigues-type Formula (31) we obtain (47). This completes the proof of the theorem. $\square$

3.4. On Some q-Analogue of Multiple Meixner Polynomials of the Second Kind

Consider the following vector measure $\vec{v}_q$ with positive q-discrete components

$$
v_i = \sum_{s=0}^{\infty} v^{\alpha,\beta_i}_q(k) \, \triangle x \, (k - 1/2) \, \delta \, (k - s), \quad i = 1, 2, \ldots, r, \tag{51}
$$

where $v_q^{\alpha,\beta_i}(s)$ is defined in (22), but here the domain for its non-identically zero part is $s \in \Omega = \mathbb{R} \setminus \{\mathbb{Z}^- \cup \{-\beta_i, -\beta_i - 1, -\beta_i - 2, \dots\}\}$, $\beta_i > 0$, $\beta_i - \beta_j \notin \mathbb{Z}$ for all $i \neq j$, and $0 < \alpha < 1$. Indeed,

$$v_q^{\alpha,\beta_i}(s) = \begin{cases} \dfrac{\alpha^x \Gamma_q(\beta_i + s)}{\Gamma_q(s+1)}, & \text{if } s \in \Omega, \\ 0, & \text{otherwise.} \end{cases}$$

Definition 3. *A polynomial $M_{q,\vec{n}}^{\alpha,\vec{\beta}}(s)$, with multi-index $\vec{n} \in \mathbb{N}^r$ and degree $|\vec{n}|$ that verifies the orthogonality conditions*

$$\sum_{s=0}^{\infty} M_{q,\vec{n}}^{\alpha,\vec{\beta}}(s) [s]_q^{(k)} v_q^{\alpha,\beta_i}(s) \triangle x(s-1/2) = 0, \qquad 0 \le k \le n_i - 1, \qquad i = 1, \dots, r, \tag{52}$$

is said to be the q-Meixner multiple orthogonal polynomial of the second kind.

The general orthogonality relations (4) have been conveniently written involving the q-analogue of the Stirling polynomials (6) as in relations (52). In Section 5 we will address the AT-property of the system of positive discrete measures (51). This fact guarantees that the q-Meixner multiple orthogonal polynomial of the second kind $M_{q,\vec{n}}^{\alpha,\vec{\beta}}(s)$ has exactly $|\vec{n}|$ different zeros on $\mathbb{R}^+$ (see [28], theorem 2.1, pp. 26–27). In Section 4, the multiple Meixner polynomials of the second kind (16) given in [28] will be recovered as q approaches 1.

To find a raising operator we substitute $[s]_q^{(k)}$ in (52) for the finite-difference expression (24) and then we use summation by parts along with conditions $v_q^{\alpha,\beta_i}(-1) = v_q^{\alpha,\beta_i}(\infty) = 0$. Thus,

$$\sum_{s=0}^{\infty} M_{q,\vec{n}}^{\alpha,\vec{\beta}}(s) \nabla [s+1]_q^{(k+1)} v_q^{\alpha,\beta_i}(s) \triangle x(s-1/2) = 0, \qquad 0 \le k \le n_i - 1, \qquad i = 1, \dots, r.$$

Using (25), one gets

$$\sum_{s=0}^{\infty} \nabla \left(M_{q,\vec{n}}^{\alpha,\vec{\beta}}(s) v_q^{\alpha,\beta_i}(s) \right) [s]_q^{(k+1)} \triangle x(s-1/2) = -\sum_{s=0}^{\infty} M_{q,\vec{n}}^{\alpha,\vec{\beta}}(s) v_q^{\alpha,\beta_i}(s) \Delta [s]_q^{(k+1)} \triangle x(s-1/2)$$

$$= -\sum_{s=0}^{\infty} M_{q,\vec{n}}^{\alpha,\vec{\beta}}(s) v_q^{\alpha,\beta_i}(s) \nabla [s+1]_q^{(k+1)} \triangle x(s-1/2).$$

Hence

$$\sum_{s=0}^{\infty} \nabla \left(M_{q,\vec{n}}^{\alpha,\vec{\beta}}(s) v_q^{\alpha,\beta_i}(s) \right) [s]_q^{(k+1)} \triangle x(s-1/2) = 0, \qquad 0 \le k \le n_i - 1, \qquad i = 1, \dots, r,$$

where

$$\nabla \left(M_{q,\vec{n}}^{\alpha,\vec{\beta}}(s) v_q^{\alpha,\beta_i}(s) \right) = \frac{q^{-|\vec{n}|+1/2} c_{q,\vec{n}}^{\alpha,\beta_i-1}}{\alpha x(\beta_i - 1)} v_q^{\alpha/q,\beta_i-1}(s) \mathcal{P}_{q,\vec{n}+\vec{e}_i}(s).$$

$\mathcal{P}_{q,\vec{n}+\vec{e}_i}(s)$ denotes a monic polynomial of degree $|\vec{n}| + 1$. Therefore, from (52) the relation

$$\sum_{s=0}^{\infty} \mathcal{P}_{q,\vec{n}+\vec{e}_i}(s) v_q^{\alpha/q,\beta_i-1}(s) [s]_q^{(k+1)} \triangle x(s-1/2) = \sum_{s=0}^{\infty} \nabla \left(M_{q,\vec{n}}^{\alpha,\vec{\beta}}(s) v_q^{\alpha,\beta_i}(s) \right) [s]_q^{(k+1)} \triangle x(s-1/2) = 0,$$

implies that $\mathcal{P}_{q,\vec{n}+\vec{e}_i}(s) = M_{q,\vec{n}+\vec{e}_i}^{\alpha/q,\vec{\beta}-\vec{e}_i}(s)$. Therefore

$$\nabla \left(M_{q,\vec{n}}^{\alpha,\vec{\beta}}(s) v_q^{\alpha,\beta_i}(s) \right) = \frac{q^{-|\vec{n}|+1/2} c_{q,\vec{n}}^{\alpha,\beta_i-1}}{\alpha x(\beta_i-1)} v_q^{\alpha/q,\beta_i-1}(s) M_{q,\vec{n}+\vec{e}_i}^{\alpha/q,\vec{\beta}-\vec{e}_i}(s),$$

which leads to the following r raising operators for the monic q-Meixner multiple orthogonal polynomials of the second kind

$$\mathcal{D}_q^{\alpha,\beta_i} M_{q,\vec{n}}^{\alpha,\vec{\beta}}(s) = -q^{1/2} M_{q,\vec{n}+\vec{e}_i}^{\alpha/q,\vec{\beta}-\vec{e}_i}(s). \tag{53}$$

The operator $\mathcal{D}_q^{\alpha,\beta_i}$ is given in (30) with the replacements: α_i by α and β by β_i, respectively. Indeed,

$$\mathcal{D}_q^{\alpha,\beta_i} f(s) = \frac{q^{|\vec{n}|+1/2}}{c_{q,\vec{n}}^{\alpha,\beta_i-1}} \left(\left(\alpha q^{\beta_i-1} \left(x(1-\beta_i) - x(s) \right) + x(s) \right) \mathcal{I} - x(s) \nabla \right) f(s), \tag{54}$$

holds for any function $f(s)$ defined on the discrete variable s.

Proposition 2. *The following finite-difference analogue of the Rodrigues-type formula holds:*

$$M_{q,\vec{n}}^{\alpha,\vec{\beta}}(s) = \mathcal{G}_q^{\vec{n},\vec{\beta},\alpha} \frac{\Gamma_q(s+1)}{\alpha^s} \mathcal{N}_{q,\vec{n}}^{\vec{\beta}} \left(\frac{\left(\alpha q^{|\vec{n}|} \right)^s}{\Gamma_q(s+1)} \right), \tag{55}$$

where

$$\mathcal{N}_{q,\vec{n}}^{\vec{\beta}} = \prod_{i=1}^r \mathcal{N}_{q,n_i}^{\beta_i}, \quad \mathcal{N}_{q,n_i}^{\beta_i} = \frac{\Gamma_q(\beta_i)}{\Gamma_q(\beta_i+s)} \nabla^{n_i} \frac{\Gamma_q(\beta_i+n_i+s)}{\Gamma_q(\beta_i+n_i)}, \tag{56}$$

and

$$\mathcal{G}_q^{\vec{n},\vec{\beta},\alpha} = (-1)^{|\vec{n}|} \left(\alpha q^{|\vec{n}|} \right)^{|\vec{n}|} q^{-\frac{|\vec{n}|}{2}} \left(\prod_{i=1}^r \frac{\prod_{j=1}^{n_i} q^{\beta_i+j-1}}{\prod_{j=1}^{n_i} c_{q,\vec{n}}^{\alpha,\beta_i+j-1}} \right) \left(\prod_{i=1}^r [-\beta_i]_q^{(n_i)} \right). \tag{57}$$

Proof. We follow the same pattern given in Proposition 1 adapted to the operator $\mathcal{N}_{q,\vec{n}}^{\vec{\beta}}$. For $i = 1, \ldots, r$, by applying k_i-times the raising operators (53) in a recursive way, the following expression holds

$$\prod_{i=1}^r \frac{\Gamma(\beta_i-k_i)}{\Gamma(\beta_i-k_i+s)} \nabla^{k_i} \frac{\Gamma_q(\beta_i+s)}{\Gamma_q(\beta_i)} \frac{(\alpha)^s}{\Gamma_q(s+1)} M_{q,\vec{n}}^{\alpha,\vec{\beta}}(s) = \prod_{i=1}^r [\beta_i-1]_q^{(k_i)} q^{|\vec{k}|/2} q^{-(|\vec{k}|)|\vec{n}|}$$

$$\times \left(\prod_{i=1}^r \alpha^{-k_i} \prod_{j=1}^{k_i} c_{q,\vec{n}}^{\alpha,\beta_i-j} \right) M_{q,\vec{n}+\vec{k}}^{\alpha/q^{|\vec{k}|},\beta_1-k_1,\ldots,\beta_r-k_r}(s) \frac{(\alpha/q^{|\vec{k}|})^s}{\Gamma_q(s+1)}.$$

Let $n_1 = n_2 = \cdots = n_r = 0$ and replace β_i by $\beta_i + k_i$ and α by $\alpha q^{|\vec{k}|}$. Finally, if we rename the new index component k_i with the old index component n_i, for $i = 1, \ldots, r$, the expression (55) holds. $\square$

3.5. q-Difference Equation for the q-Analogue of Multiple Meixner Polynomials of the Second Kind

In this section we will find the lowering operator for the q-Meixner multiple orthogonal polynomials of the second kind.

Lemma 5. *The q-Meixner multiple orthogonal polynomials of the second kind satisfy the following property*

$$\sum_{s=0}^{\infty} M_{q,\vec{n}-\vec{e}_i}^{q\alpha,\vec{\beta}+\vec{e}_i}(s) [s]_q^{(n_k-1)} v_q^{q\alpha,\beta_k+1}(s) \nabla x_1(s) = m_{k,i} \sum_{s=0}^{\infty} M_{q,\vec{n}-\vec{e}}^{q^r\alpha,\vec{\beta}+\vec{e}}(s) [s]_q^{(n_k-1)} v_q^{q^r\alpha,\beta_k+1}(s) \nabla x_1(s),$$

where

$$m_{k,i} = \frac{1 - \alpha q^{|\vec{n}|+\beta_i}}{\alpha q^{|\vec{n}|+\beta_i}} \frac{1}{x(n_k + \beta_k - \beta_i)} \prod_{j=1}^{r} \frac{\alpha q^{|\vec{n}|+\beta_j}}{1 - \alpha q^{|\vec{n}|+\beta_j}} x(n_k + \beta_k - \beta_j), \quad k, i = 1, 2, \ldots, r, \qquad (58)$$

and $\quad \vec{e} = \sum_{i=1}^{r} \vec{e}_i.$

Proof. By shifting conveniently the parameters involved in (53) and (54), respectively, one has

$$M_{q,\vec{n}}^{\alpha,\vec{\beta}}(s) = -q^{-1/2} \mathcal{D}_q^{q\alpha,\beta_i+1} \left(M_{q,\vec{n}-\vec{e}_i}^{q\alpha,\vec{\beta}+\vec{e}_i}(s) \right)$$

$$= -\frac{q^{|\vec{n}|-1}}{1 - \alpha q^{|\vec{n}|+\beta_i}} \left\{ \left(\alpha q^{\beta_i+1} \left(x(s) - x(-\beta_i) \right) - x(s) \right) M_{q,\vec{n}-\vec{e}_i}^{q\alpha,\vec{\beta}+\vec{e}_i}(s) + x(s) \triangledown M_{q,\vec{n}-\vec{e}_i}^{q\alpha,\vec{\beta}+\vec{e}_i}(s) \right\}.$$

Thus,

$$\sum_{s=0}^{\infty} M_{q,\vec{n}}^{\alpha,\vec{\beta}}(s) [s]_q^{(n_k-1)} v_q^{\alpha,\beta_k+1}(s) \triangledown x_1(s) = -\frac{q^{|\vec{n}|-1}}{1 - \alpha q^{|\vec{n}|+\beta_i}} \sum_{s=0}^{\infty} [s]_q^{(n_k-1)} v_q^{\alpha,\beta_k+1}(s) \triangledown x_1(s)$$

$$\times \left\{ \left(\alpha q^{\beta_i+1} \left(x(s) - x(-\beta_i) \right) - x(s) \right) M_{q,\vec{n}-\vec{e}_i}^{q\alpha,\vec{\beta}+\vec{e}_i}(s) + x(s) \triangledown M_{q,\vec{n}-\vec{e}_i}^{q\alpha,\vec{\beta}+\vec{e}_i}(s) \right\}.$$

Using summation by parts in the above expression we have

$$\sum_{s=0}^{\infty} M_{q,\vec{n}}^{\alpha,\vec{\beta}}(s) [s]_q^{(n_k-1)} v_q^{\alpha,\beta_k+1}(s) \triangledown x_1(s)$$

$$= \frac{\alpha q^{|\vec{n}|+\beta_i}}{1 - \alpha q^{|\vec{n}|+\beta_i}} x(n_k + \beta_k - \beta_i) \sum_{s=0}^{\infty} M_{q,\vec{n}-\vec{e}_i}^{q\alpha,\vec{\beta}+\vec{e}_i}(s) [s]_q^{(n_k-1)} v_q^{q\alpha,\beta_k+1}(s) \triangledown x_1(s)$$

$$+ \frac{\alpha q^{|\vec{n}|-n_k+2}}{1 - \alpha q^{|\vec{n}|+\beta_i}} x(\beta_k - 1) x(n_k + \beta_k - 1) \sum_{s=0}^{\infty} M_{q,\vec{n}-\vec{e}_i}^{q\alpha,\vec{\beta}+\vec{e}_i}(s) [s]_q^{(n_k-2)} v_q^{q\alpha,\beta_k+1}(s) \triangledown x_1(s).$$

From the orthogonality conditions the following relation holds:

$$\sum_{s=0}^{\infty} M_{q,\vec{n}-\vec{e}_i}^{q\alpha,\vec{\beta}+\vec{e}_i}(s) [s]_q^{(n_k-2)} v_q^{q\alpha,\beta_k+1}(s) \triangledown x_1(s) = 0.$$

Therefore,

$$\sum_{s=0}^{\infty} M_{q,\vec{n}}^{\alpha,\vec{\beta}}(s) [s]_q^{(n_k-1)} v_q^{\alpha,\beta_k+1}(s) \triangledown x_1(s) = \frac{\alpha q^{|\vec{n}|+\beta_i}}{1 - \alpha q^{|\vec{n}|+\beta_i}} x(n_k + \beta_k - \beta_i)$$

$$\times \sum_{s=0}^{\infty} M_{q,\vec{n}-\vec{e}_i}^{q\alpha,\vec{\beta}+\vec{e}_i}(s) [s]_q^{(n_k-1)} v_q^{q\alpha,\beta_k+1}(s) \triangledown x_1(s). \qquad (59)$$

Then, by iterating recursively (59), the relation (58) holds. This completes the proof of the lemma. $\quad \square$

Lemma 6. *Let $M = (m_{k,i})_{k,i=1}^{r}$ be the matrix with entries given in (58). Then, M is non-singular.*

Proof. Let us rewrite the entries in M as $m_{k,i} = c_k d_i / [n_k + \beta_k - \beta_i]_q$, where

$$c_k = q^{(1-n_k-\beta_k)/2} \prod_{j=1}^{r} \frac{\alpha q^{|\vec{n}|+\beta_j}}{1 - \alpha q^{|\vec{n}|+\beta_j}} x(n_k + \beta_k - \beta_j),$$

$$d_i = q^{\beta_i/2} \left(\frac{1 - \alpha q^{|\vec{n}|+\beta_i}}{\alpha q^{|\vec{n}|+\beta_i}} \right),$$

$$[n_k + \beta_k - \beta_i]_q = q^{(1-n_k-\beta_k+\beta_i)/2} x(n_k + \beta_k - \beta_i).$$

The matrix M is the product of three matrices; that is $M = C \cdot A \cdot D$, where $A = \left(1/[n_k + \beta_k - \beta_i]_q\right)_{k,i=1}^{r}$ and matrices C, D are the diagonal matrices $C = \operatorname{diag}(c_1, c_2, \ldots, c_r)$, $D = \operatorname{diag}(d_1, d_2, \ldots, d_r)$, respectively. □

In ([31], lemma 3.2, p. 7) it was proved that A is nonsingular. Therefore, M is also a nonsingular matrix. Indeed,

$$\det M = q^{(r-|\vec{n}|)/2} \left(\prod_{j=1}^{r} c_j d_j \right) \det A,$$

$$= \frac{\prod_{k=1}^{r-1} \prod_{l=k+1}^{r} x(\beta_l - \beta_k) q^{n_l} x(n_k - n_l + \beta_k - \beta_l)}{\prod_{k=1}^{r} \prod_{l=1}^{r} x(n_l + \beta_l - \beta_k)}. \tag{60}$$

Lemma 7. *Let $\mathbb{V}$ be the subspace of polynomials ϑ on the discrete variable $x(s)$, such that $\deg \vartheta \leq |\vec{n}| - 1$ and*

$$\sum_{s=0}^{\infty} \vartheta(s) [s]_q^{(k)} v_q^{q\alpha, \beta_j+1}(s) \triangledown x_1(s) = 0, \quad 0 \leq k \leq n_j - 2, \quad j = 1, 2, \ldots, r.$$

Then, the system $\left\{ M_{q,\vec{n}-\vec{e}_i}^{q\alpha, \vec{\beta}+\vec{e}_i}(s) \right\}_{i=1}^{r}$ is linearly independent in $\mathbb{V}$.

Proof. From orthogonality relations

$$\sum_{s=0}^{\infty} M_{q,\vec{n}-\vec{e}_j}^{q\alpha, \vec{\beta}+\vec{e}_j}(s) [s]_q^{(k)} v_q^{q\alpha, \beta_j+1}(s) \triangledown x_1(s) = 0, \quad 0 \leq k \leq n_j - 2, \quad j = 1, 2, \ldots, r,$$

we have that polynomials $M_{q,\vec{n}-\vec{e}_i}^{q\alpha, \vec{\beta}+\vec{e}_i}(s) \in \mathbb{V}$, for $i = 1, 2, \ldots, r$.

Suppose that there exist constants λ_i, $i = 1, \ldots, r$, such that

$$\sum_{i=1}^{r} \lambda_i M_{q,\vec{n}-\vec{e}_i}^{q\alpha, \vec{\beta}+\vec{e}_i}(s) = 0, \quad \text{where} \quad \sum_{i=1}^{r} |\lambda_i| > 0. \tag{61}$$

Then, multiplying the previous equation by $[s]_q^{(n_k-1)} v_q^{q\alpha, \beta_k+1}(s) \triangledown x_1(s)$ and then taking summation on s from 0 to ∞, one gets

$$\sum_{i=1}^{r} \lambda_i \sum_{s=0}^{\infty} M_{q,\vec{n}-\vec{e}_i}^{q\alpha, \vec{\beta}+\vec{e}_i}(s) [s]_q^{(n_k-1)} v_q^{q\alpha, \beta_k+1}(s) \triangledown x_1(s) = 0.$$

Using Lemma 5 and relation $\sum_{s=0}^{\infty} M_{q,\vec{n}-\vec{e}_i}^{q\alpha,\vec{\beta}+\vec{e}_i}(s)[s]_q^{(n_k-1)} v_q^{q\alpha,\beta_k+1}(s) \nabla x_1(s) \neq 0$, we obtain the following homogeneous linear system of equations

$$\sum_{i=1}^{r} m_{k,i}\lambda_i = 0, \qquad k = 1,\ldots,r,$$

or equivalently, in matrix form $M\lambda = 0$, where $\lambda = (\lambda_1,\ldots,\lambda_r)^T$. From Lemma 6, we have that M is nonsingular, which implies $\lambda_i = 0$ for $i = 1,\ldots,r$; that is, the previous assumption (61) is false. Therefore, $\left\{ M_{q,\vec{n}-\vec{e}_i}^{q\alpha,\vec{\beta}+\vec{e}_i}(s) \right\}_{i=1}^{r}$ is linearly independent in $\mathbb{V}$. Furthermore, we know that any polynomial from subspace $\mathbb{V}$ can be determined with $|\vec{n}|$ coefficients while $(|\vec{n}| - r)$ conditions are imposed on $\mathbb{V}$, consequently the dimension of $\mathbb{V}$ is at most r. Therefore, the system $\{M_{q,\vec{n}-\vec{e}_i}^{q\alpha,\vec{\beta}+\vec{e}_i}(s)\}_{i=1}^r$ spans $\mathbb{V}$. This completes the proof of the lemma. $\square$

Now we will prove that operator (8) is indeed a lowering operator for the sequence of q-Meixner multiple orthogonal polynomials of the second kind $M_{q,\vec{n}}^{\alpha,\vec{\beta}}(s)$.

Lemma 8. *The following relation holds:*

$$\Delta M_{q,\vec{n}}^{\alpha,\vec{\beta}}(s) = \sum_{i=1}^{r} \xi_i M_{q,\vec{n}-\vec{e}_i}^{q\alpha,\vec{\beta}+\vec{e}_i}(s), \tag{62}$$

where

$$\xi_i = \frac{\prod_{l=1}^{r} x(n_l + \beta_l - \beta_i)}{\prod_{k=1,k\neq i}^{r} x(\beta_i - \beta_k) \prod_{l=i+1}^{r} x(\beta_l - \beta_i)} \sum_{j=1}^{r} \frac{(1 - \alpha q^{n_j+\beta_j}) q^{|\vec{n}|-n_j+1/2}}{(1 - \alpha q^{|\vec{n}|+\beta_j}) x(n_j + \beta_j - \beta_i)}$$

$$\times \frac{(-1)^{i+j} \prod_{k=1}^{r} x(n_j + \beta_j - \beta_k)}{\prod_{k=1,k\neq j}^{r-1} q^{n_j} x(n_k - n_j + \beta_k - \beta_j) \prod_{l=j+1}^{r} q^{n_l} x(n_j - n_l + \beta_j - \beta_l)}. \tag{63}$$

Proof. Using summation by parts we have

$$\sum_{s=0}^{\infty} \Delta M_{q,\vec{n}}^{\alpha,\vec{\beta}}(s)[s]_q^{(k)} v_q^{q\alpha,\beta_j+1}(s) \nabla x_1(s) = -\sum_{s=0}^{\infty} M_{q,\vec{n}}^{\alpha,\vec{\beta}}(s) \nabla ([s]_q^{(k)} v_q^{q\alpha,\beta_j+1}(s)) \nabla x_1(s)$$

$$= -\sum_{s=0}^{\infty} M_{q,\vec{n}}^{\alpha,\vec{\beta}}(s) \varphi_{j,k}(s) v_q^{\alpha,\beta_j}(s) \nabla x_1(s), \tag{64}$$

where

$$\varphi_{j,k}(s) = q^{1/2} \left(\frac{q^{\beta_j} x(s)}{x(\beta_j)} + 1 \right) [s]_q^{(k)} - q^{-1/2} \frac{x(s)}{\alpha x(\beta_j)} [s-1]_q^{(k)},$$

is a polynomial of degree $\leq k+1$ in the variable $x(s)$. Then, from the orthogonality conditions (52) we get

$$\sum_{s=0}^{\infty} \Delta M_{q,\vec{n}}^{\alpha,\vec{\beta}}(s)[s]_q^{(k)} v_q^{q\alpha,\beta_j+1}(s) \nabla x_1(s) = 0, \qquad 0 \leq k \leq n_j - 2, \qquad j = 1,\ldots,r.$$

From Lemma 7, $\Delta M_{q,\vec{n}}^{\alpha,\vec{\beta}}(s) \in \mathbb{V}$. Moreover, $\Delta M_{q,\vec{n}}^{\alpha,\vec{\beta}}(s)$ can be expressed as a linear combination of polynomials $\{M_{q,\vec{n}-\vec{e}_i}^{q\alpha,\vec{\beta}+\vec{e}_i}(s)\}_{i=1}^r$, i.e.,

$$\Delta M_{q,\vec{n}}^{\alpha,\vec{\beta}}(s) = \sum_{i=1}^r \xi_i M_{q,\vec{n}-\vec{e}_i}^{q\alpha,\vec{\beta}+\vec{e}_i}(s), \qquad \sum_{i=1}^r |\xi_i| > 0. \tag{65}$$

Thus, for finding explicity $\xi_1, \ldots, \xi_r$ one takes into account Lemma 5 and (65) to get

$$\sum_{s=0}^\infty \Delta M_{q,\vec{n}}^{\alpha,\vec{\beta}}(s)[s]_q^{(n_k-1)} v_q^{q\alpha,\beta_k+1}(s) \, \nabla x_1(s) = \left(\sum_{i=1}^r \xi_i m_{k,i} \right) \sum_{s=0}^\infty M_{q,\vec{n}-\vec{e}}^{q^r\alpha,\vec{\beta}+\vec{e}}(s)[s]_q^{(n_k-1)}$$
$$\times v_q^{q^r\alpha,\beta_k+1}(s) \, \nabla x_1(s). \tag{66}$$

If we replace $[s]_q^{(k)}$ by $[s]_q^{(n_k-1)}$ in the left-hand side of Equation (64), then left-hand side of Equation (66) transforms into relation

$$\sum_{s=0}^\infty \Delta M_{q,\vec{n}}^{\alpha,\vec{\beta}}(s)[s]_q^{(n_k-1)} v_q^{q\alpha,\beta_k+1}(s) \, \nabla x_1(s) = -\sum_{s=0}^\infty M_{q,\vec{n}}^{\alpha,\vec{\beta}}(s)\varphi_{k,n_k-1}(s) v_q^{\alpha,\beta_k}(s) \, \nabla x_1(s)$$
$$= \frac{q^{1/2}\left(1-\alpha_k q^{n_k+\beta_k}\right)}{\alpha q^{n_k+\beta_k}} \sum_{s=0}^\infty M_{q,\vec{n}}^{\alpha,\vec{\beta}}(s)[s]_q^{(n_k)} v_q^{\alpha,\beta_k+1}(s) \, \nabla x_1(s).$$

We have used that $x(s)[s-1]_q^{(n_k-1)} = [s]_q^{(n_k)}$ to get

$$\varphi_{k,n_k-1}(s) = -\frac{q^{-1/2}\left(1-\alpha q^{n_k+\beta_k}\right)}{\alpha x(\beta_k)}[s]_q^{(n_k)} + \text{lower degree terms}.$$

Using Lemma 5, we have that

$$\sum_{s=0}^\infty \Delta M_{q,\vec{n}}^{\alpha,\vec{\beta}}(s)[s]_q^{(n_k-1)} v_q^{q\alpha,\beta_k+1}(s) \, \nabla x_1(s)$$
$$= \frac{(1-\alpha q^{n_k+\beta_k})q^{|\vec{n}|-n_k+1/2}}{1-\alpha q^{|\vec{n}|+\beta_k}} x(n_k) \sum_{s=0}^\infty M_{q,\vec{n}-\vec{e}_k}^{q\alpha,\vec{\beta}+\vec{e}_k}(s)[s]_q^{(n_k-1)} v_q^{q\alpha,\beta_k+1}(s) \, \nabla x_1(s)$$
$$= \tilde{b}_k \sum_{s=0}^\infty M_{q,\vec{n}-\vec{e}}^{q^r\alpha,\vec{\beta}+\vec{e}}(s)[s]_q^{(n_k-1)} v_q^{q^r\alpha,\beta_k+1}(s) \, \nabla x_1(s), \quad (67)$$

where

$$\tilde{b}_k = \frac{q^{1/2}(1-\alpha q^{n_k+\beta_k})}{\alpha q^{n_k+\beta_k}} \prod_{i=1}^r \frac{\alpha q^{|\vec{n}|+\beta_i}}{1-\alpha q^{|\vec{n}|+\beta_i}} x(n_k+\beta_k-\beta_i).$$

From Equations (66) and (67) we get the following linear system of equations for the unknown coefficients $\xi_1, \ldots, \xi_r$,

$$b_j = \sum_{i=1}^r \xi_i s_{j,i}, \quad k = 1, \ldots, r, \quad \Longleftrightarrow \quad S\xi = b, \quad \xi = (\xi_1, \ldots, \xi_r), \tag{68}$$

where the entries of the vector b and matrix S are as follows

$$b_j = \frac{(1-\alpha q^{n_j+\beta_j})q^{|\vec{n}|-n_j+1/2}}{(1-\alpha q^{|\vec{n}|+\beta_j})}, \quad s_{j,i} = m_{j,i}.$$

The above system (68) has a unique solution if and only if the matrix S is nonsingular. From Lemma 6, Formula (60), this condition is fulfilled. Accordingly, if $C_{j,i}$ stands for the cofactor of the entry $s_{j,i}$ and $S_i(b)$ denotes the matrix obtained from S replacing its ith column by b, then

$$\xi_i = \frac{\det S_i(b)}{\det S}, \quad i = 1, \ldots, r.$$

From Lemma 6,

$$\det S_i(b) = \sum_{j=1}^{r} b_j C_{j,i}$$

$$= \sum_{j=1}^{r} b_j (-1)^{i+j} \prod_{k=1, k\neq i}^{r-1} \prod_{l=k+1, l\neq i}^{r} x(\beta_l - \beta_k) \frac{\prod_{k=1, k\neq j}^{r-1} \prod_{l=k+1, l\neq j}^{r} q^{n_l} x(n_k - n_l + \beta_k - \beta_l)}{\prod_{k=1, k\neq i}^{r} \prod_{l=1, l\neq j}^{r} x(n_k + \beta_k - \beta_l)}.$$

Therefore, relation (62) holds. □

Theorem 3. *The q-Meixner multiple orthogonal polynomial of the second kind $M_{q,\vec{n}}^{\alpha,\vec{\beta}}(s)$ satisfies the following $(r+1)$-order q-difference equation*

$$\prod_{i=1}^{r} \mathcal{D}_{q,\vec{n}}^{q\alpha,\beta_i+1} \Delta M_{q,\vec{n}}^{\alpha,\vec{\beta}}(s) = -\sum_{i=1}^{r} q^{1/2} \xi_i \prod_{\substack{j=1 \\ j\neq i}}^{r} \mathcal{D}_{q,\vec{n}}^{q\alpha,\beta_j+1} M_{q,\vec{n}}^{\alpha,\vec{\beta}}(s), \tag{69}$$

where ξ_is are the constants in (63).

Proof. Since the operators (53) commute, we write

$$\prod_{i=1}^{r} \mathcal{D}_{q,\vec{n}}^{q\alpha,\beta_i+1} = \left(\prod_{\substack{j=1 \\ j\neq i}}^{r} \mathcal{D}_{q,\vec{n}}^{q\alpha,\beta_j+1} \right) \mathcal{D}_{q,\vec{n}}^{q\alpha,\beta_i+1}. \tag{70}$$

□

Using Formula (53) in Equation (62) by acting with the product of operators (70), we obtain the desired relation (69); that is,

$$\prod_{i=1}^{r} \mathcal{D}_{q,\vec{n}}^{q\alpha,\beta_i+1} \Delta M_{q,\vec{n}}^{\alpha,\vec{\beta}}(s) = \sum_{i=1}^{r} \xi_i \prod_{\substack{j=1 \\ j\neq i}}^{r} \mathcal{D}_{q,\vec{n}}^{q\alpha,\beta_j+1} \left(\mathcal{D}_{q,\vec{n}}^{q\alpha,\beta_i+1} M_{q,\vec{n}-\vec{e}_i}^{q\alpha,\vec{\beta}+\vec{e}_i}(s) \right)$$

$$= -\sum_{i=1}^{r} q^{1/2} \xi_i \prod_{\substack{j=1 \\ j\neq i}}^{r} \mathcal{D}_{q,\vec{n}}^{q\alpha,\beta_j+1} M_{q,\vec{n}}^{\alpha,\vec{\beta}}(s).$$

3.6. Recurrence Relation for q-Meixner Multiple Orthogonal Polynomials of the Second Kind

Theorem 4. *The q-Meixner multiple orthogonal polynomials of the second kind satisfy the following $(r+2)$-term recurrence relation*

$$x(s) M_{q,\vec{n}}^{\alpha,\vec{\beta}}(s) = M_{q,\vec{n}+\vec{e}_k}^{\alpha,\vec{\beta}}(s) + b_{\vec{n},k} M_{q,\vec{n}}^{\alpha,\vec{\beta}}(s)$$

$$+ \alpha q^{2|\vec{n}|-1} \sum_{i=1}^{r} \frac{x(n_i) x(\beta_i + n_i - 1)}{c_{q,\vec{n}+n_i\vec{e}_i}^{\alpha,\beta_i-1} c_{q,\vec{n}+n_i\vec{e}_i}^{\alpha,\beta_i-2}} \prod_{j\neq i}^{r} \frac{x(n_i + \beta_i - \beta_j)}{x(n_i + \beta_i - n_j - \beta_j)} B_{\vec{n},i} M_{q,\vec{n}-\vec{e}_i}^{\alpha,\vec{\beta}}(s), \tag{71}$$

where

$$b_{\vec{n},k} = \prod_{i=1}^{r} \frac{c_{q,\vec{n}}^{\alpha,\beta_i}}{c_{q,\vec{n}+n_i\vec{e}_i}^{\alpha,\beta_i}} \left(\sum_{i=1}^{r} \frac{-q^{|\vec{n}|} x(n_i)}{q^{n_i} c_{q,\vec{n}}^{\alpha,\beta_i}} - \frac{\alpha q^{2|\vec{n}|+1} x(\beta_k + n_k)}{c_{q,\vec{n}+n_k\vec{e}_k}^{\alpha,\beta_k+1}} \right)$$

$$+ (q-1) \left(\alpha q^{|\vec{n}|} \sum_{i=1}^{r} \frac{x(n_i) x(n_i+\beta_i-1)}{c_{q,\vec{n}+n_i\vec{e}_i}^{\alpha,\beta_i} c_{q,\vec{n}+n_i\vec{e}_i}^{\alpha,\beta_i-1}} \prod_{j\neq i}^{r} \frac{x(n_i+\beta_i-\beta_j)}{x(n_i+\beta_i-n_j-\beta_j)} + \prod_{i=1}^{r} \frac{-x(n_i)}{c_{q,\vec{n}}^{\alpha,\beta_i}} \prod_{i=1}^{r} \frac{c_{q,\vec{n}}^{\alpha,\beta_i}}{c_{q,\vec{n}+n_i\vec{e}_i}^{\alpha,\beta_i}} \right)$$

and

$$B_{\vec{n},i} = \frac{\alpha q^{|\vec{n}|} - 1}{c_{q,\vec{n}+n_i\vec{e}_i}^{\alpha,\beta_i}} \prod_{i=1}^{r} \frac{c_{q,\vec{n}}^{\alpha,\beta_i-1}}{c_{q,\vec{n}+n_i\vec{e}_i}^{\alpha,\beta_i}}.$$

Proof. Let

$$g_{\mathbf{n}}(s;\alpha) = \frac{(\alpha q^{\mathbf{n}})^s}{\Gamma_q(s+1)}, \quad \text{where} \quad \mathbf{n} = |\vec{n}|.$$

We will use Lemma 4 involving this function $g_{\mathbf{n}}(s;\alpha)$ as well as difference operator (56). Consider the following equation

$$\frac{\Gamma_q(\beta_k)}{\Gamma_q(\beta_k+s)} \nabla^{n_k+1} \frac{\Gamma_q(\beta_k+n_k+1+s)}{\Gamma_q(\beta_k+n_k+1)} \frac{(\alpha q^{|\vec{n}|+1})^s}{\Gamma_q(s+1)}$$

$$= \frac{\Gamma_q(\beta_k)}{\Gamma_q(\beta_k+s)} \nabla^{n_k} \left(q^{-s+1/2} \nabla \left(\frac{\Gamma_q(\beta_k+n_k+1+s)}{\Gamma_q(\beta_k+n_k+1)} \frac{(\alpha q^{|\vec{n}|+1})^s}{\Gamma_q(s+1)} \right) \right)$$

$$= q^{1/2} \frac{\Gamma_q(\beta_k)}{\Gamma_q(\beta_k+s)} \nabla^{n_k} \left(\frac{\Gamma_q(\beta_k+n_k+s)}{\Gamma_q(\beta_k+n_k)} \left(1 + \frac{c_{q,\vec{n}+n_k\vec{e}_k}^{\alpha,\beta_k+1}}{(\alpha_k q^{|\vec{n}|+1}) x(\beta_k+n_k)} x(s) \right) \frac{(\alpha q^{|\vec{n}|})^s}{\Gamma_q(s+1)} \right),$$

which can be rewritten as follows

$$\mathcal{N}_{q,n_k+1}^{\beta_k} \frac{(\alpha q^{|\vec{n}|+1})^s}{\Gamma_q(s+1)} = q^{1/2} \mathcal{N}_{q,n_k}^{\beta_k} \frac{(\alpha q^{|\vec{n}|})^s}{\Gamma_q(s+1)} + q^{1/2} \frac{c_{q,\vec{n}+n_k\vec{e}_k}^{\alpha,\beta_k+1}}{(\alpha q^{|\vec{n}|+1}) x(\beta_k+n_k)} \mathcal{N}_{q,n_k}^{\beta_k} x(s) \frac{(\alpha q^{|\vec{n}|})^s}{\Gamma_q(s+1)}. \tag{72}$$

Since operators (56) commute, the multiplication of Equation (72) from the left-hand side by the product $\prod_{\substack{i=1 \\ i\neq k}}^{r} \mathcal{N}_{q,n_i}^{\beta_i}$ yields

$$\mathcal{N}_{q,\vec{n}}^{\vec{\beta}} x(s) \frac{(\alpha q^{|\vec{n}|})^s}{\Gamma_q(s+1)} = \frac{\alpha x(\beta_k+n_k)}{q^{-|\vec{n}|-1/2} c_{q,\vec{n}+n_k\vec{e}_k}^{\alpha,\beta_k+1}} \left(\mathcal{N}_{q,\vec{n}+\vec{e}_k}^{\vec{\beta}} \frac{(\alpha q^{|\vec{n}|+1})^s}{\Gamma_q(s+1)} - q^{1/2} \mathcal{N}_{q,\vec{n}}^{\vec{\beta}} \frac{(\alpha q^{|\vec{n}|})^s}{\Gamma_q(s+1)} \right). \tag{73}$$

Let us recursively use Lemma 4 involving the product of r difference operators $\prod_{i=1}^{r} \mathcal{F}_{q,n_i}$ acting on the function $g_{\mathbf{n}}(s;\alpha)$, that is, the operator $\mathcal{N}_{q,\vec{n}}^{\vec{\beta}}$ (see expression (56)). Thus,

$$q^{|\vec{n}|} \mathcal{N}_{q,\vec{n}}^{\vec{\beta}} x(s) g_{\mathbf{n}}(s;\alpha) = \sum_{i=1}^{r} \prod_{j\neq i}^{r} \frac{q^{1/2} x(n_i+\beta_i-\beta_j)}{x(n_i+\beta_i-n_j-\beta_j)} \frac{x(n_i) c_{q,\vec{n}+n_j\vec{e}_j}^{\alpha,\beta_j}}{\prod_{v=1}^{r} c_{q,\vec{n}}^{\alpha,\beta_v}} \prod_{l=1}^{r} \mathcal{N}_{q,n_l-\delta_{l,i}}^{\beta_l} g_{\mathbf{n}}(s;\alpha)$$

$$+ \left(q^{|\vec{n}|} \sum_{i=1}^{r} \frac{x(n_i)}{q^{n_i} c_{q,\vec{n}}^{\alpha,\beta_i}} + (q-1) \prod_{i=1}^{r} \frac{x(n_i)}{c_{q,\vec{n}}^{\alpha,\beta_i}} \right) \mathcal{N}_{q,\vec{n}}^{\vec{\beta}} g_{\mathbf{n}}(s;\alpha) + \prod_{i=1}^{r} \frac{c_{q,\vec{n}+n_i\vec{e}_i}^{\alpha,\beta_i}}{c_{q,\vec{n}}^{\alpha,\beta_i}} x(s) \mathcal{N}_{q,\vec{n}}^{\vec{\beta}} g_{\mathbf{n}}(s;\alpha). \tag{74}$$

Hence, using expressions (73) and (74) one gets

$$
x(s)\mathcal{N}_{q,\vec{n}}^{\vec{\beta}}g_{\mathbf{n}}(s;\alpha) = q^{|\vec{n}|-1/2} \prod_{i=1}^{r} \frac{c_{q,\vec{n}}^{\alpha,\beta_i}}{c_{q,\vec{n}+n_i\vec{e}_i}^{\alpha,\beta_i}} \frac{\alpha q^{|\vec{n}|+1}x(\beta_k+n_k)}{c_{q,\vec{n}+n_k\vec{e}_k}^{\alpha,\beta_k+1}} \mathcal{N}_{q,\vec{n}+\vec{e}_k}^{\vec{\beta}}g_{\mathbf{n}+1}(s;\alpha)
$$

$$
+ \prod_{i=1}^{r} \frac{c_{q,\vec{n}}^{\alpha,\beta_i}}{c_{q,\vec{n}+n_i\vec{e}_i}^{\alpha,\beta_i}} \left(q^{|\vec{n}|} \sum_{i=1}^{r} \frac{x(n_i)}{q^{n_i}c_{q,\vec{n}}^{\alpha,\beta_i}} + (q-1)\prod_{i=1}^{r}\frac{x(n_i)}{c_{q,\vec{n}}^{\alpha,\beta_i}} - q^{|\vec{n}|}\frac{\alpha q^{|\vec{n}|+1}x(\beta_k+n_k)}{c_{q,\vec{n}+n_k\vec{e}_k}^{\alpha,\beta_k+1}} \right) \mathcal{N}_{q,\vec{n}}^{\vec{\beta}}g_{\mathbf{n}}(s;\alpha)
$$

$$
- \sum_{i=1}^{r}\prod_{j\neq i}^{r} \frac{q^{1/2}x(n_i+\beta_i-\beta_j)}{x(n_i+\beta_i-n_j-\beta_j)}\frac{-x(n_i)}{c_{q,\vec{n}+n_i\vec{e}_i}^{\alpha,\beta_i}} \prod_{l=1}^{r}\mathcal{N}_{q,n_l-\delta_{l,i}}^{\beta_l}g_{\mathbf{n}}(s;\alpha).
$$

Observe that

$$
\mathcal{N}_{q,n_i-1}^{\beta_i}g_{\mathbf{n}}(s;\alpha) = q^{-1/2}\frac{(q-1)\alpha q^{|\vec{n}|}x(n_i+\beta_i-1)}{c_{q,\vec{n}+n_i\vec{e}_i}^{\alpha,\beta_i-1}}\mathcal{N}_{q,n_i}^{\beta_i}g_{\mathbf{n}}(s;\alpha) + \frac{\alpha q^{|\vec{n}|}-1}{c_{q,\vec{n}+n_i\vec{e}_i}^{\alpha,\beta_i-1}}\mathcal{N}_{q,n_i-1}^{\beta_i}g_{\mathbf{n}-1}(s;\alpha),
$$

which is used in the previous expression when the indices l and i coincide. Therefore, the following expression holds

$$
x(s)\mathcal{N}_{q,\vec{n}}^{\vec{\beta}}g_{\mathbf{n}}(s;\alpha) = q^{|\vec{n}|-1/2} \prod_{i=1}^{r} \frac{c_{q,\vec{n}}^{\alpha,\beta_i}}{c_{q,\vec{n}+n_i\vec{e}_i}^{\alpha,\beta_i}} \frac{(\alpha q^{|\vec{n}|+1})x(\beta_k+n_k)}{c_{q,\vec{n}+n_k\vec{e}_k}^{\alpha,\beta_k+1}} \mathcal{N}_{q,\vec{n}+\vec{e}_k}^{\vec{\beta}}g_{\mathbf{n}+1}(s;\alpha)
$$

$$
+ b_{\vec{n},k}\mathcal{N}_{q,\vec{n}}^{\vec{\beta}}g_{\mathbf{n}}(s;\alpha) - q^{1/2}(1-\alpha q^{|\vec{n}|})\sum_{i=1}^{r}\prod_{j\neq i}^{r}\frac{x(n_i+\beta_i-\beta_j)}{x(n_i+\beta_i-n_j-\beta_j)}\frac{x(n_i)}{c_{q,\vec{n}+n_i\vec{e}_i}^{\alpha,\beta_i}c_{q,\vec{n}+n_i\vec{e}_i}^{\alpha,\beta_i-1}}\prod_{l=1}^{r}\mathcal{N}_{q,n_l-\delta_{l,i}}^{\beta_l}g_{\mathbf{n}-1}(s;\alpha).
$$

Finally, multiplying from the left both sides of the previous expression by $\mathcal{G}_q^{\vec{n},\vec{\beta},\alpha}\Gamma_q(\beta_i)/\Gamma_q(\beta_i+s)$ and using Rodrigues-type Formula (55), we obtain (71). This completes the proof of the theorem. $\square$

4. Limit Relations as q Approaches 1

The lattice $x(s) = (q^s-1)/(q-1)$ allows to transit from the non-uniform distribution of points $(q^s-1)/(q-1)$, $s = 0,1,\dots$, to the uniform distribution s, as q approaches 1. Under this limiting process one expects that the q-algebraic relations studied in this paper transform into the corresponding relations for discrete multiple orthogonal polynomials [28]. Indeed, the q-analogue of Rodrigues-type Formulas (31) and (55) will be transformed into their discrete counterparts (15) and (16), respectively. As a consequence, the recurrence relations (19) and (20) can be derived from (47) and (71), respectively.

We begin by analyzing the Rodrigues-type formulas, which then can be used for addressing the limit relations involving other algebraic properties.

Proposition 3. *The following limiting relations for q-Meixner multiple orthogonal polynomials of the first kind* (31) *and second kind* (55) *hold:*

$$
\lim_{q\to 1} M_{q,\vec{n}}^{\vec{\alpha},\beta}(s) = (\beta)_{|\vec{n}|}\prod_{i=1}^{r}\left(\frac{\alpha_i}{\alpha_i-1}\right)^{n_i}\frac{\Gamma(\beta)\Gamma(s+1)}{\Gamma(\beta+s)}\prod_{i=1}^{r}\alpha_i^{-s}\nabla^{n_i}\alpha_i^{s}\left(\frac{\Gamma(\beta+|\vec{n}|+s)}{\Gamma(\beta+|\vec{n}|)\Gamma(s+1)}\right), \tag{75}
$$

$$
\lim_{q\to 1} M_{q,\vec{n}}^{\alpha,\vec{\beta}}(s) = \left(\frac{\alpha}{\alpha-1}\right)^{|\vec{n}|}\left(\prod_{i=1}^{r}(\beta_i)_{n_i}\right)\frac{\Gamma(s+1)}{\alpha^s}\prod_{i=1}^{r}\frac{\Gamma(\beta_i)}{\Gamma(\beta_i+s)}\nabla^{n_i}\frac{\Gamma(\beta_i+n_i+s)}{\Gamma(\beta_i+n_i)}\left(\frac{\alpha^s}{\Gamma(s+1)}\right). \tag{76}
$$

The right-hand side limiting results are the corresponding discrete multiple orthogonal polynomials $M_{\vec{n}}^{\vec{\alpha},\beta}(s)$ and $M_{\vec{n}}^{\alpha,\vec{\beta}}(s)$ given in (15) *and* (16), *respectively.*

Proof. We begin by proving (75). Let us rewrite the m-th action of the difference operator ∇ on a function $f(s)$ defined on the q-lattice $x(s)$ as follows (see formula (3.2.29) from [38])

$$\nabla^m f(s) = q^{\binom{m+1}{2}/2 - ms} \sum_{k=0}^{m} \begin{bmatrix} m \\ k \end{bmatrix} (-1)^k q^{\binom{m-k}{2}} f(s-k), \tag{77}$$

where

$$\begin{bmatrix} m \\ k \end{bmatrix} = \frac{(q;q)_m}{(q;q)_k (q;q)_{m-k}}, \qquad m = 1,2,\ldots,$$

$$(a;q)_k = \prod_{j=0}^{k-1}(1 - aq^j) \quad \text{for} \quad k > 0, \quad \text{and} \quad (a;q)_0 = 1.$$

Here the expression $(a;q)_k$ denotes the q-analogue of the Pochhammer symbol [37,38,43,44]. Moreover, expression (77) is a q-analogue of (11).

In (31) we have the following expression

$$M_{q,\vec{n}}^{\vec{\alpha},\beta}(s) = G_q^{\vec{n},\vec{\alpha},\beta} \frac{\Gamma_q(\beta)\Gamma_q(s+1)}{\Gamma_q(\beta+s)} \prod_{i=1}^{r} (\alpha_i)^{-s} \nabla^{n_i} (\alpha_i q^{n_i})^s \left(\frac{\Gamma_q(\beta + |\vec{n}| + s)}{\Gamma_q(\beta + |\vec{n}|)\Gamma_q(s+1)} \right),$$

where the normalizing coefficient $G_q^{\vec{n},\vec{\alpha},\beta}$ is given in (33) and it tends to the following expression, as q approaches to 1

$$(\beta)_{|\vec{n}|} \left(\prod_{i=1}^{r} \left(\frac{\alpha_i}{\alpha_i - 1} \right)^{n_i} \right).$$

Without loss of generality, let us consider a multi-index $\vec{n} = (n_1, n_2)$ and rewrite the above expression in accordance with Formula (77); that is, we first need to express $\nabla^{n_1}(\alpha_1 q^{n_1})^s \Gamma_q(\beta + |\vec{n}| + s)/(\Gamma_q(\beta + |\vec{n}|)\Gamma_q(s+1))$ in terms of a finite sum and then compute the action of ∇^{n_2} on the product formed by this resulting expression and $(\alpha_2 q^{n_2})^s$. Namely,

$$M_{q,n_1,n_2}^{\alpha_1,\alpha_2,\beta}(s) = G_q^{n_1,n_2,\alpha_1,\alpha_2,\beta} \frac{\Gamma_q(\beta)\Gamma_q(s+1)}{\Gamma_q(\beta+s)} \left(\alpha_2^{-s} \nabla^{n_2}(\alpha_2 q^{n_2})^s \right) \left(\alpha_1^{-s} \nabla^{n_1}(\alpha_1 q^{n_1})^s \right) \frac{\Gamma_q(\beta + |\vec{n}| + s)}{\Gamma_q(\beta + |\vec{n}|)\Gamma_q(s+1)}$$

$$= G_q^{n_1,n_2,\alpha_1,\alpha_2,\beta} q^{(\binom{n_1+1}{2}+\binom{n_2+1}{2})/2} \frac{\Gamma_q(\beta)\Gamma_q(s+1)}{\Gamma_q(\beta+s)\Gamma_q(\beta+n_1+n_2)}$$

$$\times \sum_{k=0}^{n_1} \sum_{l=0}^{n_2} (-1)^{l+k} \begin{bmatrix} n_2 \\ l \end{bmatrix} \begin{bmatrix} n_1 \\ k \end{bmatrix} \frac{q^{\binom{n_2-l}{2}-ln_2+\binom{n_1-k}{2}-kn_1}}{\alpha_2^l \alpha_1^k} \frac{\Gamma_q(\beta + n_1 + n_2 - k - l + s)}{\Gamma_q(s-k-l+1)}. \tag{78}$$

Applying limit in the above expression as q approaches to 1 yields

$$\lim_{q \to 1} M_{q,n_1,n_2}^{\alpha_1,\alpha_2,\beta}(s) = (\beta)_{n_1+n_2} \left(\frac{\alpha_1}{\alpha_1 - 1} \right)^{n_1} \left(\frac{\alpha_2}{\alpha_2 - 1} \right)^{n_2} \frac{\Gamma(\beta)\Gamma(s+1)}{\Gamma(\beta+s)}$$

$$= \sum_{k=0}^{n_1} \sum_{l=0}^{n_2} (-1)^{l+k} \binom{n_2}{l} \binom{n_1}{k} \frac{1}{\alpha_2^l \alpha_1^k} \frac{\Gamma(\beta + n_1 + n_2 - k - l + s)}{\Gamma(s-k-l+1)}. \tag{79}$$

Using (11), one rewrites Equation (79) such that it involves the product of raising operators as in (13) to obtain

$$
\lim_{q \to 1} M_{q,n_1,n_2}^{\alpha_1,\alpha_2,\beta}(s) = (\beta)_{n_1+n_2} \left(\frac{\alpha_1}{\alpha_1 - 1} \right)^{n_1} \left(\frac{\alpha_2}{\alpha_2 - 1} \right)^{n_2} \frac{\Gamma(\beta)\Gamma(s+1)}{\Gamma(\beta + s)}
$$

$$
\times \left(\alpha_2^{-s} \, \nabla^{n_2} \, \alpha_2^s \right) \left(\alpha_1^{-s} \, \nabla^{n_1} \, \alpha_1^s \right) \frac{\Gamma(\beta + n_1 + n_2 + s)}{\Gamma(\beta + n_1 + n_2)\Gamma(s+1)}
$$

$$
= M_{n_1,n_2}^{\alpha_1,\alpha_2,\beta}(s),
$$

which coincides with (15) for $\vec{n} = (n_1, n_2)$. Observe that repeating the aforementioned procedure for a multi-index $\vec{n}$ of dimension r, we obtain for the polynomial

$$
M_{q,\vec{n}}^{\vec{\alpha},\beta}(s) = \mathcal{G}_q^{\vec{n},\vec{\alpha},\beta} q^{\sum_{i=1}^{r} \binom{n_i+1}{2}/2} \frac{\Gamma_q(\beta)\Gamma_q(s+1)}{\Gamma_q(\beta + s)\Gamma_q(\beta + |\vec{n}|)}
$$

$$
\times \sum_{k_1=0}^{n_1} \cdots \sum_{k_r=0}^{n_r} (-1)^{|\vec{k}|} \begin{bmatrix} n_r \\ k_r \end{bmatrix} \cdots \begin{bmatrix} n_1 \\ k_1 \end{bmatrix} \frac{q^{\binom{n_r-k_r}{2}-k_r n_r + \cdots + \binom{n_1-k_1}{2}-k_1 n_1}}{\alpha_r^{k_r} \cdots \alpha_1^{k_1}} \frac{\Gamma_q(\beta + |\vec{n} - \vec{k}| + s)}{\Gamma_q(s - |\vec{k}| + 1)},
$$

where $\vec{k} = (k_1, \ldots, k_r)$, the following relation

$$
\lim_{q \to 1} M_{q,\vec{n}}^{\vec{\alpha},\beta}(s) = (\beta)_{|\vec{n}|} \left(\prod_{i=1}^{r} \left(\frac{\alpha_i}{\alpha_i - 1} \right)^{n_i} \right) \frac{\Gamma(\beta)\Gamma(s+1)}{\Gamma(\beta + s)} \prod_{i=1}^{r} \alpha_i^{-s} \, \nabla^{n_i} \, \alpha_i^s \left(\frac{\Gamma(\beta + |\vec{n}| + s)}{\Gamma(\beta + |\vec{n}|)\Gamma(s+1)} \right),
$$

$$
= M_{\vec{n}}^{\vec{\alpha},\beta}(s).
$$

This proves the expression (75).

Next, we will prove the second limiting relation (76). Notice that the normalizing coefficient $\mathcal{G}_q^{\vec{n},\vec{\beta},\alpha}$ given in (57) has the following limit expression, as q approaches 1,

$$
\lim_{q \to 1} \mathcal{G}_q^{\vec{n},\vec{\beta},\alpha} = \lim_{q \to 1} (-1)^{|\vec{n}|} \left(\alpha q^{|\vec{n}|} \right)^{|\vec{n}|} q^{-\frac{|\vec{n}|}{2}} \left(\prod_{i=1}^{r} \frac{\prod_{j=1}^{n_i} q^{\beta_i+j-1}}{\prod_{j=1}^{n_i} \left(\alpha q^{|\vec{n}|+\beta_i+j-1} - 1 \right)} \right) \left(\prod_{i=1}^{r} [-\beta_i]_q^{(n_i)} \right)
$$

$$
= \left(\frac{\alpha}{\alpha - 1} \right)^{|\vec{n}|} \left(\prod_{i=1}^{r} (\beta_i)_{n_i} \right).
$$

From (55) and (77) we have

$$
M_{q,\vec{n}}^{\vec{\alpha},\beta}(s) = \mathcal{G}_q^{\vec{n},\vec{\beta},\alpha} q^{\sum_{i=1}^{r} \binom{n_i+1}{2}/2} \prod_{i=1}^{r} \frac{\Gamma_q(\beta_i)}{\Gamma_q(\beta_i + n_i)} \frac{\Gamma_q(s+1)}{\alpha^s}
$$

$$
\times \sum_{k_r=0}^{n_r} \cdots \sum_{k_1=0}^{n_1} (-1)^{|\vec{k}|} \begin{bmatrix} n_r \\ k_r \end{bmatrix} \cdots \begin{bmatrix} n_1 \\ k_1 \end{bmatrix} \frac{q^{\binom{n_r-k_r}{2}-k_r n_r + \cdots + \binom{n_1-k_1}{2}-k_1 n_1}}{\Gamma_q(\beta_r + s)\Gamma_q(\beta_{r-1} + s - k_r) \cdots \Gamma_q(\beta_1 + s - k_r - \cdots - k_2)}
$$

$$
\times \frac{\left(\alpha q^{|\vec{n}|} \right)^{s-|\vec{k}|} \Gamma_q(\beta_r + n_r + s - k_r) \cdots \Gamma_q(\beta_2 + n_2 + s - k_r - \cdots - k_2)\Gamma_q(\beta_1 + n_1 + s - |\vec{k}|)}{\Gamma_q(s - |\vec{k}| + 1)}.
$$

Therefore, we evaluate the following limit:

$$\lim_{q \to 1} M_{q,\vec{n}}^{\vec{\alpha},\vec{\beta}}(s) = \left(\frac{\alpha}{\alpha-1}\right)^{|\vec{n}|} \left(\prod_{i=1}^{r}(\beta_i)_{n_i}\right) \frac{\Gamma(s+1)}{\alpha^s}$$

$$\times \sum_{k_r=0}^{n_r} \cdots \sum_{k_1=0}^{n_1} (-1)^{|\vec{k}|} \binom{n_r}{k_r} \cdots \binom{n_1}{k_1} \frac{1}{\Gamma(\beta_r+s)\Gamma(\beta_{r-1}+s-k_r)\cdots\Gamma(\beta_1+s-k_r-\cdots-k_2)}$$

$$\times \frac{\alpha^{s-|\vec{k}|}\Gamma(\beta_r+n_r+s-k_r)\cdots\Gamma(\beta_2+n_2+s-k_r-\cdots-k_2)\Gamma(\beta_1+n_1+s-|\vec{k}|)}{\Gamma(s-|\vec{k}|+1)}.$$

Finally, using (11) one rewrites the right-hand side as follows

$$\lim_{q \to 1} M_{q,\vec{n}}^{\alpha,\vec{\beta}}(s) = \left(\frac{\alpha}{\alpha-1}\right)^{|\vec{n}|} \left(\prod_{i=1}^{r}(\beta_i)_{n_i}\right) \frac{\Gamma(s+1)}{\alpha^s} \prod_{i=1}^{r} \frac{\Gamma(\beta_i)}{\Gamma(\beta_i+s)} \nabla^{n_i} \frac{\Gamma(\beta_i+n_i+s)}{\Gamma(\beta_i+n_i)} \left(\frac{\alpha^s}{\Gamma(s+1)}\right)$$

$$= M_{\vec{n}}^{\alpha,\vec{\beta}}(s).$$

This completes the proof of expression (76). □

5. Appendix: AT-Property for the Studied Discrete Measures

Lemma 9. *The system of functions*

$$\alpha_1^s, x(s)\alpha_1^s, \ldots, x(s)^{n_1-1}\alpha_1^s, \ldots, \alpha_r^s, x(s)\alpha_r^s, \ldots, x(s)^{n_r-1}\alpha_r^s, \tag{80}$$

with $\alpha_i > 0$, $i = 1, 2, \ldots, r$, with all the α_i different, and $(\alpha_i/\alpha_j) \neq q^k$, $k \in \mathbb{Z}$, $i, j = 1, \ldots, r$, $i \neq j$, forms a Chebyshev system on $\mathbb{R}^+$ for every $\vec{n} = (n_1, \ldots, n_r) \in \mathbb{N}^r$.

Proof. For a Chebyshev system every linear combination $\sum_{i=1}^{r} Q_{n_i-1}(x(s))\alpha_i^s$ has at most $|\vec{n}|-1$ zeros on $\mathbb{R}^+$ for every $Q_{n_i-1}(x(s)) \in \mathbb{P}_{n_i-1} \setminus \{0\}$. Since $x(s) = c_1 q^s + c_3$, where c_1, c_3 are constants, we consider $\sum_{i=1}^{r} Q_{n_i-1}(q^s)\alpha_i^s$, instead. Thus, the system (80) transforms into

$$a_{1,0}^s, a_{1,1}^s, \ldots, a_{1,n_1-1}^s, \ldots, a_{r,0}^s, a_{r,1}^s, \ldots, a_{r,n_r-1}^s,$$

where $a_{i,k} = (q^k \alpha_i)$, with $k = 0, \ldots, n_i-1$, $i = 1, \ldots, r$. Observe that $a_{j,m} \neq a_{l,p}$ for $j \neq l$, $m \neq p$. Hence, identity $a_{i,k} = e^{\log a_{i,k}}$ yields the well-known Chebyshev system (see [34], p. 138)

$$e^{s\log a_{1,0}}, e^{s\log a_{1,1}}, \ldots, e^{s\log a_{1,n_1-1}}, \ldots, e^{s\log a_{r,0}}, e^{s\log a_{r,1}}, \ldots, e^{s\log a_{r,n_r-1}}.$$

Then, we conclude that the functions (80) form a Chebyshev system on $\mathbb{R}^+$. □

Lemma 10. *Let $\beta_i > 0$ and $\beta_i - \beta_j \notin \mathbb{Z}$ whenever $i \neq j$. Assume $v(s)$ is a continuous function with no zeros on $\mathbb{R}^+$, then the functions*

$$v(s)\Gamma_q(s+\beta_1), v(s)x(s)\Gamma_q(s+\beta_1), \ldots, v(s)x(s)^{n_1-1}\Gamma_q(s+\beta_1),$$

$$\vdots \tag{81}$$

$$v(s)\Gamma_q(s+\beta_r), v(s)x(s)\Gamma_q(s+\beta_r), \ldots, v(s)x(s)^{n_r-1}\Gamma_q(s+\beta_r),$$

form a Chebyshev system on Ω for every $\vec{n} \in \mathbb{N}^r$.

Proof. For the system of functions (81) we have a Chebyshev system on Ω for every $\vec{n} \in \mathbb{N}^r$ if and only if every linear combination of these functions (except the one with each coefficient equals 0) has at most $|\vec{n}| - 1$ zeros. This linear combination can be rewritten as a function of the system

$$v(s)\Gamma_q\left(s + \beta_1\right), v(s)\left[s + \beta_1\right]_q^{(1)} \Gamma_q\left(s + \beta_1\right), \ldots,$$
$$v(s)\left[s + \beta_1 + n_1 - 2\right]_q^{(n_1-1)} \Gamma_q\left(s + \beta_1\right),$$
$$v(s)\Gamma_q\left(s + \beta_r\right), v(s)\left[s + \beta_r\right]_q^{(1)} \Gamma_q\left(s + \beta_r\right), \ldots,$$
$$v(s)\left[s + \beta_1 + n_r - 2\right]_q^{(n_r-1)} \Gamma_q\left(s + \beta_r\right),$$

where $[s + \beta_i]_q^{(n_i)}$, $i = 1, \ldots, r$, is given in (6).

Observe that

$$[s + k - 1]_q^{(k)} \Gamma_q(s) = \Gamma_q\left(s + k\right),$$

holds. Therefore, the above system transforms into

$$v(s)\Gamma_q\left(s + \beta_1\right), v(s)\Gamma_q\left(s + \beta_1 + 1\right), \ldots, v(s)\Gamma_q\left(s + \beta_1 + n_1 - 1\right),$$
$$\vdots \tag{82}$$
$$v(s)\Gamma_q\left(s + \beta_r\right), v(s)\Gamma_q\left(s + \beta_r + 1\right), \ldots, v(s)\Gamma_q\left(s + \beta_r + n_r - 1\right).$$

Thus, it is sufficient to prove that these systems (82) form a Chebyshev system on Ω for every $\vec{n} \in \mathbb{N}^r$. If we define the matrix $\mathcal{A}\left(\vec{n}, s_1, \ldots, s_{|\vec{n}|}\right)$ by

$$\begin{pmatrix} \Gamma_q\left(s_1 + \beta_1\right) & \Gamma_q\left(s_2 + \beta_1\right) & \cdots & \Gamma_q\left(s_{|\vec{n}|} + \beta_1\right) \\ \vdots & \vdots & & \vdots \\ \Gamma_q\left(s_1 + \beta_1 + n_1 - 1\right) & \Gamma_q\left(s_2 + \beta_1 + n_1 - 1\right) & \cdots & \Gamma_q\left(s_{|\vec{n}|} + \beta_1 + n_1 - 1\right) \\ \vdots & \vdots & & \vdots \\ \Gamma_q\left(s_1 + \beta_r\right) & \Gamma_q\left(s_2 + \beta_r\right) & \cdots & \Gamma_q\left(s_{|\vec{n}|} + \beta_r\right) \\ \vdots & \vdots & & \vdots \\ \Gamma_q\left(s_1 + \beta_r + n_1 - 1\right) & \Gamma_q\left(s_2 + \beta_r + n_1 - 1\right) & \cdots & \Gamma_q\left(s_{|\vec{n}|} + \beta_r + n_1 - 1\right) \end{pmatrix},$$

the proof is reduced to showing that $\det \mathcal{A}\left(\vec{n}, s_1, \ldots, s_{|\vec{n}|}\right) \neq 0$, for every $|\vec{n}|$, and different points $s_1, \ldots, s_{|\vec{n}|}$ in Ω, because $|v| > 0$ on Ω. Now we replace the q-gamma function in $\mathcal{A}\left(\vec{n}, s_1, \ldots, s_{|\vec{n}|}\right)$ by the integral representation

$$\Gamma_q(s) = \int_0^{\frac{1}{1-q}} t^{s-1} E_q^{-qt} d_q t = \int_0^{x(\infty)} t^{s-1} E_q^{-qt} d_q t, \quad s > 0, \tag{83}$$

where

$$E_q^z = {}_0\varphi_0\left(-; -; q, -\left(1 - q\right) z\right)$$

denotes the q-analogue of the exponential function. From multilinearity of the determinant we take $|\vec{n}|$ integrations out of $|\vec{n}|$ rows to obtain

$$\det \mathcal{A}\left(\vec{n}, s_1, \ldots, s_{|\vec{n}|}\right) = \underbrace{\int_0^{x(\infty)} \cdots \int_0^{x(\infty)}}_{|\vec{n}| \text{ times}} \prod_{1 \leq i \leq |\vec{n}|} E_q^{-qt_i} t_i^{s_i-1}$$

$$\times \det \mathcal{B}\left(\vec{n}, t_1, \ldots, t_{|\vec{n}|}\right) d_q t_1 \ldots d_q t_{|\vec{n}|}, \qquad (84)$$

where

$$\mathcal{B}\left(\vec{n}, t_1, \ldots, t_{|\vec{n}|}\right) = \begin{pmatrix} t_1^{\beta_1} & t_2^{\beta_1} & \cdots & t_{|\vec{n}|}^{\beta_1} \\ \vdots & \vdots & & \vdots \\ t_1^{\beta_1+n_1-1} & t_2^{\beta_1+n_1-1} & \cdots & t_{|\vec{n}|}^{\beta_1+n_1-1} \\ \vdots & \vdots & & \vdots \\ t_1^{\beta_r} & t_2^{\beta_r} & \cdots & t_{|\vec{n}|}^{\beta_r} \\ \vdots & \vdots & & \vdots \\ t_1^{\beta_r+n_r-1} & t_2^{\beta_r+n_r-1} & \cdots & t_{|\vec{n}|}^{\beta_r+n_r-1} \end{pmatrix}.$$

Notice that, from ([34], p. 138, example 4) we know that the functions

$$t^{\beta_1}, \ldots, t^{\beta_1+n_1-1}, \ldots, t^{\beta_r}, \ldots, t^{\beta_r+n_r-1},$$

form a Chebyshev system on $\mathbb{R}^+$ if all the exponents are different, which is in accordance with our choice $\beta_i - \beta_j \notin \mathbb{Z}$ whenever $i \neq j$. Moreover, if all $n_i < N+1$, then the exponents involved in the above matrix are different for $\beta_i - \beta_j \notin \{0, 1, \ldots, N\}$ whenever $i \neq j$. Hence, $\det \mathcal{B}\left(\vec{n}, t_1, \ldots, t_{|\vec{n}|}\right)$ does not vanish for distinct $t_1, \ldots, t_{|\vec{n}|}$. Now, for a permutation σ of $\{1, \ldots, |\vec{n}|\}$ we make a change of variables $t_i \mapsto t_{\sigma(i)}$ in the integral (84). Thus, we have

$$\det \mathcal{A}\left(\vec{n}, t_1, \ldots, t_{|\vec{n}|}\right) = \underbrace{\int_0^{x(\infty)} \cdots \int_0^{x(\infty)}}_{|\vec{n}| \text{ times}} \prod_{1 \leq i \leq |\vec{n}|} E_q^{-qt_i} \det \mathcal{B}\left(\vec{n}, t_1, \ldots, t_{|\vec{n}|}\right)$$

$$\times \operatorname{sgn}(\sigma) \prod_{1 \leq j \leq |\vec{n}|} t_{\sigma(j)}^{s_j-1} d_q t_1 \ldots d_q t_{|\vec{n}|}. \qquad (85)$$

We average (85) over all permutation σ, i.e.,

$$\det \mathcal{A}\left(\vec{n}, s_1, \ldots, s_{|\vec{n}|}\right) = \frac{1}{n!} \sum_{\sigma \in S_{|\vec{n}|}} \underbrace{\int_0^{x(\infty)} \cdots \int_0^{x(\infty)}}_{|\vec{n}| \text{ times}} \prod_{1 \leq i \leq |\vec{n}|} E_q^{-qt_i}$$

$$\times \det \mathcal{B}\left(\vec{n}, t_1, \ldots, t_{|\vec{n}|}\right) \operatorname{sgn}(\sigma) \prod_{1 \leq j \leq |\vec{n}|} t_{\sigma(j)}^{s_j-1} d_q t_1 \ldots d_q t_{|\vec{n}|},$$

being $S_{|\vec{n}|}$ the permutation group. Now, relabeling the choice of points, i.e., $t_1, \ldots, t_{|\vec{n}|}$, where $0 < t_1 < \cdots < t_{|\vec{n}|}$, we have

$$\det \mathcal{A}\left(\vec{n}, t_1, \ldots, t_{|\vec{n}|}\right) = \frac{1}{n!} \underbrace{\int_0^{x(\infty)} \cdots \int_0^{x(\infty)}}_{0 < t_1 < \cdots < t_{|\vec{n}|}} \prod_{1 \leq i \leq |\vec{n}|} E_q^{-qt_i} \det \mathcal{B}\left(\vec{n}, t_1, \ldots, t_{|\vec{n}|}\right)$$

$$\times \sum_{\sigma \in S_{|\vec{n}|}} \mathrm{sgn}\,(\sigma) \prod_{1 \leq j \leq |\vec{n}|} t_{\sigma(j)}^{s_j - 1} d_q t_1 \ldots d_q t_{|\vec{n}|}. \tag{86}$$

As a result, from the definition of determinant we have

$$\sum_{\sigma \in S_{|\vec{n}|}} \mathrm{sgn}\,(\sigma) \prod_{1 \leq j \leq |\vec{n}|} t_{\sigma(j)}^{s_j - 1} = \begin{vmatrix} t_1^{s_1 - 1} & t_1^{s_2 - 1} & \cdots & t_1^{s_{|\vec{n}|} - 1} \\ t_2^{s_1 - 1} & t_2^{s_2 - 1} & \cdots & t_2^{s_{|\vec{n}|} - 1} \\ \vdots & \vdots & & \vdots \\ t_{|\vec{n}|}^{s_1 - 1} & t_{|\vec{n}|}^{s_2 - 1} & \cdots & t_{|\vec{n}|}^{s_{|\vec{n}|} - 1} \end{vmatrix}. \tag{87}$$

Taking into account that $t_1, \ldots, t_{|\vec{n}|}$ are strictly positive and different, then using the result in ([34], p. 138, example 3) with multi-index $(1, \ldots, 1)$, will imply that (87) is different from zero if all the $s_1, \ldots, s_{|\vec{n}|}$ are different. Accordingly, for distinct $s_1, \ldots, s_{|\vec{n}|}$, the integrand of Equation (86) has a constant sign in the region of integration and hence $\det \mathcal{A}\left(\vec{n}, s_1, \ldots, s_{|\vec{n}|}\right)$ does not vanish. $\square$

As a consequence of Lemma 10 the system of measures $\mu_1, \mu_2, \ldots, \mu_r$ given in (51) forms an AT system on Ω.

6. Concluding Remarks

We have studied two families of multiple orthogonal polynomials on a non-uniform lattice, i.e., q-Meixner multiple orthogonal polynomials of the first and second kind, respectively. They are derived from two systems of q-discrete measures. Each system forms an AT-system. For these families of multiple q-orthogonal polynomials we have obtained the Rodrigues-type Formulas (31) and (55) as well as the recurrence relations (47) and (71), and the q-difference equations (41) and (69). The use of some q-difference operators has played an important role in deriving the aforementioned algebraic properties. Finally, in the limit situation $q \to 1$, we have obtained the multiple Meixner polynomials given in [28].

In closing, we address some research directions and open problems:

Problem 1. *A description of the main term of the logarithm asymptotics of the q-analogues of multiple Meixner polynomials deserves special attention. For such a purpose, we will use an algebraic function formulation for the solution of the equilibrium problem with constraints [45–47] to describe the zero distribution of multiple orthogonal polynomials [48]. This approach has been recently developed for multiple Meixner polynomials in [21] (see [49] as well as [17,50] for other approaches). Moreover, by analyzing the limiting behavior of the coefficients of the recurrence relations for such polynomials we expect to obtain the main term of their asymptotics.*

Problem 2. *In [51] the authors use the annihilation and creation operators a_i, $a_i^\dagger$ ($i = 1, \ldots, r$) satisfying the commutation relations*

$$[a_i, a_j^\dagger] = \delta_{i,j}, \qquad [a_i^\dagger, a_j^\dagger] = [a_i, a_j] = 0, \quad i, j = 1, \ldots, r.$$

The generated Lie algebra is formed by r copies of the Heisenberg–Weyl algebra $W_i = \mathrm{span}\{a_i, a_i^\dagger, 1\}$. For a more detailed and technical information about orthogonal polynomials in the Lie algebras see [52] as well as [53] for quantum mechanics and polynomials of a discrete variable.

The normalized simultaneous eigenvectors of the r number operators $N_i = a_i^\dagger a_i$ are denoted by

$$|n_1, n_2, \ldots, n_r\rangle = |n_1\rangle |n_2\rangle \cdots |n_r\rangle,$$

Indeed,

$$N_i |n_1, n_2, \ldots, n_r\rangle = n_i |n_1, n_2, \ldots, n_r\rangle,$$
$$\langle m_1, m_2, \ldots, m_r | n_1, n_2, \ldots, n_r\rangle = \delta_{m_1,n_1} \cdots \delta_{m_r,n_r}.$$

Moreover,

$$a_i^\dagger |n_1, n_2, \ldots, n_r\rangle = \sqrt{n_i + 1} |n_1, \ldots, n_i + 1, \ldots, n_r\rangle,$$
$$a_i |n_1, n_2, \ldots, n_r\rangle = \sqrt{n_i} |n_1, \ldots, n_i - 1, \ldots, n_r\rangle,$$

The Bargmann realization in terms of coordinates z_i, $i = 1, \ldots, r$, in $\mathbb{C}^r$ has

$$a_i = \frac{\partial}{\partial z_i}, \quad a_i^\dagger = z_i,$$
$$\langle z_1, z_2, \ldots, z_r | n_1, n_2, \ldots, n_r\rangle = \frac{z_1^{n_1} \cdots z_r^{n_r}}{\sqrt{n_1! \cdots n_r!}}.$$

For the model in [51]

$$H_i^{\vec{\alpha},\beta} = a_i + \sum_{k=1}^{r} \frac{N_k}{1 - \alpha_k} + \left(\frac{\alpha_i}{1 - \alpha_i} + \sum_{j=1}^{r} \frac{\alpha_j}{(1 - \alpha_j)^2} a_j^\dagger \right) \left(\sum_{k=1}^{r} N_k + \beta \right), \quad i = 1, \ldots, r,$$

represent the set of non-Hermitian operators defined in the universal enveloping algebra formed by the r copies W_i.

The operators making up the H_i generate an isomorphic Lie algebra to that of the diffeomorphisms in $\mathbb{C}^r$ spanned by vector fields of the form

$$Z = \sum_{i=1}^{r} f_i(\vec{z}) \frac{\partial}{\partial z_i} + g(\vec{z}), \quad \vec{z} = (z_1, \ldots, z_r).$$

The authors indicated that although in the coordinate realization where

$$a_i = \frac{1}{\sqrt{2}} \left(x_i + \frac{\partial}{\partial x_i} \right), \quad a^\dagger = \frac{1}{\sqrt{2}} \left(x_i - \frac{\partial}{\partial x_i} \right),$$

the operators H_i are third order differential operators, they can be considered as Hamiltonians and are simultaneously diagonalized by the multiple Meixner polynomials of the first kind.

Consider the states $|x, \vec{\alpha}, \beta\rangle$ defined by means of the combination of states $|n_1, \ldots, n_r\rangle$ as:

$$|x, \vec{\alpha}, \beta\rangle = N_{x,\vec{\alpha},\beta}^r \sum_{\vec{n}} \frac{M_{\vec{n}}^{\vec{\alpha},\beta}(x)}{\sqrt{n_1! \cdots n_r!}} |n_1, n_2, \ldots, n_r\rangle, \quad x \in \mathbb{N}.$$

Thus,

$$H_i^{\vec{\alpha},\beta}|x,\vec{\alpha},\beta\rangle = N_{x,\vec{\alpha},\beta}^r \sum_{\vec{n}} \frac{1}{\sqrt{n_1!\cdots n_r!}} \Big[M_{\vec{n}+\vec{e}_i}^{\vec{\alpha},\beta}(x)$$
$$+ \left((\beta+|\vec{n}|)\left(\frac{\alpha_i}{1-\alpha_i}\right) + \sum_{k=1}^r \frac{n_k}{1-\alpha_k}\right) M_{\vec{n}}^{\vec{\alpha},\beta}(x)$$
$$+ \sum_{j=1}^r \frac{\alpha_j n_j\,(\beta+|\vec{n}|-1)}{(\alpha_j-1)^2} M_{\vec{n}-\vec{e}_j}^{\vec{\alpha},\beta}(x)\Big]|n_1,n_2,\ldots,n_r\rangle.$$

In [51], by using the recurrence relation (19) for multiple Meixner polynomials of the first kind, the following relation

$$H_i^{\vec{\alpha},\beta}|x,\vec{\alpha},\beta\rangle = x|x,\vec{\alpha},\beta\rangle,$$

holds.

Despite the fact the operators are non-Hermitian, they have a real spectrum given by the lattice, i.e., the non-negative integers. The states $|x,\vec{\alpha},\beta\rangle$ are uniquely defined as the joint eigenstates of the Hamiltonian operators with eigenvalues equal to x. Moreover,

$$[H_i^{\vec{\alpha},\beta}, H_j^{\vec{\alpha},\beta}]|x,\vec{\alpha},\beta\rangle = 0.$$

However, these Hamiltonians do not commute pairwise. Indeed,

$$[H_i^{\vec{\alpha},\beta}, H_j^{\vec{\alpha},\beta}] = a_i - a_j + \frac{\alpha_i-\alpha_j}{(1-\alpha_i)(1-\alpha_j)}\left(\beta+\sum_{k=1}^r N_k\right).$$

Finally, because they do not commute and yet have common eigenvectors, the authors in [51] say that they form a 'weakly' integrable system.

The physical model described above motivates the study of a q-deformed model, which is currently being considered by using the results of the present paper involving the q-analogue of multiple Meixner polynomials of the first kind. In particular, the recurrence relation (47).

Author Contributions: Conceptualization, J.A. and A.M.R.-A.; methodology, J.A.; formal analysis, J.A.; investigation, J.A. and A.M.R.-A.; resources, J.A. and A.M.R.-A.; writing—original draft preparation, J.A. and A.M.R.-A.; writing—review and editing, J.A.; visualization, J.A. and A.M.R.-A.; supervision, J.A.; project administration, J.A.; funding acquisition, J.A. All authors have read and agreed to the published version of the manuscript.

References

1. Hermite, C. *Sur la Fonction Exponentielle*; Gauthier-Villars: Paris, France, 1874; pp. 1–33. Available online: https://archive.org/details/surlafonctionexp00hermuoft/page/n1 (accessed on 6 July 2020).
2. Aptekarev, A.I. Multiple orthogonal polynomials. *J. Comput. Appl. Math.* **1998**, *99*, 423–447. [CrossRef]
3. Aptekarev, A.I.; Branquinho, A.; Van Assche, W. Multiple orthogonal polynomials for classical weights. *Trans. Am. Math. Soc.* **2003**, *335*, 3887–3914. [CrossRef]
4. Arvesú, J.; Soria-Lorente, A. On Infinitely Many Rational Approximants to $\zeta(3)$. *Mathematics* **2019**, *7*, 1176. [CrossRef]
5. Kalyagin, V.A. Higher order difference operator's spectra characteristics and the convergence of the joint rational approximations. *Dokl. Akad. Nauk* **1995**, *340*, 15–17.
6. Nikishin, E.M. On simultaneous Padé approximations. *Mat. Sb.* **1980**, *113*, 499–519; English Transl.: *Math. USSR Sb.* **1982**, *41*.

7. Prévost, M.; Rivoal, T. Remainder Padé approximants for the exponential function. *Constr. Approx.* **2007**, *25*, 109–123. [CrossRef]
8. Sorokin, V.N. Hermite-Padé approximations of polylogarithms. *Izv. Vyssh. Uchebn. Zaved. Mat.* **1994**, *5*, 49–59.
9. Angelesco, A. Sur l'approximation simultanée de plusieurs intégrales définies. *CR Acad. Sci. Paris* **1918**, *167*, 629–631.
10. Brezinski, C.; Van Iseghem, J. Vector orthogonal polynomials of dimension *d*. In *Approximation and Computation: A Festschrift in Honor of Walter Gautschi*; Zahar, R.V.M., Ed.; ISNM International Series of Numerical Mathematics; Birkhäuser: Boston, MA, USA, 1994; Volume 119, pp. 29–39.
11. Bustamante, J.; Lagomasino, G.L. Hermite-Padé approximants for Nikishin systems of analytic functions. *Mat. Sb.* **1992**, *183*, 117–138. [CrossRef]
12. De Bruin, M.G. Simultaneous Padé approximation and orthogonality. In *Polynômes Orthogonaux et Applications*; Lecture Notes in Mathematics 1171; Brezinski, C., Eds.; Springer: Berlin/Heidelberg, Germany, 1985; pp. 74–83.
13. De Bruin, M.G. Some aspects of simultaneous rational approximation. In *Numerical Analysis and Mathematical Modeling*; Banach Center Publications 24; PWN-Polish Scientific Publishers: Warsaw, Poland, 1990; pp. 51–84.
14. Gonchar, A.A.; Rakhmanov, E.A.; Sorokin, V.N. Hermite-Padé approximants for systems of Markov-type functions. *Mat. Sb.* **1997**, *188*, 33–58. [CrossRef]
15. Kalyagin, V.A. On a class of polynomials defined by two orthogonality relations. *Mat. Sb.* **1979**, *110*, 609–627. [CrossRef]
16. Mahler, K. Perfect systems. *Compos. Math.* **1968**, *19*, 95–166.
17. Sorokin, V.N. A generalization of classical orthogonal polynomials and the convergence of simultaneous Padé approximants. *J. Soviet Math.* **1989**, *45*, 1461–1499. [CrossRef]
18. Kaliaguine, V.A.; Ronveaux A. On a system of classical polynomials of simultaneous orthogonality. *J. Comput. Appl. Math.* **1996**, *67*, 207–217. [CrossRef]
19. Sorokin, V.N. Simultaneous Padé approximants for finite and infinite intervals. *Izv. Vyssh. Uchebn. Zaved. Mat.* **1984**, *8*, 45–52.
20. Álvarez-Fernández, C.; Fidalgo Prieto, U.; Mañas, M. Multiple orthogonal polynomials of mixed type: Gauss-Borel factorization and the multi-component 2D toda hierarchy. *Adv. Math.* **2011**, *227*, 1451–1525. [CrossRef]
21. Aptekarev, A.I.; Arvesú, J. Asymptotics for multiple Meixner polynomials. *J. Math. Anal. Appl.* **2014**, *411*, 485–505. [CrossRef]
22. Schweiger, F. *Multidimensional Continued Fractions*; Oxford Science Publications; Oxford University Press: Oxford, UK, 2000.
23. Bleher, P.M.; Kuijlaars, A.B.J. Random matrices with external source and multiple orthogonal polynomials. *Int. Math. Res. Not. IMRN* **2004**, *3*, 109–129. [CrossRef]
24. Borodin, A.; Ferrari, P.L.; Sasamoto, T. Two speed TASEP. *J. Stat. Phys.* **2009**, *137*, 936–977. [CrossRef]
25. Daems, E.; Kuijlaars, A.B.J. A Christoffel–Darboux formula for multiple orthogonal polyno-mials. *J. Approx. Theory* **2004**, *130*, 190–202. [CrossRef]
26. Johansson, K. Discrete orthogonal polynomial ensembles and the Plancherel measure. *Ann. Math.* **2001**, *153*, 259–296. [CrossRef]
27. Kuijlaars, A.B.J. Multiple orthogonal polynomials in random matrix theory. In Proceedings of the International Congress of Mathematicians, Hyderabad, India, 19–27 August 2010; Bhatia, R., Ed.; Hindustan Book Agency: New Delhi, India, 2010; Volume 3, pp. 1417–1432.
28. Arvesú, J.; Coussement, J.; Van Assche, W. Some discrete multiple orthogonal polynomials. *J. Comput. Appl. Math.* **2003**, *153*, 19–45. [CrossRef]
29. Van Assche, W.; Coussement, E. Some classical multiple orthogonal polynomials. *J. Comput. Appl. Math.* **2001**, *127*, 317–347. [CrossRef]
30. Van Assche, W.; Yakubovich, S.B. Multiple orthogonal polynomials associated with Macdonald functions. *Integral Transform. Spec. Funct.* **2007**, *9*, 229–244. [CrossRef]
31. Arvesú, J.; Esposito, C. A high-order *q*-difference equation for *q*-Hahn multiple orthogonal polynomials. *J. Differ. Equ. Appl.* **2012**, *18*, 833–847. [CrossRef]
32. Arvesú, J.; Ramírez-Aberasturis, A.M. On the *q*-Charlier multiple orthogonal polynomials. *SIGMA Symmetry Integr. Geom. Methods Appl.* **2015**, *11*, 026. [CrossRef]

33. Postelmans, K.; Van Assche, W. Multiple little q-Jacobi polynomials. *J. Comput. Appl. Math.* **2005**, *178*, 361–375. [CrossRef]

34. Nikishin, E.M.; Sorokin, V.N. Rational approximations and orthogonality. In *Translations of Mathematical Monographs*; American Mathematical Society: Providence, RI, USA, 1991; Volume 92.

35. Álvarez-Nodarse, R.; Arvesú, J. On the q-polynomials in the exponential lattice $x(s) = c_1 q^s + c_3$. *Integral Transform. Spec. Funct.* **1999**, *8*, 299–324. [CrossRef]

36. Arvesú, J. On some properties of q-Hahn multiple orthogonal polynomials. *J. Comput. Appl. Math.* **2010**, *233*, 1462–1469. [CrossRef]

37. Gasper, G.; Rahman, M. Basic hypergeometric series. In *Encyclopedia of Mathematics and its Applications*, 2nd ed.; Cambridge University Press: Cambridge, UK, 2004; Volume 96. [CrossRef]

38. Nikiforov, A.F.; Suslov, S.K.; Uvarov, V.B. *Classical Orthogonal Polynomials of A Discrete Variable*; Springer Series in Computational Physics; Springer: Berlin/Heidelberg, Germany, 1991. [CrossRef]

39. Lee, D.W. Difference equations for discrete classical multiple orthogonal polynomials. *J. Approx. Theory* **2008**, *150*, 132–152. [CrossRef]

40. Coussement, J.; Van Assche, W. Differential equations for multiple orthogonal polynomials with respect to classical weights. *J. Phys. A Math. Gen.* **2006**, *39*, 3311–3318. [CrossRef]

41. Van Assche, W. Non-symmetric linear difference equations for multiple orthogonal polynomials. *CRM Proc. Lect. Notes* **2000**, *25*, 391–405.

42. Van Assche, W. Difference equations for multiple Charlier and Meixner polynomials. In Proceedings of the Sixth International Conference on Difference Equations, Augsburg, Germany, 30 July–3 August 2001; CRC: Boca Raton, FL, USA, 2004; pp. 549–557.

43. Koekoek, R.; Lesky, P.A.; Swarttouw, R.F. *Hypergeometric Orthogonal Polynomials and Their q-Analogues*; Springer Monographs in Mathematics; Springer: Berlin/Heidelberg, Germany, 2010. [CrossRef]

44. Nikiforov, A.F.; Uvarov, V.B. Polynomial solutions of hypergeometric type difference equations and their classification. *Integral Transform. Spec. Funct.* **1993**, *1*, 223–249. [CrossRef]

45. Bleher, P.M.; Delvaux, S.; Kuijlaars, A.B.J. Random matrix model with external source and a constrained vector equilibrium problem. *Commun. Pure Appl. Math.* **2011**, *64*, 116–160. [CrossRef]

46. Dragnev, P.D.; Saff, E.B. Constrained energy problems with applications to orthogonal polynomials of a discrete variable. *J. Anal. Math.* **1997**, *72*, 223–259. [CrossRef]

47. Gonchar, A.A.; Rakhmanov, E.A. On the equilibrium problem for vector potentials. *Uspekhi Mat. Nauk* **1985**, *40*, 155–156; English Transl.: *Russ. Math. Surv.* **1985**, *40*, 183–184. [CrossRef]

48. Rakhmanov, E.A. Equilibrium measure and the distribution of zeros of the extremal polynomials of a discrete variable. *Mat. Sb.* **1996**, *187*, 109–124; English Transl.: *Sb. Math.* **1996**, *187*, 1213–1228. [CrossRef]

49. Aptekarev, A.I.; Kalyagin, V.A.; Lysov, V.G.; Tulyakov, D.N. Equilibrium of vector potentials and uniformization of the algebraic curves of genus 0. *J. Comput. Appl. Math.* **2009**, *233*, 602–616. [CrossRef]

50. Lysov, V.G. Strong asymptotics of the Hermite-Padé approximants for a system of Stieltjes functions with Laguerre weight. *Mat. Sb.* **2005**, *196*, 99–122; English Transl.: *Sb. Math.* **2005**, *196*, 1815–1840. [CrossRef]

51. Miki, H.; Tsujimoto, S.; Vinet, L.; Zhedanov, A. An algebraic model for the multiple Meixner polynomials of the first kind. *J. Phys. A Math. Theor.* **2012**, *45*, 325205. [CrossRef]

52. Granovskii, Y.I.; Zhedanov, A. Orthogonal polynomials in the Lie algebras. *Sov. Phys. J.* **1986**, *29*, 387–393. [CrossRef]

53. Floreanini, R.; LeTourneux, J.; Vinet, L. Quantum mechanics and polynomials of a discrete variable. *Ann. Phys.* **1993**, *226*, 331–349. [CrossRef]

3

Exceptional Set for Sums of Symmetric Mixed Powers of Primes

Jinjiang Li [1], Chao Liu [1], Zhuo Zhang [1] and Min Zhang [2,*]

[1] Department of Mathematics, China University of Mining and Technology, Beijing 100083, China; jinjiang.li.math@gmail.com (J.L.); chao.liu@student.cumtb.edu.cn (C.L.); zhuo.zhang.math@foxmail.com (Z.Z.)

[2] School of Applied Science, Beijing Information Science and Technology University, Beijing 100192, China

* Correspondence: min.zhang.math@gmail.com

Abstract: The main purpose of this paper is to use the Hardy–Littlewood method to study the solvability of mixed powers of primes. To be specific, we consider the even integers represented as the sum of one prime, one square of prime, one cube of prime, and one biquadrate of prime. However, this representation can not be realized for all even integers. In this paper, we establish the exceptional set of this kind of representation and give an upper bound estimate.

Keywords: Waring–Goldbach problem; circle method; exceptional set; symmetric form

MSC: 11P05, 11P32, 11P55

1. Introduction and Main Result

Let $N, k_1, k_2, \ldots, k_s$ be natural numbers which satisfy $2 \leqslant k_1 \leqslant k_2 \leqslant \cdots \leqslant k_s$, $N > s$. Waring's problem of unlike powers concerns the possibility of representation of N in the form

$$N = x_1^{k_1} + x_2^{k_2} + \cdots + x_s^{k_s}. \tag{1}$$

For previous literature, the reader could refer to section P12 of LeVeque's *Reviews in number theory* and the bibliography of Vaughan [1]. For the special case, $k_1 = k_2 = \cdots = k_s$, an interesting problem is to determine the value for $k \geqslant 2$, called Waring's problem, of the function $G(k)$, the least positive number s such that every *sufficiently large* number can be represented the sum of at most s k-th powers of natural numbers. For this problem, there are only two values of the function $G(k)$ determined exactly. To be specific, $G(2) = 4$, by Lagrange in 1770, and $G(4) = 16$, by Davenport [2]. The majority of information for $G(k)$ has been derived from the Hardy–Littlewood method. This method has arised from a celebrated paper of Hardy and Ramanujan [3], which focused on the partition function.

There are many authors who devoted to establish many kinds of generalisations of this classical version of Waring's problem. Among these results, it is necessary to illustrate some of the majority variants. We begin with the most famous Waring–Goldbach problem, for which one devotes to investigate the possibility of the representation of integers as sums of k-th powers of prime numbers. In order to explain the associated congruence conditions, we denote by k a natural number and p a prime number. We write $\theta = \theta(k; p)$ as the integer with the properties $p^\theta | k$ and $p^\theta \nmid k$, and then define $\gamma = \gamma(k, p)$ by

$$\gamma(k, p) = \begin{cases} \theta + 2, & \text{when } p = 2 \text{ and } \theta > 0, \\ \theta + 1, & \text{otherwise.} \end{cases}$$

Also, we set

$$K(k) = \prod_{(p-1)|k} p^\gamma.$$

Denote by $H(k)$ the smallest integer s, which satisfies every sufficiently large integer congruent to s modulo $K(k)$ can be represented as the sum of s k-th powers of primes . By noting the fact that for $(p-1)|k$, we have $p^\theta(p-1)|k$, provided that $a^k \equiv 1 \pmod{p^\gamma}$ and $(p,a) = 1$. This states the seemingly awkward definition of $H(k)$, because if n is the sum of s k-th powers of primes exceeding $k+1$, then it must satisfy $n \equiv s \pmod{K(k)}$. Trivially, further congruence conditions could arise from the primes p which satisfy $(p-1) \nmid k$. Following the previous investigations of Vinogradov [4,5], Hua systematically considered and investigated the additive problems involving prime variables in his famous book (see Hua [6,7]).

For the nonhomogeneous case, the most optimistic conjecture suggests that, for each prime p, if the Equation (1) has p-adic solutions and satisfies

$$k_1^{-1} + k_2^{-1} + \cdots + k_s^{-1} > 1, \tag{2}$$

then n can be written as the sum of unlike powers of positive integers (1) provided that n is sufficiently large in terms of k. For $s = 3$, such an claim maybe not true in certain situations (see Jagy and Kaplansky [8], or Exercise 5 of Chapter 8 of Vaughan [1]). However, a guide of application for the Hardy–Littlewood method suggests that the condition (2) should ensure at least that *almost all* integers satisfying the expected congruence conditions can be represented. Moreover, once subject to the following condition

$$k_1^{-1} + k_2^{-1} + \cdots + k_s^{-1} > 2, \tag{3}$$

a standard application of the Hardy–Littlewood method suggests that all the integers, which satisfy necessary congruence conditions, could be written in the form (1). Meanwhile, a conventional argument of the circle method shows that in situations in which the condition (2) does not hold, then every sufficiently large integer can not be represented in the expected form.

Since the Hardy–Littlewood method, the investigation of Waring's problem for unlike powers has produced splendid progress in circle method, especially for the classical version of Waring's problem. Additive Waring's problems of unlike powers involving squares, cubes or biquadrates offen attract greater interest of many mathematicians than those cases with higher mixed powers, and the current circumstance is quite satisfactory. For example, the reader can refer to references [9–19].

The Waring–Goldbach problem of mixed powers concerns the representation of N which satisfying some necessary congruence conditions as the form

$$N = p_1^{k_1} + p_2^{k_2} + \cdots + p_s^{k_s},$$

where $p_1, p_2, \ldots, p_s$ are prime variables.

In 2002, Brüdern and Kawada [20] proved that for every sufficiently large even integer N, the equation

$$N = x + p_2^2 + p_3^3 + p_4^4$$

is solvable with x being an almost–prime $\mathcal{P}_2$ and the p_j $(j = 2,3,4)$ primes. As usual, $\mathcal{P}_r$ denotes an almost–prime with at most r prime factors, counted according to multiplicity. On the other hand, in 2015, Zhao [21] established that, for $k = 3$ or 4, every sufficiently large even integer N can be represented as the form

$$N = p_1 + p_2^2 + p_3^3 + p_4^k + 2^{v_1} + 2^{v_2} + \cdots + 2^{v_{t(k)}},$$

where $p_1, \ldots, p_4$ are primes, $v_1, v_2, \ldots, v_{t(k)}$ are natural numbers, and $t(3) = 16$, $t(4) = 18$, which is an improvement result of Liu and Lü [22]. Afterwards, Lü [23] improved the result of Zhao [21] and showed that every sufficiently large even integer N can be represented as a sum of one prime, one square of prime, one cube of prime, one biquadrate of prime and 16 powers of 2.

In view of the results of Brüdern and Kawada [20], Zhao [21], Liu and Lü [22] and Lü [23], it is reasonable to conjecture that, for sufficiently large integer N satisfying $N \equiv 0 \,(\mathrm{mod}\, 2)$, the following Diophantine equation

$$N = p_1 + p_2^2 + p_3^3 + p_4^4$$

is solvable, here and below the letter p, with or without subscript, always denotes a prime number. However, this conjecture may be out of reach at present with the known methods and techniques.

In this paper, we shall consider the exceptional set of the problem (4) and establish the following result.

Theorem 1. *Let $E(N)$ denote the number of positive integers n, which satisfy $n \equiv 0 \,(\mathrm{mod}\, 2)$, up to N, which can not be represented as*

$$n = p_1 + p_2^2 + p_3^3 + p_4^4. \tag{4}$$

Then, for any $\varepsilon > 0$, we have

$$E(N) \ll N^{\frac{61}{144}+\varepsilon}.$$

We will establish Theorem 1 by using a pruning process into the Hardy–Littlewood circle method. For the treatment on minor arcs, we will employ the argument developed by Wooley in [24] combined with the new estimates for exponential sum over primes developed by Zhao [25]. For the treatment on major arcs, we shall prune the major arcs further and deal with them respectively. The explicit details will be given in the related sections.

Notation. In this paper, let p, with or without subscripts, always denote a prime number; ε always denotes a sufficiently small positive constant, which may not be the same at different occurrences. The letter c always denotes a positive constant. As usual, we use $\chi \bmod q$ to denote a Dirichlet character modulo q, and $\chi^0 \bmod q$ the principal character. Moreover, we use $\varphi(n)$ and $d(n)$ to denote the Euler's function and Dirichlet's divisor function, respectively. $e(x) = e^{2\pi i x}$; $f(x) \ll g(x)$ means that $f(x) = O(g(x))$; $f(x) \asymp g(x)$ means that $f(x) \ll g(x) \ll f(x)$. N is a sufficiently large integer and $n \in (N/2, N]$, and hence $\log N \asymp \log n$.

2. Outline of the Proof of Theorem 1

Let N be a sufficiently large positive integer. By a splitting argument, it is sufficient to consider the even integers $n \in (N/2, N]$. For the application of the Hardy–Littlewood method, it is necessary to define the Farey dissection. For this purpose, we set the parameters as follows

$$A = 100^{100}, \quad Q_0 = \log^A N, \quad Q_1 = N^{\frac{1}{6}}, \quad Q_2 = N^{\frac{5}{6}}, \quad \mathfrak{I}_0 = \left[-\frac{1}{Q_2}, 1 - \frac{1}{Q_2} \right].$$

By Dirichlet's rational approximation lemma (for instance, see Lemma 12 on p.104 of [26], or Lemma 2.1 of [1]), each $\alpha \in (-1/Q_2, 1 - 1/Q_2]$ can be represented in the form

$$\alpha = \frac{a}{q} + \lambda, \qquad |\lambda| \leqslant \frac{1}{qQ_2},$$

for some integers a, q with $1 \leqslant a \leqslant q \leqslant Q_2$ and $(a, q) = 1$. Define

$$\mathfrak{M}(q, a) = \left[\frac{a}{q} - \frac{1}{qQ_2}, \frac{a}{q} + \frac{1}{qQ_2}\right], \qquad \mathfrak{M} = \bigcup_{1 \leqslant q \leqslant Q_1} \bigcup_{\substack{1 \leqslant a \leqslant q \\ (a,q)=1}} \mathfrak{M}(q, a),$$

$$\mathfrak{M}_0(q, a) = \left[\frac{a}{q} - \frac{Q_0^{100}}{qN}, \frac{a}{q} + \frac{Q_0^{100}}{qN}\right], \qquad \mathfrak{M}_0 = \bigcup_{1 \leqslant q \leqslant Q_0^{100}} \bigcup_{\substack{1 \leqslant a \leqslant q \\ (a,q)=1}} \mathfrak{M}_0(q, a),$$

$$\mathfrak{m}_1 = \mathfrak{I}_0 \setminus \mathfrak{M}, \qquad \mathfrak{m}_2 = \mathfrak{M} \setminus \mathfrak{M}_0.$$

Then we obtain the Farey dissection

$$\mathfrak{I}_0 = \mathfrak{M}_0 \cup \mathfrak{m}_1 \cup \mathfrak{m}_2. \tag{5}$$

For $k = 1, 2, 3, 4$, we define

$$f_k(\alpha) = \sum_{X_k < p \leqslant 2X_k} e(p^k \alpha),$$

where $X_k = (N/16)^{\frac{1}{k}}$. Let

$$\mathscr{R}(n) = \sum_{\substack{n = p_1 + p_2^2 + p_3^3 + p_4^4 \\ X_i < p_i \leqslant 2X_i \\ i=1,2,3,4}} 1.$$

From (5), one has

$$\mathscr{R}(n) = \int_0^1 \left(\prod_{k=1}^4 f_k(\alpha)\right) e(-n\alpha) d\alpha = \int_{-\frac{1}{Q_2}}^{1-\frac{1}{Q_2}} \left(\prod_{k=1}^4 f_k(\alpha)\right) e(-n\alpha) d\alpha$$

$$= \left\{\int_{\mathfrak{M}_0} + \int_{\mathfrak{m}_1} + \int_{\mathfrak{m}_2}\right\} \left(\prod_{k=1}^4 f_k(\alpha)\right) e(-n\alpha) d\alpha.$$

In order to prove Theroem 1, we need the two following propositions:

Proposition 1. *For $n \in (N/2, N]$, there holds*

$$\int_{\mathfrak{M}_0} \left(\prod_{k=1}^4 f_k(\alpha)\right) e(-n\alpha) d\alpha = \frac{\Gamma(2)\Gamma(\frac{3}{2})\Gamma(\frac{4}{3})\Gamma(\frac{5}{4})}{\Gamma(\frac{25}{12})} \mathfrak{S}(n) \frac{n^{\frac{13}{12}}}{\log^4 n} + O\left(\frac{n^{\frac{13}{12}}}{\log^5 n}\right), \tag{6}$$

where $\mathfrak{S}(n)$ is the singular series defined in (10), which is absolutely convergent and satisfies

$$(\log \log n)^{-c^*} \ll \mathfrak{S}(n) \ll d(n) \tag{7}$$

for any integer n satisfying $n \equiv 0 \,(\mathrm{mod}\, 2)$ and some fixed constant $c^* > 0$.

The proof of (6) in Proposition 1 follows from the well–know standard technique in the Hardy–Littlewood method. For more information, one can see pp. 90–99 of Hua [7], so we omit the details herein. For the properties (7) of singular series, we shall give the proof in Section 4.

Proposition 2. *Let $\mathscr{Z}(N)$ denote the number of integers $n \in (N/2, N]$ satisfying $n \equiv 0 \,(\mathrm{mod}\, 2)$ such that*

$$\sum_{j=1}^2 \left|\int_{\mathfrak{m}_j} \left(\prod_{k=1}^4 f_k(\alpha)\right) e(-n\alpha) d\alpha\right| \gg \frac{n^{\frac{13}{12}}}{\log^5 n}.$$

Then we have

$$\mathcal{Z}(N) \ll N^{\frac{61}{144}+\varepsilon}.$$

The proof of Proposition 2 will be given in Section 5. The remaining part of this section is devoted to establishing Theorem 1 by using Proposition 1 and Proposition 2.

Proof of Theorem 1. From Proposition 2, we deduce that, with at most $O(N^{\frac{61}{144}+\varepsilon})$ exceptions, all even integers $n \in (N/2, N]$ satisfy

$$\sum_{j=1}^{2} \left| \int_{\mathfrak{m}_j} \left(\prod_{k=1}^{4} f_k(\alpha) \right) e(-n\alpha) \mathrm{d}\alpha \right| \ll \frac{n^{\frac{13}{12}}}{\log^5 n},$$

from which and Proposition 1, we conclude that, with at most $O(N^{\frac{61}{144}+\varepsilon})$ exceptions, for all even integers $n \in (N/2, N]$, $\mathscr{R}(n)$ holds the asymptotic formula

$$\mathscr{R}(n) = \frac{\Gamma(2)\Gamma(\frac{3}{2})\Gamma(\frac{4}{3})\Gamma(\frac{5}{4})}{\Gamma(\frac{25}{12})} \mathfrak{S}(n) \frac{n^{\frac{13}{12}}}{\log^4 n} + O\left(\frac{n^{\frac{13}{12}}}{\log^5 n} \right).$$

In other words, all even integers $n \in (N/2, N]$ can be represented in the form $p_1 + p_2^2 + p_3^3 + p_4^4$ with at most $O(N^{\frac{61}{144}+\varepsilon})$ exceptions, where p_1, p_2, p_3, p_4 are prime numbers. By a splitting argument, we get

$$E(N) \ll \sum_{0 \leqslant \ell \ll \log N} \mathcal{Z}\left(\frac{N}{2^\ell} \right) \ll \sum_{0 \leqslant \ell \ll \log N} \left(\frac{N}{2^\ell} \right)^{\frac{61}{144}+\varepsilon} \ll N^{\frac{61}{144}+\varepsilon}.$$

This completes the proof of Theorem 1.

3. Some Auxiliary Lemmas

In this section, we shall list some necessary lemmas which will be used in proving Proposition 2.

Lemma 1. *Suppose that α is a real number, and that $|\alpha - a/q| \leqslant q^{-2}$ with $(a, q) = 1$. Let $\beta = \alpha - a/q$. Then we have*

$$f_k(\alpha) \ll d^{\delta_k}(q)(\log x)^c \left(X_k^{1/2} \sqrt{q(1+N|\beta|)} + X_k^{4/5} + \frac{X_k}{\sqrt{q(1+N|\beta|)}} \right),$$

where $\delta_k = \frac{1}{2} + \frac{\log k}{\log 2}$ and c is a constant.

Proof. See Theorem 1.1 of Ren [27]. $\square$

Lemma 2. *Suppose that α is a real number, and that there exist $a \in \mathbb{Z}$ and $q \in \mathbb{N}$ with*

$$(a, q) = 1, \qquad 1 \leqslant q \leqslant X \qquad and \qquad |q\alpha - a| \leqslant X^{-1}.$$

If $P^{2\delta 2^{1-k}} \leqslant X \leqslant P^{k-2\delta 2^{1-k}}$, then one has

$$\sum_{P < p \leqslant 2P} e(p^k \alpha) \ll P^{1-\delta 2^{1-k}+\varepsilon} + \frac{P^{1+\varepsilon}}{q^{1/2}\left(1 + P^k|\alpha - a/q|\right)^{1/2}},$$

where $\delta = 1/3$ for $k \geqslant 4$.

Proof. See Lemma 2.4 of Zhao [25]. $\square$

Lemma 3. *Suppose that α is a real number, and that there are $a \in \mathbb{Z}$ and $q \in \mathbb{N}$ with*

$$(a, q) = 1, \qquad 1 \leqslant q \leqslant Q \qquad and \qquad |q\alpha - a| \leqslant Q^{-1}.$$

If $P^{\frac{1}{2}} \leqslant Q \leqslant P^{\frac{5}{2}}$, then one has

$$\sum_{P < p \leqslant 2P} e(p^3 \alpha) \ll P^{1 - \frac{1}{12} + \varepsilon} + \frac{q^{-\frac{1}{6}} P^{1+\varepsilon}}{\left(1 + P^3 |\alpha - a/q|\right)^{1/2}}.$$

Proof. See Lemma 8.5 of Zhao [25]. $\qquad\qquad\qquad\qquad\qquad\qquad\qquad\qquad\qquad\qquad\qquad$ $\square$

Lemma 4. *For $\alpha \in \mathfrak{m}_1$, we have*

$$f_3(\alpha) \ll N^{\frac{11}{36} + \varepsilon} \qquad and \qquad f_4(\alpha) \ll N^{\frac{23}{96} + \varepsilon}.$$

Proof. For $\alpha \in \mathfrak{m}_1$, we have $Q_1 \leqslant q \leqslant Q_2$. By Lemma 3, we get

$$f_3(\alpha) \ll X_3^{\frac{11}{12} + \varepsilon} + X_3^{1+\varepsilon} Q_1^{-\frac{1}{6}} \ll N^{\frac{11}{36} + \varepsilon}.$$

From Lemma 2, we obtain

$$f_4(\alpha) \ll X_4^{\frac{23}{24} + \varepsilon} + X_4^{1+\varepsilon} Q_1^{-\frac{1}{2}} \ll N^{\frac{23}{96} + \varepsilon}.$$

This completes the proof of Lemma 4.

$\qquad$ $\square$

For $1 \leqslant a \leqslant q$ with $(a, q) = 1$, set

$$\mathcal{I}(q, a) = \left[\frac{a}{q} - \frac{1}{qQ_0}, \frac{a}{q} + \frac{1}{qQ_0} \right], \qquad \mathcal{I} = \bigcup_{\substack{1 \leqslant q \leqslant Q_0}} \bigcup_{\substack{a = -q \\ (a,q)=1}}^{2q} \mathcal{I}(q, a). \qquad (8)$$

For $\alpha \in \mathfrak{m}_2$, by Lemma 1, we have

$$f_3(\alpha) \ll \frac{N^{\frac{1}{3}} \log^c N}{q^{\frac{1}{2} - \varepsilon} (1 + N|\lambda|)^{1/2}} + N^{\frac{4}{15} + \varepsilon} = V_3(\alpha) + N^{\frac{4}{15} + \varepsilon}, \qquad (9)$$

say. Then we obtain the following Lemma.

Lemma 5. *We have*

$$\int_{\mathcal{I}} |V_3(\alpha)|^4 d\alpha = \sum_{\substack{1 \leqslant q \leqslant Q_0}} \sum_{\substack{a = -q \\ (a,q)=1}}^{2q} \int_{\mathcal{I}(q,a)} |V_3(\alpha)|^4 d\alpha \ll N^{\frac{1}{3}} \log^c N.$$

Proof. We have

$$\sum_{\substack{1\leqslant q\leqslant Q_0}} \sum_{\substack{a=-q\\(a,q)=1}}^{2q} \int_{\mathcal{I}(q,a)} |V_3(\alpha)|^4 d\alpha$$

$$\ll \sum_{\substack{1\leqslant q\leqslant Q_0}} q^{-2+\varepsilon} \sum_{\substack{a=-q\\(a,q)=1}}^{2q} \int_{|\lambda|\leqslant\frac{1}{Q_0}} \frac{N^{\frac{4}{3}}\log^c N}{(1+N|\lambda|)^2} d\lambda$$

$$\ll \sum_{\substack{1\leqslant q\leqslant Q_0}} q^{-2+\varepsilon} \sum_{\substack{a=-q\\(a,q)=1}}^{2q} \left(\int_{|\lambda|\leqslant\frac{1}{N}} N^{\frac{4}{3}}\log^c N d\lambda + \int_{\frac{1}{N}\leqslant|\lambda|\leqslant\frac{1}{Q_0}} \frac{N^{\frac{4}{3}}\log^c N}{N^2\lambda^2} d\lambda \right)$$

$$\ll N^{\frac{1}{3}}\log^c N \sum_{1\leqslant q\leqslant Q_0} q^{-2+\varepsilon}\varphi(q) \ll N^{\frac{1}{3}}Q_0^{\varepsilon}\log^c N \ll N^{\frac{1}{3}}\log^c N.$$

This completes the proof of Lemma 5. □

4. The Singular Series

In this section, we shall concentrate on investigating the properties of the singular series which appear in Proposition 1. First, we illustrate some notations. For $k \in \{1,2,3,4\}$ and a Dirichlet character χ mod q, we define

$$C_k(\chi,a) = \sum_{h=1}^{q} \overline{\chi(h)}e\left(\frac{ah^k}{q}\right), \qquad C_k(q,a) = C_k(\chi^0,a),$$

where χ^0 is the principal character modulo q. Let $\chi_1, \chi_2, \chi_3, \chi_4$ be Dirichlet characters modulo q. Set

$$B(n,q,\chi_1,\chi_2,\chi_3,\chi_4) = \sum_{\substack{a=1\\(a,q)=1}}^{q} C_1(\chi_1,a)C_2(\chi_2,a)C_3(\chi_3,a)C_4(\chi_4,a)e\left(-\frac{an}{q}\right),$$

$$B(n,q) = B(n,q,\chi^0,\chi^0,\chi^0,\chi^0),$$

and write

$$A(n,q) = \frac{B(n,q)}{\varphi^4(q)}, \qquad \mathfrak{S}(n) = \sum_{q=1}^{\infty} A(n,q). \tag{10}$$

Lemma 6. For $(a,q)=1$ and any Dirichlet character χ mod q, there holds

$$|C_k(\chi,a)| \leqslant 2q^{1/2}d^{\beta_k}(q)$$

with $\beta_k = (\log k)/\log 2$.

Proof. See the Problem 14 of Chapter VI of Vinogradov [28]. □

Lemma 7. Let p be a prime and $p^\alpha \| k$. For $(a,p)=1$, if $\ell \geqslant \gamma(p)$, we have $C_k(p^\ell,a) = 0$, where

$$\gamma(p) = \begin{cases} \alpha+2, & \text{if } p \neq 2 \text{ or } p = 2, \alpha = 0; \\ \alpha+3, & \text{if } p = 2, \alpha > 0. \end{cases}$$

Proof. See Lemma 8.3 of Hua [7]. □

For $k \geqslant 1$, we define

$$S_k(q,a) = \sum_{m=1}^{q} e\left(\frac{am^k}{q}\right).$$

Lemma 8. *Suppose that $(p,a) = 1$. Then*

$$S_k(p,a) = \sum_{\chi \in \mathscr{A}_k} \overline{\chi(a)}\tau(\chi),$$

where $\mathscr{A}_k$ denotes the set of non–principal characters χ modulo p for which χ^k is principal, and $\tau(\chi)$ denotes the Gauss sum

$$\sum_{m=1}^{p} \chi(m)e\left(\frac{m}{p}\right).$$

Also, there hold $|\tau(\chi)| = p^{1/2}$ and $|\mathscr{A}_k| = (k, p-1) - 1$.

Proof. See Lemma 4.3 of Vaughan [1]. □

Lemma 9. *For $(p,n) = 1$, we have*

$$\left| \sum_{a=1}^{p-1} \left(\prod_{k=1}^{4} \frac{S_k(p,a)}{p} \right) e\left(-\frac{an}{p}\right) \right| \leqslant 24p^{-\frac{3}{2}}. \tag{11}$$

Proof. We denote by $\mathcal{S}$ the left-hand side of (11). It follows from Lemma 8 that

$$\mathcal{S} = \frac{1}{p^4} \sum_{a=1}^{p-1} \left(\prod_{k=1}^{4} \left(\sum_{\chi_k \in \mathscr{A}_k} \overline{\chi_k(a)}\tau(\chi_k) \right) \right) e\left(-\frac{an}{p}\right).$$

If $|\mathscr{A}_k| = 0$ for some $k \in \{1,2,3,4\}$, then $\mathcal{S} = 0$. If this is not the case, then

$$\mathcal{S} = \frac{1}{p^4} \sum_{\chi_1 \in \mathscr{A}_1} \sum_{\chi_2 \in \mathscr{A}_2} \sum_{\chi_3 \in \mathscr{A}_3} \sum_{\chi_4 \in \mathscr{A}_4} \tau(\chi_1)\tau(\chi_2)\tau(\chi_3)\tau(\chi_4)$$

$$\times \sum_{a=1}^{p-1} \overline{\chi_1(a)\chi_2(a)\chi_3(a)\chi_4(a)}e\left(-\frac{an}{p}\right).$$

From Lemma 8, the quadruple outer sums have no more than $4! = 24$ terms. For each of these terms, there holds

$$|\tau(\chi_1)\tau(\chi_2)\tau(\chi_3)\tau(\chi_4)| = p^2.$$

Since in any one of these terms $\overline{\chi_1(a)\chi_2(a)\chi_3(a)\chi_4(a)}$ is a Dirichlet character $\chi \pmod p$, the inner sum is

$$\sum_{a=1}^{p-1} \chi(a)e\left(-\frac{an}{p}\right) = \overline{\chi(-n)} \sum_{a=1}^{p-1} \chi(-an)e\left(-\frac{an}{p}\right) = \overline{\chi(-n)}\tau(\chi).$$

By noting the fact that $\tau(\chi^0) = -1$ for principal character $\chi^0 \bmod p$, we derive that

$$|\overline{\chi(-n)}\tau(\chi)| \leqslant p^{\frac{1}{2}}.$$

From the above arguments, we deduce that

$$|\mathcal{S}| \leqslant \frac{1}{p^4} \cdot 24 \cdot p^2 \cdot p^{\frac{1}{2}} = 24p^{-\frac{3}{2}},$$

which completes the proof of Lemma 9. □

Lemma 10. *Let $\mathcal{L}(p,n)$ denote the number of solutions of the congruence*

$$x_1 + x_2^2 + x_3^3 + x_4^4 \equiv n \pmod p, \qquad 1 \leqslant x_1, x_2, x_3, x_4 \leqslant p-1.$$

Then, for $n \equiv 0 \,(\mathrm{mod}\,2)$, we have $\mathcal{L}(p,n) > 0$.

Proof. We have

$$p \cdot \mathcal{L}(p,n) = \sum_{a=1}^{p} C_1(p,a)C_2(p,a)C_3(p,a)C_4(p,a)e\left(-\frac{an}{p}\right) = (p-1)^4 + E_p,$$

where

$$E_p = \sum_{a=1}^{p-1} C_1(p,a)C_2(p,a)C_3(p,a)C_4(p,a)e\left(-\frac{an}{p}\right).$$

By Lemma 8, we obtain

$$|E_p| \leqslant (p-1)(\sqrt{p}+1)(2\sqrt{p}+1)(3\sqrt{p}+1).$$

It is easy to check that $|E_p| < (p-1)^4$ for $p \geqslant 7$. Therefore, we obtain $\mathcal{L}(p,n) > 0$ for $p \geqslant 7$. For $p = 2,3,5$, we can check $\mathcal{L}(p,n) > 0$ one by one. This completes the proof of Lemma 10. $\square$

Lemma 11. *$A(n,q)$ is multiplicative in q.*

Proof. From the definition of $A(n,q)$ in (10), it is sufficient to show that $B(n,q)$ is multiplicative in q. Suppose $q = q_1 q_2$ with $(q_1, q_2) = 1$. Then we obtain

$$B(n,q_1q_2) = \sum_{\substack{a=1 \\ (a,q_1q_2)=1}}^{q_1q_2} \left(\prod_{k=1}^{4} C_k(q_1q_2,a)\right)e\left(-\frac{an}{q_1q_2}\right)$$

$$= \sum_{\substack{a_1=1 \\ (a_1,q_1)=1}}^{q_1} \sum_{\substack{a_2=1 \\ (a_2,q_2)=1}}^{q_2} \left(\prod_{k=1}^{4} C_k(q_1q_2, a_1q_2 + a_2q_1)\right)e\left(-\frac{a_1 n}{q_1}\right)e\left(-\frac{a_2 n}{q_2}\right). \tag{12}$$

For $(q_1, q_2) = 1$, there holds

$$C_k(q_1q_2, a_1q_2 + a_2q_1) = \sum_{\substack{m=1 \\ (m,q_1q_2)=1}}^{q_1q_2} e\left(\frac{(a_1q_2 + a_2q_1)m^k}{q_1q_2}\right)$$

$$= \sum_{\substack{m_1=1 \\ (m_1,q_1)=1}}^{q_1} \sum_{\substack{m_2=1 \\ (m_2,q_2)=1}}^{q_2} e\left(\frac{(a_1q_2 + a_2q_1)(m_1q_2 + m_2q_1)^k}{q_1q_2}\right)$$

$$= \sum_{\substack{m_1=1 \\ (m_1,q_1)=1}}^{q_1} e\left(\frac{a_1(m_1q_2)^k}{q_1}\right) \sum_{\substack{m_2=1 \\ (m_2,q_2)=1}}^{q_2} e\left(\frac{a_2(m_2q_1)^k}{q_2}\right)$$

$$= C_k(q_1,a_1)C_k(q_2,a_2). \tag{13}$$

Putting (13) into (12), we deduce that

$$B(n,q_1q_2) = \sum_{\substack{a_1=1 \\ (a_1,q_1)=1}}^{q_1} \left(\prod_{k=1}^{4} C_k(q_1,a_1)\right)e\left(-\frac{a_1 n}{q_1}\right) \sum_{\substack{a_2=1 \\ (a_2,q_2)=1}}^{q_2} \left(\prod_{k=1}^{4} C_k(q_2,a_2)\right)e\left(-\frac{a_2 n}{q_2}\right)$$

$$= B(n,q_1)B(n,q_2).$$

This completes the proof of Lemma 11. $\square$

Lemma 12. *Let $A(n,q)$ be as defined in (10). Then*

(i) we have

$$\sum_{q>Z} |A(n,q)| \ll Z^{-\frac{1}{2}+\varepsilon} d(n),$$

and thus the singular series $\mathfrak{S}(n)$ is absolutely convergent and satisfies $\mathfrak{S}(n) \ll d(n)$.

(ii) there exists an absolute positive constant $c^* > 0$, such that, for $n \equiv 0 \,(\mathrm{mod}\ 2)$,

$$\mathfrak{S}(n) \gg (\log\log n)^{-c^*}.$$

Proof. From Lemma 11, we know that $B(n,q)$ is multiplicative in q. Therefore, there holds

$$B(n,q) = \prod_{p^t \| q} B(n,p^t) = \prod_{p^t \| q} \sum_{\substack{a=1 \\ (a,p)=1}}^{p^t} \left(\prod_{k=1}^{4} C_k(p^t,a) \right) e\left(-\frac{an}{p^t} \right). \tag{14}$$

From (14) and Lemma 7, we deduce that $B(n,q) = \prod_{p\|q} B(n,p)$ or 0 according to q is square–free or not. Thus, one has

$$\sum_{q=1}^{\infty} A(n,q) = \sum_{\substack{q=1 \\ q \text{ square–free}}}^{\infty} A(n,q). \tag{15}$$

Write

$$\mathcal{R}(p,a) := \prod_{k=1}^{4} C_k(p,a) - \prod_{k=1}^{4} S_k(p,a).$$

Then

$$A(n,p) = \frac{1}{(p-1)^4} \sum_{a=1}^{p-1} \left(\prod_{k=1}^{4} S_k(p,a) \right) e\left(-\frac{an}{p} \right) + \frac{1}{(p-1)^4} \sum_{a=1}^{p-1} \mathcal{R}(p,a) e\left(-\frac{an}{p} \right). \tag{16}$$

Applying Lemma 6 and noticing that $S_k(p,a) = C_k(p,a) + 1$, we get $S_k(p,a) \ll p^{\frac{1}{2}}$, and thus $\mathcal{R}(p,a) \ll p^{\frac{3}{2}}$. Therefore, the second term in (16) is $\leqslant c_1 p^{-\frac{3}{2}}$. On the other hand, from Lemma 9, we can see that the first term in (16) is $\leqslant 2^4 \cdot 24 p^{-\frac{3}{2}} = 384 p^{-\frac{3}{2}}$. Let $c_2 = \max(c_1, 384)$. Then we have proved that, for $p \nmid n$, there holds

$$|A(n,p)| \leqslant c_2 p^{-\frac{3}{2}}. \tag{17}$$

Moreover, if we use Lemma 6 directly, it follows that

$$|B(n,p)| = \left| \sum_{a=1}^{p-1} \left(\prod_{k=1}^{4} C_k(p,a) \right) e\left(-\frac{an}{p} \right) \right| \leqslant \sum_{a=1}^{p-1} \prod_{k=1}^{4} |C_k(p,a)|$$
$$\leqslant (p-1) \cdot 2^4 \cdot p^2 \cdot 24 = 384 p^2 (p-1),$$

and therefore

$$|A(n,p)| = \frac{|B(n,p)|}{\varphi^4(p)} \leqslant \frac{384 p^2}{(p-1)^3} \leqslant \frac{2^3 \cdot 384 p^2}{p^3} = \frac{3072}{p}. \tag{18}$$

Let $c_3 = \max(c_2, 3072)$. Then, for square–free q, we have

$$|A(n,q)| = \left(\prod_{\substack{p|q \\ p \nmid n}} |A(n,p)| \right) \left(\prod_{\substack{p|q \\ p|n}} |A(n,p)| \right) \leqslant \left(\prod_{\substack{p|q \\ p \nmid n}} (c_3 p^{-\frac{3}{2}}) \right) \left(\prod_{\substack{p|q \\ p|n}} (c_3 p^{-1}) \right)$$
$$= c_3^{\omega(q)} \left(\prod_{p|q} p^{-\frac{3}{2}} \right) \left(\prod_{p|(n,q)} p^{\frac{1}{2}} \right) \ll q^{-\frac{3}{2}+\varepsilon} (n,q)^{\frac{1}{2}}.$$

Hence, by (15), we obtain

$$
\sum_{q>Z} |A(n,q)| \ll \sum_{q>Z} q^{-\frac{3}{2}+\varepsilon}(n,q)^{\frac{1}{2}} = \sum_{d|n}\sum_{q>\frac{Z}{d}} (dq)^{-\frac{3}{2}+\varepsilon}d^{\frac{1}{2}} = \sum_{d|n} d^{-1+\varepsilon}\sum_{q>\frac{Z}{d}} q^{-\frac{3}{2}+\varepsilon}
$$

$$
\ll \sum_{d|n} d^{-1+\varepsilon}\left(\frac{Z}{d}\right)^{-\frac{1}{2}+\varepsilon} = Z^{-\frac{1}{2}+\varepsilon}\sum_{d|n} d^{-\frac{1}{2}+\varepsilon} \ll Z^{-\frac{1}{2}+\varepsilon}d(n).
$$

This proves (i) of Lemma 12.

To prove (ii) of Lemma 12, by Lemma 11, we first note that

$$
\mathfrak{S}(n) = \prod_p \left(1 + \sum_{t=1}^{\infty} A(n,p^t)\right) = \prod_p (1 + A(n,p))
$$

$$
= \left(\prod_{p\le c_3} (1 + A(n,p))\right)\left(\prod_{\substack{p>c_3\\p\nmid n}} (1 + A(n,p))\right)\left(\prod_{\substack{p>c_3\\p|n}} (1 + A(n,p))\right). \tag{19}
$$

From (17), we have

$$
\prod_{\substack{p>c_3\\p\nmid n}} (1 + A(n,p)) \ge \prod_{p>c_3}\left(1 - \frac{c_3}{p^{3/2}}\right) \ge c_4 > 0. \tag{20}
$$

By (18), we know that there are $c_5 > 0$ such that

$$
\prod_{\substack{p>c_3\\p|n}} (1 + A(n,p)) \ge \prod_{\substack{p>c_3\\p|n}}\left(1 - \frac{c_3}{p}\right) \ge \prod_{p|n}\left(1 - \frac{c_3}{p}\right) \gg (\log\log n)^{-c_5}. \tag{21}
$$

On the other hand, it is easy to see that

$$
1 + A(n,p) = \frac{p\cdot\mathcal{L}(p,n)}{\varphi^4(p)}.
$$

By Lemma 10, we know that $\mathcal{L}(p,n) > 0$ for all p with $n \equiv 0\,(\mathrm{mod}\,2)$, and thus $1 + A(n,p) > 0$. Therefore, there holds

$$
\prod_{p\le c_3} (1 + A(n,p)) \ge c_6 > 0. \tag{22}
$$

Combining the estimates (19)–(22), and taking $c^* = c_5 > 0$, we derive that

$$
\mathfrak{S}(n) \gg (\log\log n)^{-c^*}.
$$

This completes the proof Lemma 12. $\qquad\square$

5. Proof of Proposition 2

In this section, we shall give the proof of Proposition 2. We denote by $\mathcal{Z}_j(N)$ the set of integers n satisfying $n \in [N/2, N]$ and $n \equiv 0\,(\mathrm{mod}\,2)$ for which the following estimate

$$
\left|\int_{\mathfrak{m}_j}\left(\prod_{k=1}^{4} f_k(\alpha)\right)e(-n\alpha)\mathrm{d}\alpha\right| \gg \frac{n^{\frac{13}{12}}}{\log^5 n} \tag{23}
$$

holds. For convenience, we use $\mathcal{Z}_j$ to denote the cardinality of $\mathcal{Z}_j(N)$ for abbreviation. Also, we define the complex number $\xi_j(n)$ by taking $\xi_j(n) = 0$ for $n \notin \mathcal{Z}_j(N)$, and when $n \in \mathcal{Z}_j(N)$ by means of the equation

$$\left| \int_{\mathfrak{m}_j} \left(\prod_{k=1}^{4} f_k(\alpha) \right) e(-n\alpha) d\alpha \right| = \xi_j(n) \int_{\mathfrak{m}_j} \left(\prod_{k=1}^{4} f_k(\alpha) \right) e(-n\alpha) d\alpha. \tag{24}$$

Plainly, one has $|\xi_j(n)| = 1$ whenever $\xi_j(n)$ is nonzero. Therefore, we obtain

$$\sum_{n \in \mathcal{Z}_j(N)} \xi_j(n) \int_{\mathfrak{m}_j} \left(\prod_{k=1}^{4} f_k(\alpha) \right) e(-n\alpha) d\alpha = \int_{\mathfrak{m}_j} \left(\prod_{k=1}^{4} f_k(\alpha) \right) \mathcal{K}_j(\alpha) d\alpha, \tag{25}$$

where the exponential sum $\mathcal{K}_j(\alpha)$ is defined by

$$\mathcal{K}_j(\alpha) = \sum_{n \in \mathcal{Z}_j(N)} \xi_j(n) e(-n\alpha).$$

For $j = 1, 2$, set

$$I_j = \int_{\mathfrak{m}_j} \left(\prod_{k=1}^{4} f_k(\alpha) \right) \mathcal{K}_j(\alpha) d\alpha.$$

By (23)–(25), we derive that

$$I_j \gg \sum_{n \in \mathcal{Z}_j(N)} \frac{n^{\frac{13}{12}}}{\log^5 n} \gg \frac{\mathcal{Z}_j N^{\frac{13}{12}}}{\log^5 N}, \qquad j = 1, 2. \tag{26}$$

By Lemma 2.1 of Wooley [24] with $k = 2$, we know that, for $j = 1, 2$, there holds

$$\int_0^1 |f_2(\alpha) \mathcal{K}_j(\alpha)|^2 d\alpha \ll N^\varepsilon (\mathcal{Z}_j N^{\frac{1}{2}} + \mathcal{Z}_j^2). \tag{27}$$

It follows from Cauchy's inequality, Lemma 4 and (27) that

$$I_1 \ll \left(\sup_{\alpha \in \mathfrak{m}_1} |f_3(\alpha)| \right) \left(\sup_{\alpha \in \mathfrak{m}_1} |f_4(\alpha)| \right) \left(\int_0^1 |f_2(\alpha) \mathcal{K}_1(\alpha)|^2 d\alpha \right)^{\frac{1}{2}} \left(\int_0^1 |f_1(\alpha)|^2 d\alpha \right)^{\frac{1}{2}}$$

$$\ll N^{\frac{11}{36} + \varepsilon} \cdot N^{\frac{23}{96} + \varepsilon} \cdot \left(N^\varepsilon (\mathcal{Z}_1 N^{\frac{1}{2}} + \mathcal{Z}_1^2) \right)^{\frac{1}{2}} \cdot N^{\frac{1}{2}}$$

$$\ll N^{\frac{301}{288} + \varepsilon} \left(\mathcal{Z}_1^{\frac{1}{2}} N^{\frac{1}{4}} + \mathcal{Z}_1 \right) \ll \mathcal{Z}_1^{\frac{1}{2}} N^{\frac{373}{288} + \varepsilon} + \mathcal{Z}_1 N^{\frac{301}{288} + \varepsilon}. \tag{28}$$

Combining (26) and (28), we get

$$\mathcal{Z}_1 N^{\frac{13}{12}} \log^{-5} N \ll I_1 \ll \mathcal{Z}_1^{\frac{1}{2}} N^{\frac{373}{288} + \varepsilon} + \mathcal{Z}_1 N^{\frac{301}{288} + \varepsilon},$$

which implies

$$\mathcal{Z}_1 \ll N^{\frac{61}{144} + \varepsilon}. \tag{29}$$

Next, we give the upper bound for $\mathcal{Z}_2$. By (9), we obtain

$$I_2 \ll \int_{\mathfrak{m}_2} |f_1(\alpha) f_2(\alpha) V_3(\alpha) f_4(\alpha) \mathcal{K}_2(\alpha)| d\alpha$$

$$+ N^{\frac{4}{15} + \varepsilon} \cdot \int_{\mathfrak{m}_2} |f_1(\alpha) f_2(\alpha) f_4(\alpha) \mathcal{K}_2(\alpha)| d\alpha$$

$$= I_{21} + I_{22}, \tag{30}$$

say. For $\alpha \in \mathfrak{m}_2$, we have either $Q_0^{100} < q < Q_1$ or $Q_0^{100} < N|q\alpha - a| < NQ_2^{-1} = Q_1$. Therefore, by Lemma 1, we get

$$\sup_{\alpha \in \mathfrak{m}_2} |f_4(\alpha)| \ll \frac{N^{\frac{1}{4}}}{\log^{40A} N}. \tag{31}$$

In view of the fact that $\mathfrak{m}_2 \subseteq \mathcal{I}$, where $\mathcal{I}$ is defined by (8), Hölder's inequality, the trivial estimate $\mathcal{K}_2(\alpha) \ll \mathcal{Z}_2$ and Theorem 4 of Hua (See [7], p. 19), we obtain

$$I_{21} \ll \mathcal{Z}_2 \sup_{\alpha \in \mathfrak{m}_2} |f_4(\alpha)| \times \left(\int_0^1 |f_1(\alpha)|^2 d\alpha \right)^{\frac{1}{2}} \left(\int_0^1 |f_2(\alpha)|^4 d\alpha \right)^{\frac{1}{4}} \left(\int_{\mathcal{I}} |V_3(\alpha)|^4 d\alpha \right)^{\frac{1}{4}}$$

$$\ll \mathcal{Z}_2 \cdot \frac{N^{\frac{1}{4}}}{\log^{40A} N} \cdot N^{\frac{1}{2}} \cdot (N \log^c N)^{\frac{1}{4}} \cdot (N^{\frac{1}{3}} \log^c N)^{\frac{1}{4}} \ll \frac{\mathcal{Z}_2 N^{\frac{13}{12}}}{\log^{30A} N}. \tag{32}$$

Moreover, it follows from (27), (31) and Cauchy's inequality that

$$I_{22} \ll N^{\frac{4}{15}+\varepsilon} \cdot \sup_{\alpha \in \mathfrak{m}_2} |f_4(\alpha)| \times \left(\int_0^1 |f_1(\alpha)|^2 d\alpha \right)^{\frac{1}{2}} \left(\int_0^1 |f_2(\alpha)\mathcal{K}_2(\alpha)|^2 d\alpha \right)^{\frac{1}{2}}$$

$$\ll N^{\frac{4}{15}+\varepsilon} \cdot \frac{N^{\frac{1}{4}}}{\log^{40A} N} \cdot N^{\frac{1}{2}} \cdot \left(N^\varepsilon (\mathcal{Z}_2 N^{\frac{1}{2}} + \mathcal{Z}_2^2) \right)^{\frac{1}{2}}$$

$$\ll N^{\frac{61}{60}+\varepsilon} (\mathcal{Z}_2^{\frac{1}{2}} N^{\frac{1}{4}} + \mathcal{Z}_2) \ll \mathcal{Z}_2^{\frac{1}{2}} N^{\frac{19}{15}+\varepsilon} + \mathcal{Z}_2 N^{\frac{61}{60}+\varepsilon}. \tag{33}$$

Combining (26), (30), (32) and (33), we deduce that

$$\frac{\mathcal{Z}_2 N^{\frac{13}{12}}}{\log^5 N} \ll I_2 = I_{21} + I_{22} \ll \frac{\mathcal{Z}_2 N^{\frac{13}{12}}}{\log^{30A} N} + \mathcal{Z}_2^{\frac{1}{2}} N^{\frac{19}{15}+\varepsilon} + \mathcal{Z}_2 N^{\frac{61}{60}+\varepsilon},$$

which implies

$$\mathcal{Z}_2 \ll N^{\frac{11}{30}+\varepsilon}. \tag{34}$$

From (29) and (34), we have

$$\mathcal{Z}(N) \ll \mathcal{Z}_1 + \mathcal{Z}_2 \ll N^{\frac{61}{144}+\varepsilon},$$

which completes the proof of Proposition 2.

Author Contributions: All authors contributed equally to this work. All authors have read and agreed to the published version of the manuscript.

Acknowledgments: The authors would like to express the most sincere gratitude to the referee for his/her patience in refereeing this paper.

References

1. Vaughan, R.C. *The Hardy-Littlewood Method*; Cambridge University Press: Cambridge, UK, 1997.
2. Davenport, H. On Waring's problem for fourth powers. *Ann. Math.* **1939**, *40*, 731–747.
3. Hardy, G.H.; Ramanujan, S. Asymptotic formulae in combinatory analysis. *Proc. Lond. Math. Soc.* **1918**, *17*, 75–115.
4. Vinogradov, I.M. Representation of an odd number as a sum of three primes. *Dokl. Akad. Nauk SSSR* **1937**, *15*, 6–7.
5. Vinogradov, I.M. Some theorems concerning the theory of primes. *Mat. Sb.* **1937**, *44*, 179–195.

6.   Hua, L.K. *Additive Primzahltheorie*; B. G. Teubner Verlagsgesellschaft: Leipzig, Germany, 1959.

7.   Hua, L.K. *Additive Theory of Prime Numbers*; American Mathematical Society: Providence, RI, USA, 1965.

8.   Jagy, W.C.; Kaplansky, I. Sums of squares, cubes, and higher powers. *Exp. Math.* **1995**, *4*, 169–173.

9.   Brüdern, J. On Waring's problem for cubes and biquadrates. *J. Lond. Math. Soc.* **1988**, *37*, 25–42.

10.   Brüdern, J.; Wooley, T.D. On Waring's problem: Three cubes and a sixth power. *Nagoya Math. J.* **2001**, *163*, 13–53.

11.   Davenport, H.; Heilbronn, H. Note on a result in the additive theory of numbers. *Proc. Lond. Math. Soc.* **1937**, *2*, 142–151.

12.   Davenport, H.; Heilbronn, H. On Waring's problem: Two cubes and one square. *Proc. Lond. Math. Soc.* **1937**, *2*, 73–104.

13.   Roth, K.F. Proof that almost all positive integers are sums of a square, a positive cube and a fourth power. *J. Lond. Math. Soc.* **1949**, *24*, 4–13.

14.   Vaughan, R.C. A ternary additive problem. *Proc. Lond. Math. Soc.* **1980**, *41*, 516–532.

15.   Hooley, C. On a new approach to various problems of Waring's type. In *Recent Progress in Analytic Number Theory*; Academic Press: London, UK, 1981; Volume 1, pp. 127–191.

16.   Davenport, H. On Waring's problem for cubes. *Acta Math.* **1939**, *71*, 123–143.

17.   Lu, M.G. On Waring's problem for cubes and fifth power. *Sci. China Ser. A* **1993**, *36*, 641–662.

18.   Kawada, K.; Wooley, T.D. Sums of fourth powers and related topics. *J. Reine Angew. Math.* **1999**, *512*, 173–223.

19.   Vaughan, R.C. A new iterative method in Waring's problem. *Acta Math.* **1989**, *162*, 1–71.

20.   Brüdern, J.; Kawada, K. Ternary problems in additive prime number theory. In *Analytic Number Theory*; Jia, C., Matsumoto, K., Eds.; Kluwer: Dordrecht, The Netherlands, 2002; pp. 39–91.

21.   Zhao, L. On unequal powers of primes and powers of 2. *Acta Math. Hungar.* **2015**, *146*, 405–420.

22.   Liu, Z.; Lü, G. On unlike powers of primes and powers of 2. *Acta Math. Hung.* **2011**, *132*, 125–139.

23.   Lü, X.D. On unequal powers of primes and powers of 2. *Ramanujan J.* **2019**, *50*, 111–121.

24.   Wooley, T.D. Slim exceptional sets and the asymptotic formula in Waring's problem. *Math. Proc. Camb. Philos. Soc.* **2003**, *134*, 193–206.

25.   Zhao, L. On the Waring-Goldbach problem for fourth and sixth powers. *Proc. Lond. Math. Soc.* **2014**, *108*, 1593–1622.

26.   Pan, C.D.; Pan, C.B. *Goldbach Conjecture*; Science Press: Beijing, China, 1981.

27.   Ren, X.M. On exponential sums over primes and application in Waring–Goldbach problem. *Sci. China Ser. A* **2005**, *48*, 785–797.

28.   Vinogradov, I.M. *Elements of Number Theory*; Dover Publications, New York, USA, 1954.

4

Symmetry Identities of Changhee Polynomials of type Two

Joohee Jeong [1,†,‡]**, Dong-Jin Kang** [2,‡] **and Seog-Hoon Rim** [1,*,‡]

1   Department of Mathematics Education, Kyungpook National University, Daegu 41566, Korea; jhjeong@knu.ac.kr
2   Department of Computer Engineering, Information Technology Services, Kyungpook National University, Daegu 41566, Korea; djkang@knu.ac.kr
*   Correspondence: shrim@knu.ac.kr.
†   Current address: 80 Daehakro, Bukgu, Daegu 41566, Korea.
‡   These authors contributed equally to this work.

Abstract: In this paper, we consider Changhee polynomials of type two, which are motivated from the recent work of D. Kim and T. Kim. We investigate some symmetry identities for the Changhee polynomials of type two which are derived from the properties of symmetry for the fermionic p-adic integral on $\mathbb{Z}_p$.

Keywords: Changhee polynomials; Changhee polynomials of type two; fermionic p-adic integral on $\mathbb{Z}_p$

1. Introduction

Let p be a fixed odd prime number. Throughout this paper, $\mathbb{Z}_p$, $\mathbb{Q}_p$ and $\mathbb{C}_p$ will denote the ring of p-adic integers, the field of p-adic rational numbers and the completion of the algebraic closure of $\mathbb{Q}_p$.

The p-adic norm $|\cdot|_p$ is normalized as $|p|_p = \frac{1}{p}$.

Let $f(x)$ be a continulus funciton on $\mathbb{Z}_p$. Then the fermionic p-adic integral on $\mathbb{Z}_p$ is defined by Kim in [1] as

$$\int_{\mathbb{Z}_p} f(x) d\mu_{-1}(x) = \lim_{N \to \infty} \sum_{x=0}^{p^N - 1} f(x) \mu_{-1}(x) = \lim_{x \to \infty} \sum_{x=0}^{p^N - 1} f(x)(-1)^x. \tag{1}$$

For $n \in \mathbb{N}$, by (1), we get

$$\int_{\mathbb{Z}_p} f(x+n) d\mu_{-1}(x) + (-1)^{n-1} \int_{\mathbb{Z}_p} f(x) d\mu_{-1}(x)$$
$$= 2 \sum_{\ell=0}^{n-1} f(\ell)(-1)^{n-1-\ell} \tag{2}$$

as shown in [2–5]. In particular, if we take $n = 1$, then we have

$$\int_{\mathbb{Z}_p} f(x+1) d\mu_{-1}(x) + \int_{\mathbb{Z}_p} f(x) d\mu_{-1}(x) = 2f(0), \tag{3}$$

which is noted in [6,7].

In the previous paper [8], D. Kim and T. Kim introduced the Changhee polynomials $\widetilde{Ch}_n(x)$ of type two by the generating function

$$\sum_{n=0}^{\infty} \widetilde{Ch}_n(x) \frac{t^n}{n!} = \frac{2}{(1+t) + (1+t)^{-1}} (1+t)^x. \tag{4}$$

By exploiting the method of fermionic p-adic integral on $\mathbb{Z}_p$, the Changhee polynomials of type two can be represented by the fermionic p-adic integrals of $\mathbb{Z}_p$: for $t \in \mathbb{C}_p$ with $|t|_p < p^{-\frac{1}{p-1}}$,

$$\int_{\mathbb{Z}_p} (1+t)^{2y+1+2x} d\mu_{-1}(y) = \frac{2}{(1+t)^2 + 1}(1+t)^{2x+1}$$
$$= \sum_{n=0}^{\infty} \widetilde{Ch}_n(x)\frac{t^n}{n!} \tag{5}$$

When $x = 0$, $\widetilde{Ch}_n = \widetilde{Ch}_n(0)$ are called the Changhee numbers of type two.

In this paper, we will introduce further generalization of Changhee polynomials of type two, by using again fermionic p-adic integration on $\mathbb{Z}_p$.

We investigate some symmetry identities for the w-Changhee polynomials of type two which are derived from the properties of symmetry for the fermionic p-adic integral on $\mathbb{Z}_p$. Many authors investigated symmetric properties of special polynomials and numbers. See [9–12] and their references.

We introduce w-Changhee polynomials of type two in Section 3.

2. Changhee Polynomials and Numbers of Type Two

In this section, we use the techniques presented in the articles of C. Cesarano, C. Fornaro [13] and C. Cesarno [14], in particular the similarity of Chebyshev polynomials.

By using the generating functions of Changhee numbers and polynomials of type two, we have the following result.

Proposition 1. *For $n \in \mathbb{N}$ and $1 \le k \le n$, we have*

$$\widetilde{Ch}_n(x) = \sum_{m=0}^{n} \binom{n}{m}(2x)_m \widetilde{Ch}_{n-m}, \tag{6}$$

where $(x)_n = x(x-1)\cdots(x-n+1)$, $(n \ge 1)$, $(x)_0 = 1$.

Proof of Proposition 1.

$$\sum_{n=0}^{\infty} \widetilde{Ch}_n(x)\frac{t^n}{n!} = \frac{2}{(1+t) + (1+t)^{-1}}(1+t)^{2x}$$
$$= \sum_{m=0}^{\infty} \widetilde{Ch}_m \frac{t^m}{m!} \sum_{\ell=0}^{\infty}(2x)_\ell \frac{t^n}{\ell!}$$
$$= \sum_{n=0}^{\infty}\left(\sum_{m=0}^{n}\binom{n}{m}\widetilde{Ch}_m(2x)_{n-m}\right)\frac{t^n}{n!}$$

□

The Stirling number $S_1(\ell, n)$ of the first kind is defined in [2–5,15] by the generating function

$$(\log(1+t))^n = n!\sum_{\ell=n}^{\infty} S_1(\ell, n),$$

and the Stirling number $S_2(m, n)$ of the second kind is given in [4] by the generating function

$$(e^t - 1)^n = n!\sum_{m=n}^{\infty} S_2(m, n)\frac{t^m}{m!}.$$

As is well known, the Euler polynomials $E_n(x)$ are defined in [16–18] by the generating function

$$\frac{2}{e^t+1}e^{xt} = \sum_{n=0}^{\infty} E_n(x)\frac{t^n}{n!}. \tag{7}$$

When $x = 0$, $E_n = E_n(0)$, $(n \geq 0)$, are called the n-th Euler numbers, whereas the Euler numbers E_n^* of the second kind are given by the generating function

$$\text{sech}(t) = \frac{2}{e^t+e^{-t}} = \sum_{n=0}^{\infty} E_n^*\frac{t^n}{n!} \tag{8}$$

as noted in [16,19].

Before we proceed, we study some relevant relations between the Changhee numbers of type two and the Euler numbers of the second kind.

Proposition 2. For $n \in \mathbb{N}$ and $0 \leq k \leq n$, we have

$$\widetilde{Ch}_n = \sum_{k=0}^{n} E_k^* S_1(n,k). \tag{9}$$

Proof of Proposition 2. From the generating functions of Changhee numbers of type two shown in (8), we have

$$\sum_{n=0}^{\infty} \widetilde{Ch}_n \frac{t^n}{n!} = \frac{2}{(1+t)+(1+t)^{-1}} = \frac{2}{e^{\log(1+t)}+e^{-\log(1+t)}}$$

$$= \text{sech}(\log(1+t)) \tag{10}$$

$$= \sum_{n=0}^{\infty} E_n^* \frac{(\log(1+t))^n}{n!} = \sum_{n=0}^{\infty} \left(\sum_{k=0}^{n} E_k^* S_1(n,k)\right) \frac{t^n}{n!}.$$

Thus we have the result. □

The result above helps us to derive some values of Changhee numbers of type two $\widetilde{Ch}_n$'s as follows: from $E_0^* = 1$, $E_1^* = 0$, $E_2^* = -1$, $E_3^* = 0$, $E_4^* = 5$, $E_5^* = 0$ and $S_1(n,n) = 0$ for $n \geq 0$, $S_1(n,0) = 0$ for $n \geq 1$, $S_1(2,1) = 1$, $S_1(3,1) = 2$, $S_1(4,1) = 6$, $S_1(5,1) = 24$, $S_1(3,2) = 3$, $S_1(4,2) = 11$, $S_1(5,2) = 50$, $S_1(4,3) = 6$, $S_1(5,3) = 35$, $S_1(5,4) = 10$,

$$\widetilde{Ch}_0 = E_0^* S_1(0,0) = 1,$$
$$\widetilde{Ch}_1 = E_0^* S_1(1,0) + E_1^* S_1(1,1) = 0 + 0 = 0,$$
$$\widetilde{Ch}_2 = E_0^* S_1(2,0) + E_1^* S_1(2,1) + E_2^* S_1(2,2) = 0 + 0 - 1 = -1.$$
$$\widetilde{Ch}_3 = E_0^* S_1(3,0) + E_1^* S_1(3,1) + E_2^* S_1(3,2) + E_3^* S_1(3,3)$$
$$= 0 + 0 - 3 + 0 = -3,$$
$$\widetilde{Ch}_4 = E_0^* S_1(4,0) + E_1^* S_1(4,1) + E_2^* S_1(4,2) + E_3^* S_1(4,3) + E_4^* S_1(4,4)$$
$$= 0 + 0 - 11 + 0 + 5 = -6,$$
$$\widetilde{Ch}_5 = E_0^* S_1(5,0) + E_1^* S_1(5,1) + E_2^* S_1(5,2) + E_3^* S_1(5,3) + E_4^* S_1(5,4)$$
$$+ E_5^* S_1(5,5) = 0 + 0 - 50 + 0 + 50 + 0 = 0.$$

For the inversion formulas for Proposition 2, we have the following.

Proposition 3. *For $n \in \mathbb{N}$ and $0 \le k \le n$, we have*

$$E_n^* = \sum_{k=0}^{n} \widetilde{Ch}_k S_2(n,k).$$

Proof of Proposition 3. From (6) and (8), we get the following, by replacing t by $e^t - 1$:

$$\frac{2}{(1+t)^2+1}(1+t) = \sum_{n=0}^{\infty} \widetilde{Ch}_n \frac{t^n}{n!}$$

$$\frac{2}{e^{2t}+1}e^t = \sum_{k=0}^{\infty} \widetilde{Ch}_k \frac{1}{k!}(e^t-1)^k$$

$$= \sum_{n=0}^{\infty} \left(\sum_{k=0}^{n} \widetilde{Ch}_k S_2(n,k) \right) \frac{t^n}{n!}$$

$$= \frac{2}{e^t+e^{-t}} = \sum_{n=0}^{\infty} E_n^* \frac{t^n}{n!}. \tag{11}$$

Now (11) gives us the desired result $E_n^* = \sum_{k=0}^{n} \widetilde{Ch}_k S_2(n,k)$. $\square$

Also by using the fermionic p-adic integration on $\mathbb{Z}_p$, we can represent Changhee numbers of type two as follows.

Proposition 4 (Witt's formula for Changhee numbers of type two).

For $n \in \mathbb{N}$, we have

$$\widetilde{Ch}_n = \int_{\mathbb{Z}_p} (2x+1)_n d\mu_{-1}(x). \tag{12}$$

Proof of Proposition 4. First, we observe

$$\int_{\mathbb{Z}_p} (1+t)^{2x+1} d\mu_{-1}(x) = \int_{\mathbb{Z}_p} \sum_{n=0}^{\infty} (2x+1)_n \frac{t^n}{n!} d\mu_{-1}(x)$$

$$= \sum_{n=0}^{\infty} \int_{\mathbb{Z}_p} (2x+1)_n d\mu_{-1} \frac{t^n}{n!}, \tag{13}$$

On the other hand, by the definition of fermionic p-adic integration on $\mathbb{Z}_p$,

$$\int_{\mathbb{Z}_p} (1+t)^{2x+1} d\mu_{-1}(x) = \frac{2}{(1+t)^2+1}(1+t) = \sum_{n=0}^{\infty} \widetilde{Ch}_n \frac{t^n}{n!}. \tag{14}$$

Thus, by comparing the coefficients of both sides of (13) and (14), we have the desired result. $\square$

3. Symmetry of w-Changhee Polynomials of Type Two

Motivated from D. Kim and T. Kim [20], for $w \in \mathbb{N}$, we define w-Changhee polynomials of type two by the following generating function

$$\frac{2}{(1+t)^{2w}+1}(1+t)^{2wx+1} = \sum_{n=0}^{\infty} \widetilde{Ch}_{n,w}(x) \frac{t^n}{n!}. \tag{15}$$

When $x = 0$, $\widetilde{Ch}_{n,w} = \widetilde{Ch}_{n,w}(0)$ are called the w-Changhee numbers of type two. When $w = 1$, $\widetilde{Ch}_{n,1}(x) = \widetilde{Ch}_n(x)$ are just the Changhee polynomials of type two in (4). For the case of $w = \frac{1}{2}$, the $\frac{1}{2}$-Changhee polynomials of type two are related to the well-known Changhee polynomials of type two, i.e., $\widetilde{Ch}_{n,\frac{1}{2}}(x) = \widetilde{Ch}_n(x+1)$.

The generating function of w-Changhee polynomials of type two can be related with Changhee polynomials of type two or Changhee numbers of type two as follows.

Proposition 5. *For $n, w, \ell \in \mathbb{N}$ and $1 \le \ell \le n$, we have*

$$(1) \ \widetilde{Ch}_{n,w}(x) = \sum_{\ell=0}^{n} \widetilde{Ch}_{\ell}(2wx), \ and$$

$$(2) \ \widetilde{Ch}_{n,w}(x) = \sum_{\ell=0}^{n} \binom{n}{\ell} (2wx)_{\ell} \widetilde{Ch}_{n-\ell}.$$

Proof of Proposition 5. (1) is immediate from the definition. For (2), we have

$$\sum_{n=0}^{\infty} \widetilde{Ch}_{n,w}(x) \frac{t^n}{n!} = \left(\sum_{\ell=0}^{\infty} \widetilde{Ch}_{\ell} \frac{t^{\ell}}{\ell!} \right) (1+t)^{2wx}$$

$$= \left(\sum_{\ell=0}^{\infty} \widetilde{Ch}_{\ell} \frac{t^{\ell}}{\ell!} \right) \left(\sum_{m=0}^{\infty} (2wx)_m \frac{t^m}{m!} \right)$$

$$= \sum_{n=0}^{\infty} \left(\sum_{\ell=0}^{n} \binom{n}{\ell} (2wx)_{\ell} \widetilde{Ch}_{n-\ell} \right\} \frac{t^n}{n!}.$$

$\square$

From (3), we can easily derive the following:

$$2 \sum_{\ell=0}^{n} (-1)^{\ell} (1+t)^{2\ell} = \frac{2\{1 + (-1)^{n+1}(1+t)^{2(n+1)}\}}{(1+t)^2 + 1} \tag{16}$$

The left hand side of (16) can be written as

$$2 \sum_{\ell=0}^{n} (-1)^{\ell} (1+t)^{2\ell} = \sum_{n=0}^{\infty} \left(\sum_{\ell=0}^{n-1} (-1)^{\ell} (2\ell)_n \right) \frac{t^n}{n!} \tag{17}$$

We use the notation of λ-falling factorial in [12,21] for $\lambda \in \mathbb{R}$,

$$(\ell \mid \lambda)_n = \begin{cases} \ell(\ell - \lambda) \cdots (\ell - \lambda(n-1)), & (\text{if } n \ge 1) \\ 1, & (\text{if } n = 0). \end{cases}$$

Then the right hand side of (17) can be written as

$$2 \sum_{\ell=0}^{n-1} (-1)^{\ell} (1+t)^{2\ell} = \sum_{n=0}^{\infty} T_m(n; (\ell \mid \tfrac{1}{2})) \frac{t^n}{n!}. \tag{18}$$

where we denote, for $\lambda \in \mathbb{R}$,

$$T_m(n; (\ell \mid \lambda)) = \sum_{\ell=0}^{n} (-1)^{\ell} (\ell \mid \lambda)_m.$$

For $n \in \mathbb{N}, n \equiv 1 \pmod 2, m \ge 0$ we have

$$\sum_{m=0}^{\infty} 2 \left(\sum_{\ell=0}^{n} (-1)^{\ell} (-2\ell)_m \right) \frac{t^m}{m!} = \frac{2(1 + (1+t)^{2(n+1)})}{(1+t)^2 + 1}. \tag{19}$$

On the other hand, by (4) and (18), we have

$$\sum_{m=0}^{\infty} \left(\widetilde{Ch}_m + \widetilde{Ch}_m(n+1) \right) \frac{t^m}{m!} = \frac{2(1+t)}{(1+t)^2+1} + \frac{2(1+t)^{2(n+1)}(1+t)}{(1+t)^2+1}$$

$$= 2 \sum_{\ell=0}^{n} (-1)^\ell (1+t)^{2\ell+1} \tag{20}$$

$$= 2 T_m(n; (\ell + \tfrac{1}{2} \,|\, \tfrac{1}{2})).$$

Now we consider a quotient of fermionic p-adic integrals on $\mathbb{Z}_p$,

$$\frac{2 \int_{\mathbb{Z}_p} (1+t)^{2w_2 x_2} d\mu_{-1}(x_2)}{\int_{\mathbb{Z}_p} (1+t)^{2w_1 w_2 x_1} d\mu_{-1}(x_1)} = \sum_{\ell=0}^{w_1-1} (-1)(1+t)^{2w_2\ell}$$

$$= \sum_{m=0}^{\infty} \sum_{\ell=0}^{w_1-1} (-1)^\ell (2w_2\ell)_m$$

$$= \sum_{m=0}^{\infty} \sum_{\ell=0}^{w_1-1} (2w_2)^m (-1)^\ell \left(\ell \,|\, \tfrac{1}{2w_2} \right)_m \tag{21}$$

$$= \sum_{m=0}^{\infty} (2w_2)^m T_m(w_1 - 1 \,|\, (\ell \,|\, \tfrac{1}{2w_2})),$$

where $T_m(n \,|\, (\ell \,|\, \lambda)) = \sum_{\ell=0}^{n} (-1)^\ell (\ell \,|\, \lambda)_m$ for $\lambda \in \mathbb{R}$.

For the symmetry of w-Changhee polynomials of type two, we consider the following quotient form of fermionic p-adic integration on $\mathbb{Z}_p$.

$$T(w_1, w_2) = \frac{2 \int_{\mathbb{Z}_p} \int_{\mathbb{Z}_p} (1+t)^{2w_1 x_1 + 2w_2 x_2 + 2} d\mu_{-1}(x_1)\, d\mu_{-1}(x_2)}{\int_{\mathbb{Z}_p} (1+t)^{2w_1 w_2 x_1 + 1} d\mu_{-1}(x_1)} (1+t)^{2w_1 w_2 x}$$

$$= \int_{\mathbb{Z}_p} (1+t)^{2w_1 x_1 + 1} d\mu_{-1}(x_1) (1+t)^{2w_1 w_2 x}$$

$$\times \frac{\int_{\mathbb{Z}_p} (1+t)^{2w_2 x_2} d\mu_{-1}(x_2)}{\int_{\mathbb{Z}_p} (1+t)^{2w_1 w_2 x_1} d\mu_{-1}(x_1)} \tag{22}$$

$$= \left(\sum_{\ell=0}^{\infty} \widetilde{Ch}_{\ell,w_1}(w_2 x) \frac{t^\ell}{\ell!} \right) \left(\sum_{k=0}^{\infty} (2w_2)^k T_k(w_1 - 1 \,|\, (k \,|\, \tfrac{1}{2w_2})) \right)$$

$$= \sum_{n=0}^{\infty} \left(\sum_{k=0}^{n} \binom{n}{k} \widetilde{Ch}_{n-k,w_1}(w_2 x)(2w_k)^k T_k(w_1 - 1 \,|\, (k \,|\, \tfrac{1}{2w_2})) \right) \frac{t^n}{n!}.$$

Similarly we have the following identity for $T(w_1, w_2)$ because $T(w_1, w_2)$ is symmetric on w_1 and w_2.

$$T(w_1, w_2) = \sum_{n=0}^{\infty} \left(\sum_{k=0}^{n} \binom{n}{k} \widetilde{Ch}_{n-k,w_2}(w_1 x)(2w_1)^k T_k(w_2 - 1 \,|\, (k \,|\, \tfrac{1}{2w_1})) \right) \frac{t^n}{n!}. \tag{23}$$

Thus, by (22) and (23), we have the following theorem.

Theorem 1. *For $w_1, w_2 \in \mathbb{N}$ with $w_1 \equiv 1 \pmod 2$, $w_2 \equiv 1 \pmod 2$ and $n \geq 0$, we have*

$$\sum_{k=0}^{n} \binom{n}{k} \widetilde{Ch}_{n-k,w_2}(w_1 x)(2w_1)^k T_k(w_2 - 1 \,|\, (k \,|\, \tfrac{1}{2w_1}))$$

$$= \sum_{k=0}^{n} \binom{n}{k} \widetilde{Ch}_{n-k,w_1}(w_2 x)(2w_2)^k T_k(w_1 - 1 \,|\, (k \,|\, \tfrac{1}{2w_2})).$$

If we take $w_2 = 1$ in Theorem 1, we have the following

Corollary 1. *For $w_1 \in \mathbb{N}$ with $w_1 \equiv 1 \pmod 2$ and $n \geq 0$, we have*

$$\widetilde{Ch}_n(w_1 x) = \sum_{k=0}^{n} \binom{n}{k} \widetilde{Ch}_{n-k,w_1}(x) 2^k T_k\left(w_1 - 1 \mid (k \mid \tfrac{1}{2})\right).$$

From (22), we rewrite $T(w_1, w_2)$ as follows:

$$
\begin{aligned}
T(w_1, w_2) &= \int_{\mathbb{Z}_p} (1+t)^{2w_1 x_1} d\mu_{-1}(x_1) (1+t)^{2w_1 w_2 x} \\
&\quad \times \frac{2 \int_{\mathbb{Z}_p} (1+t)^{2w_2 x_2} d\mu_{-1}(x_2)}{\int_{\mathbb{Z}_p} (1+t)^{2w_1 w_2 x_1} d\mu_{-1}(x_1)} \\
&= \int_{\mathbb{Z}_p} (1+t)^{2w_1 x_1} d\mu_{-1}(x_1) (1+t)^{2w_1 w_2 x} \\
&\quad \times 2 \sum_{\ell=0}^{w_1-1} (1+t)^{2w_2 \ell} (-1)^\ell \\
&= 2 \sum_{\ell=0}^{w_1-1} (-1)^\ell \int_{\mathbb{Z}_p} (1+t)^{2w_1 x_1 + 2w_1 w_2 x + 2w_2 \ell} d\mu_{-1}(x_1) \\
&= 2 \sum_{\ell=0}^{w_1-1} (-1)^\ell \int_{\mathbb{Z}_p} (1+t)^{2w_1 x_1 + 2w_1 w_2 x + \frac{w_2}{w_1} \ell} d\mu_{-1}(x_1) \\
&= 2 \sum_{\ell=0}^{w_1-1} (-1)^\ell \sum_{k=0}^{\infty} \widetilde{Ch}_{k,w_1}\left(w_2 x + \frac{w_2}{w_1} \ell\right) \frac{t^k}{k!} \\
&= \sum_{n=0}^{\infty} \left(2 \sum_{\ell=0}^{w_1-1} (-1)^\ell \widetilde{Ch}_{n,w_1}\left(\frac{w_2}{w_1} \ell + w_2 x\right)\right) \frac{t^n}{n!}
\end{aligned}
\tag{24}
$$

Similarly, by the symmetry of $T(w_1, w_2)$, we have the following identity

$$T(w_1, w_2) = \sum_{n=0}^{\infty} \left(2 \sum_{\ell=0}^{w_2-1} (-1)^\ell \widetilde{Ch}_{n,w_2}\left(\frac{w_1}{w_2} \ell + w_1 x\right)\right) \frac{t^n}{n!}. \tag{25}$$

Now from (24) and (25), we have the following theorem.

Theorem 2. *For $w_1, w_2 \in \mathbb{N}$ with $w_1 \equiv 1 \pmod 2$, $w_2 \equiv 1 \pmod 2$ and $n \geq 0$, we have*

$$\sum_{\ell=0}^{w_1-1} (-1)^\ell \widetilde{Ch}_{n,w_1}\left(\frac{w_2}{w_1} \ell + w_2 x\right) = \sum_{\ell=0}^{w_2-1} (-1)^\ell \widetilde{Ch}_{n,w_2}\left(\frac{w_1}{w_2} \ell + w_1 x\right).$$

When we take $w_2 = 1$, we have

$$\widetilde{Ch}_n(w_1 \ell + w_1 x) = \sum_{\ell=0}^{w_1-1} (-1)^\ell \widetilde{Ch}_{n,w_1}\left(\frac{\ell}{w_1} + x\right).$$

4. Conclusions

The Changhee polynomials of type two are considered by D. Kim and T. Kim (see [8]) and various properties on their polynomials and numbers are investigated.

In this paper, we investigate some symmetry identities for the Changhee polynomials of type two which are derived from the properties of symmetry for the fermionic p-adic integrals on $\mathbb{Z}_p$. The techniques presented in the articles by Cesarano and Fornaro [13,14], paticularly the Chebyshev polynomials, are used.

Especially we introduce w-Changhee polynomials of type two and investigate interesting symmetry identities.

For the cases of $w = 1$, $w = \frac{1}{2}$ and $w = \frac{1}{4}$, the symmetry of the w-Changhee polynomials of type two are related to the works of Changhee polynomials of type two, those of well-known Changhee polynomials (see [4,22]), and those of the Catalan polynomials (see [20]) respectively.

Recently, many works are done on some identities of special polynomials in the view point of degenerate sense (see [15,20,21]). Our result could be developed in that direction also: i.e., on the symmetry of the degenerate w-Changhee polynomials of type two.

Finally, we remark that our results on symmetry of two variables could be extended to the three variables case.

Author Contributions: All authors contributed equally to this work. All authors read and approved the final manuscript.

Acknowledgments: The authors would like to thank the referees for their valuable comments which improved the original manuscript in its present form.

References

1.   Kim, T. q-Volkenborn integration. *Russ. J. Math. Phys.* **2002**, *9*, 288–299.
2.   Kim, D.S. Identities associated with generalized twisted Euler polynomials twisted by ramified roots of unity. *Adv. Stud. Contemp. Math. (Kyungshang)* **2012**, *22*, 363–377.
3.   Kim, D.S.; Kim, T.; Kim, Y.H.; Lee, S.H. Some arithmetic properties of Bernoulli and Euler numbers. *Adv. Stud. Contemp. Math. (Kyungshang)* **2012**, *22*, 467–480.
4.   Kim, D.S.; Kim, T.; Seo, J.J. A note on Changhee polynomials and numbers. *Adv. Stud. Theor. Phys.* **2013**, *7*, 993–1003. [CrossRef]
5.   Kim, T.; Rim, S.-H. New Changhee q-Euler numbers and polynomials associated with p-adic q-integerals. *Comput. Math. Appl.* **2007**, *54*, 484–489. [CrossRef]
6.   Kim, T. Non Archmedean q-integrals associated with multiple Changhee q-Bernoulli polynomials. *Russ. Math. Phys.* **2003**, *10*, 91–98.
7.   Kim, T. p-adic q-integrals associated with the Changhee-Barnes' q-Bernoulli polynomials. *Integr. Transf. Spec. Funct.* **2004**, *15*, 415–420. [CrossRef]
8.   Kim, D.S.; Kim, T. A note on type 2 Changhee and Daehee polynomials. *arXiv* **2018**, arXiv:1809.05217.
9.   Kim, D.S.; Lee, N.; Na, H.; Park, K.H. Abundant symmetry for higher-order Bernoulli polynomials (I). *Adv. Stud. Contemp. Math.* **2013**, *23*, 461–482.
10.   Kim, D.S.; Lee, N.; Na, J.; Park, K.H. Abundant symmetry for higher-order Bernoulli polynomials (II). *Proc. Jangjeon Math. Soc.* **2013**, *16*, 359–378.
11.   Kim, T. Symmetry p-adic invariant integral on $\mathbb{Z}_p$ for Bernoulli and Euler polynomials. *J. Differ. Equ. Appl.* **2008**, *14*, 1267–1277. [CrossRef]
12.   Kim, T.; Dolgy, D. On the identities of symmetry for degenerate Bernoulli polynomials of order r. *Adv. Stud. Contemp. Math.* **2015**, *25*, 457–462.
13.   Cesarano, C.; Fornaro, C. A note on two-variable Chebyshev polynomials. *Georgian Math. J.* **2017**, *24*, 339–349. [CrossRef]
14.   Cesarno, C. Generalized Chebyshev polynomials. *Hacet. J. Math. Stat.* **2014**, *43*, 731–740.
15.   Kim, T.; Kim, D.S. Identities for degenerate Bernoulli polynomials and Korobov polynomials of the first kind. *Sci. China Math.* **2018**. [CrossRef]
16.   Kim, D.S.; Kim, T. Some p-adic integrals on $\mathbb{Z}_p$ associated with trigonometric functions. *Russ. J. Math. Phys.* **2018**, *25*, 300–308. [CrossRef]
17.   Simsek, Y. Identities on Changhee numbers and Apostol-type Daehee polynomials. *Adv. Stud. Contemp. Math. (Kyungshang)* **2017**, *27*, 199–212.
18.   Zhang, W.P. Number of solutions to a congruence equations mod p. *J. Northwest Univ. Natl. Sci.* **2016**, *46*, 313–316. (In Chinese)
19.   Knuth, D.E.; Buckholtz, T.J. Computation of Tangent, Euler, and Bernoulli Numbers. *Math. Comput.* **1967**, *21*, 663–688. [CrossRef]

20. Kim, D.S.; Kim, T. Triple symmetric identities for w-Caralan polynomials. *J. Korean Math. Soc.* **2017**, *54*, 1243–1264.

21. Kim, T.; Kim, D.S. Identities of symmetry for degenerate Euler polynomials and alternating generalized falling factorial sums. *Iran. J. Sci. Technol. Trans. A Sci.* **2017**, *41*, 939–949. [CrossRef]

22. Kim, T. Symmetry of power sum polynomials and multivariate fermionic p-adic invariant integral on $\mathbb{Z}_p$. *Russ. J. Math. Phys.* **2009**, *16*, 93–96. [CrossRef]

5

On Classical Gauss Sums and Some of their Properties

Li Chen

School of Mathematics, Northwest University, Xi'an 710127, China; cl1228@stumail.nwu.edu.cn

Abstract: The goal of this paper is to solve the computational problem of one kind rational polynomials of classical Gauss sums, applying the analytic means and the properties of the character sums. Finally, we will calculate a meaningful recursive formula for it.

Keywords: third-order character; classical Gauss sums; rational polynomials; analytic method; recursive formula

2010 Mathematics Subject Classification: 11L05, 11L07

1. Introduction

Let $q \geq 3$ be an integer. For any Dirichlet character χ mod q, according to the definition of classical Gauss sums $\tau(\chi)$, we can write

$$\tau(\chi) = \sum_{a=1}^{q} \chi(a) e\left(\frac{a}{q}\right),$$

where $e(y) = e^{2\pi i y}$.

Since this sum appears in numerous classical number theory problems, and it has a close connection with the trigonometric sums, we believe that classical Gauss sums play a crucial part in analytic number theory. Because of this phenomenon, plenty of experts have researched Gauss sums. Meanwhile, more conclusions have been obtained as regards their arithmetic properties. Such as the following results provided by Chen and Zhang [1]:

Let p be an odd prime with $p \equiv 1$ mod 4, λ be any fourth-order character mod p. Then one has the identity

$$\tau^2(\lambda) + \tau^2\left(\overline{\lambda}\right) = \sqrt{p} \cdot \sum_{a=1}^{p-1} \left(\frac{a + \overline{a}}{p}\right) = 2\sqrt{p} \cdot \alpha,$$

where $\left(\frac{*}{p}\right) = \chi_2$ denotes the the Legendre's symbol mod p (please see Reference [1,2] for its definition and related properties), and $\alpha = \sum_{a=1}^{\frac{p-1}{2}} \left(\frac{a + \overline{a}}{p}\right)$.

If p is a prime with $p \equiv 1$ mod 3, ψ is any third-order character mod p, then Zhang and Hu [3] had already obtained an analogous result (see Lemma 1). However, perhaps the most beautiful and important property of Gauss sums $\tau(\chi)$ is that $|\tau(\chi)| = \sqrt{q}$, for any primitive character χ mod q.

Reference [2] and References [4–13] have a good deal of various elementary properties of Gauss sums. In this paper, the following rational polynomials of Gauss sums attract our attention.

$$U_k(p, \chi) = \frac{\tau^{3k}(\chi)}{\tau^{3k}(\overline{\chi})} + \frac{\tau^{3k}(\overline{\chi})}{\tau^{3k}(\chi)}, \tag{1}$$

where p is an odd prime, k is a non-negative integer, χ is any non-principal character mod p.

Observing the basic properties of Equation (1), we noticed that hardly anyone had published research in any academic papers to date. We consider that the question is significant. In addition, the regularity of the value distribution of classical Gauss sums could be better revealed. Presently, we will explain certain properties discovered in our investigation. See that $U_k(p, \chi)$ has some good properties. In fact, for some special character χ mod p, the second-order linear recurrence formula for $U_k(p, \chi)$ for all integers $k \geq 0$ may be found similarly.

The goal of this paper is to use the analytic method and the properties of the character sums to solve the computational problem of $U_k(p, \chi)$, and to calculate two recursive formulae, which are listed hereafter:

Theorem 1. *Let p be a prime with $p \equiv 1$ mod 12, ψ be any third-order character mod p. Then, for any positive integer k, we can deduce the following second-order linear recursive formulae*

$$U_{k+1}(p, \psi) = \frac{d^2 - 2p}{p} \cdot U_k(p, \psi) - U_{k-1}(p, \psi),$$

where the initial values $U_0(p, \psi) = 2$ and $U_1(p, \psi) = \frac{d^2 - 2p}{p}$, d is uniquely determined by $4p = d^2 + 27b^2$ and $d \equiv 1$ mod 3.

So we can deduce the general term

$$U_k(p, \psi) = \left(\frac{d^2 - 2p + 3dbi\sqrt{3}}{2p} \right)^k + \left(\frac{d^2 - 2p - 3dbi\sqrt{3}}{2p} \right)^k, \quad i^2 = -1.$$

Theorem 2. *Let p be a prime with $p \equiv 7$ mod 12, ψ be any third-order character mod p. Then, for any positive integer k, we will obtain the second-order linear recursive formulae*

$$U_{k+1}(p, \psi) = \frac{i(2p - d^2)}{p} \cdot U_k(p, \psi) - U_{k-1}(p, \psi),$$

where the initial values $U_0(p, \psi) = 2$, $U_1(p, \psi) = \frac{i(2p - d^2)}{p}$ and $i^2 = -1$.

Similarly, we can also deduce the general term

$$U_k(p, \psi) = i^k \left(\frac{2p - d^2 + \sqrt{8p^2 - 4pd^2 + d^4}}{2p} \right)^k + i^k \left(\frac{2p - d^2 - \sqrt{8p^2 - 4pd^2 + d^4}}{2p} \right)^k.$$

2. Several Lemmas

We have used five simple and necessary lemmas to prove our theorems. Hereafter, we will apply relevant properties of classical Gauss sums and the third-order character mod p, all of which can be found in books concerning elementary and analytic number theory, such as in References [2,10], so we will not duplicate the related contents.

Lemma 1. *If p is any prime with $p \equiv 1$ mod 3, ψ is any third-order character mod p, then, we have the equation*

$$\tau^3(\psi) + \tau^3(\overline{\psi}) = dp,$$

where $\tau(\psi)$ denotes the classical Gauss sums, d is uniquely determined by $4p = d^2 + 27b^2$ and $d \equiv 1$ mod 3.

Proof. See References [3] or [8]. □

Lemma 2. *Let p be a prime with $p \equiv 1 \bmod 3$, ψ be any third-order character $\bmod\ p$, $\chi_2 = \left(\frac{*}{p}\right)$ denotes the Legendre's symbol $\bmod\ p$. The following identity holds*

$$\tau^2\left(\overline{\psi}\right) = \left(\frac{-1}{p}\right)\psi(4)\tau(\chi_2)\tau(\psi\chi_2).$$

Proof. Firstly, using the properties of Gauss sums, we get

$$\sum_{a=1}^{p-1}\overline{\psi}\left(a(a+1)\right) = \frac{1}{\tau(\psi)}\sum_{b=1}^{p-1}\psi(b)\sum_{a=1}^{p-1}\overline{\psi}(a)e\left(\frac{b(a+1)}{p}\right)$$

$$= \frac{\tau^2(\overline{\psi})}{\tau(\psi)} = \frac{\tau^3(\overline{\psi})}{p}.$$

$$(2)$$

On the other side, we get the sums

$$\sum_{a=1}^{p-1}\overline{\psi}\left(a(a+1)\right) = \psi(4)\sum_{a=0}^{p-1}\overline{\psi}\left(4a^2+4a\right)$$

$$= \psi(4)\sum_{a=0}^{p-1}\overline{\psi}\left((2a+1)^2-1\right) = \psi(4)\sum_{a=0}^{p-1}\overline{\psi}\left(a^2-1\right)$$

$$= \frac{\psi(4)}{\tau(\psi)}\sum_{b=1}^{p-1}\psi(b)\sum_{a=0}^{p-1}e\left(\frac{b(a^2-1)}{p}\right) = \frac{\psi(4)}{\tau(\psi)}\sum_{b=1}^{p-1}\psi(b)e\left(\frac{-b}{p}\right)\sum_{a=0}^{p-1}e\left(\frac{ba^2}{p}\right)$$

$$= \frac{\psi(4)\tau(\chi_2)}{\tau(\psi)}\sum_{b=1}^{p-1}\psi(b)\chi_2(b)e\left(\frac{-b}{p}\right) = \frac{\psi(4)\chi_2(-1)\tau(\chi_2)\tau(\psi\chi_2)}{\tau(\psi)}.$$

$$(3)$$

Combining Equations (2) and (3), we obtain

$$\tau^2\left(\overline{\psi}\right) = \left(\frac{-1}{p}\right)\psi(4)\tau(\chi_2)\tau(\psi\chi_2).$$

Now, Lemma 2 has been proved. $\square$

Lemma 3. *Let p be a prime with $p \equiv 1 \bmod 6$, χ be any sixth-order character $\bmod\ p$. Then, about classical Gauss sums $\tau(\chi)$, the following holds:*

$$\tau^3(\chi) + \tau^3\left(\overline{\chi}\right) = \begin{cases} p^{\frac{1}{2}}\left(d^2-2p\right) & \text{if } p = 12h+1, \\ -i\cdot p^{\frac{1}{2}}\left(d^2-2p\right) & \text{if } p = 12h+7, \end{cases}$$

where $i^2 = -1$, d is uniquely determined by $4p = d^2 + 27b^2$ and $d \equiv 1 \bmod 3$.

Proof. Since $p \equiv 1 \bmod 6$, ψ is a third-order character $\bmod\ p$. Any sixth-order character χ $\bmod\ p$ can be denoted as $\chi = \psi\chi_2$ or $\chi = \overline{\psi}\chi_2$. Note that $\psi^3(4) = 1$, $\overline{\psi}^3(4) = 1$ and $\chi_2^3 = \chi_2$, from Lemma 2 we deduce

$$\tau^6\left(\overline{\psi}\right) = \left(\frac{-1}{p}\right)\tau^3(\chi_2)\tau^3(\psi\chi_2)$$

$$(4)$$

and

$$\tau^6\left(\psi\right) = \left(\frac{-1}{p}\right)\tau^3(\chi_2)\tau^3\left(\overline{\psi}\chi_2\right).$$

$$(5)$$

Adding Equations (4) and (5), and then applying Lemma 1 we have

$$\left(\frac{-1}{p}\right)\tau^3(\chi_2)\left(\tau^3(\psi\chi_2)+\tau^3\left(\overline{\psi}\chi_2\right)\right) = \tau^6\left(\overline{\psi}\right) + \tau^6\left(\psi\right)$$

$$= \left(\tau^3\left(\overline{\psi}\right)+\tau^3\left(\psi\right)\right)^2 - 2p^3 = d^2p^2 - 2p^3.$$

$$(6)$$

Note that χ_2 is a real character mod p, $\overline{\psi}\chi_2 = \overline{\psi\chi_2}$, and $\tau(\chi_2) = \sqrt{p}$. If $p \equiv 1 \bmod 4$; $\tau(\chi_2) = i \cdot \sqrt{p}$, $i^2 = -1$, if $p \equiv 3 \bmod 4$. From Equation (6) we may immediately prove the sum

$$\tau^3(\psi\chi_2) + \tau^3(\overline{\psi}\chi_2) = \begin{cases} p^{\frac{1}{2}}\left(d^2 - 2p\right) & \text{if } p = 12h+1, \\ -i \cdot p^{\frac{1}{2}}\left(d^2 - 2p\right) & \text{if } p = 12h+7. \end{cases} \tag{7}$$

Let $\chi = \psi\chi_2$, then χ is a sixth-order character mod p and $\overline{\psi}\chi_2 = \overline{\chi}$. From Equation (7) we can deduce the sum term

$$\tau^3(\chi) + \tau^3(\overline{\chi}) = \begin{cases} p^{\frac{1}{2}}\left(d^2 - 2p\right) & \text{if } p = 12h+1, \\ -i \cdot p^{\frac{1}{2}}\left(d^2 - 2p\right) & \text{if } p = 12h+7. \end{cases}$$

The proof of Lemma 3 has been completed. $\square$

Lemma 4. *Let p be a prime with $p \equiv 7 \bmod 12$, ψ be any three-order character mod p. Then, we compute the sum term*

$$\frac{\tau^3(\overline{\psi})}{\tau^3(\psi)} + \frac{\tau^3(\psi)}{\tau^3(\overline{\psi})} = \frac{i \cdot (2p - d^2)}{p}.$$

Proof. Let ψ be a three-order character mod p. Then, for any six-order character χ mod p, we must have $\chi = \psi\chi_2$ or $\chi = \overline{\chi}\chi_2$. Without loss of generality we suppose that $\chi = \psi\chi_2$, then note that $\psi(-1) = 1$, $\chi_2(-1) = -1$ and Theorem 7.5.4 in Reference [10], we acquire

$$\sum_{a=0}^{p-1} e\left(\frac{ba^2}{p}\right) = \chi_2(b) \cdot \sqrt{p}, \ (p,b) = 1.$$

Using the properties of Gauss sums we can write

$$\sum_{a=0}^{p-1}\chi(a^2-1) = \frac{1}{\tau(\overline{\chi})}\sum_{b=1}^{p-1}\overline{\chi}(b)\sum_{a=0}^{p-1}e\left(\frac{b(a^2-1)}{p}\right)$$
$$= \frac{1}{\tau(\overline{\chi})}\sum_{b=1}^{p-1}\overline{\chi}(b)e\left(\frac{-b}{p}\right)\sum_{a=0}^{p-1}e\left(\frac{ba^2}{p}\right) = \frac{\sqrt{p}}{\tau(\overline{\chi})}\sum_{b=1}^{p-1}\overline{\chi}(b)\chi_2(b)e\left(\frac{-b}{p}\right) \tag{8}$$
$$= \frac{\overline{\chi}(-1)\chi_2(-1)\sqrt{p}\ \tau(\overline{\chi}\chi_2)}{\tau(\overline{\chi})} = \frac{\sqrt{p}\ \tau(\overline{\chi}\chi_2)}{\tau(\overline{\chi})}.$$

Noting that $\overline{\chi}^2 = \overline{\psi}^2 = \psi$, we can deduce

$$\sum_{a=0}^{p-1}\chi(a^2-1) = \sum_{a=0}^{p-1}\chi((a+1)^2-1) = \sum_{a=1}^{p-1}\chi(a)\chi(a+2)$$
$$= \frac{1}{\tau(\overline{\chi})}\sum_{b=1}^{p-1}\overline{\chi}(b)\sum_{a=1}^{p-1}\chi(a)e\left(\frac{b(a+2)}{p}\right) = \frac{\tau(\chi)}{\tau(\overline{\chi})}\sum_{b=1}^{p-1}\overline{\chi}^2(b)e\left(\frac{2b}{p}\right) \tag{9}$$
$$= \frac{\overline{\psi}(2)\tau(\chi)\tau(\psi)}{\tau(\overline{\chi})}.$$

Obviously, $\overline{\chi}\chi_2 = \overline{\psi}$ and $\psi^3(2) = 1$, applying Equations (8) and (9) we have

$$\tau^3(\chi) = p^{\frac{3}{2}} \cdot \frac{\tau^3(\overline{\psi})}{\tau^3(\psi)}. \tag{10}$$

Similarly, we can see

$$\tau^3(\overline{\chi}) = p^{\frac{3}{2}} \cdot \frac{\tau^3(\psi)}{\tau^3(\overline{\psi})}. \tag{11}$$

Combining Equation (10), Equation (11) and Lemma 3 we compute

$$\frac{\tau^3(\psi)}{\tau^3(\overline{\psi})} + \frac{\tau^3(\overline{\psi})}{\tau^3(\psi)} = \frac{1}{p^{\frac{3}{2}}}\left(\tau^3(\chi) + \tau^3(\overline{\chi})\right) = \frac{i\cdot(2p-d^2)}{p}.$$

This completes the proof of Lemma 4. $\square$

Lemma 5. *Let p be a prime with $p \equiv 1 \bmod 12$, ψ be any three-order character mod p. Then, we obtain the sum term*

$$\frac{\tau^3(\overline{\psi})}{\tau^3(\psi)} + \frac{\tau^3(\psi)}{\tau^3(\overline{\psi})} = \frac{d^2-2p}{p}.$$

Proof. From Lemma 3 and the method of proving Lemma 4 we can easily deduce Lemma 5. $\square$

3. Proofs of the Theorems

In this section, we prove our two theorems. For Theorem 1, since $p \equiv 1 \bmod 12$, ψ is a third-order character mod p, then for any positive integer k, let

$$U_k(p) = \frac{\tau^{3k}(\psi)}{\tau^{3k}(\overline{\psi})} + \frac{\tau^{3k}(\overline{\psi})}{\tau^{3k}(\psi)}.$$

From Lemma 5 we have

$$U_1(p) = \frac{\tau^3(\overline{\psi})}{\tau^3(\psi)} + \frac{\tau^3(\psi)}{\tau^3(\overline{\psi})} = \frac{d^2-2p}{p} \tag{12}$$

and

$$\frac{d^2-2p}{p}\cdot U_k(p) = U_k(p)U_1(p) = \left(\frac{\tau^{3k}(\psi)}{\tau^{3k}(\overline{\psi})} + \frac{\tau^{3k}(\overline{\psi})}{\tau^{3k}(\psi)}\right)\cdot\left(\frac{\tau^3(\overline{\psi})}{\tau^3(\psi)} + \frac{\tau^3(\psi)}{\tau^3(\overline{\psi})}\right)$$

$$= \frac{\tau^{3k+3}(\overline{\psi})}{\tau^{3k+3}(\psi)} + \frac{\tau^{3k+3}(\psi)}{\tau^{3k+3}(\overline{\psi})} + \frac{\tau^{3k-3}(\overline{\psi})}{\tau^{3k-3}(\psi)} + \frac{\tau^{3k-3}(\psi)}{\tau^{3k-3}(\overline{\psi})} = U_{k+1}(p) + U_{k-1}(p). \tag{13}$$

Combining Equations (12) and (13) we may immediately compute the second-order linear recursive formula

$$U_{k+1}(p) = \frac{d^2-2p}{p}\cdot U_k(p) - U_{k-1}(p) \tag{14}$$

with initial values $U_0(p) = 2$ and $U_1(p) = \frac{d^2-2p}{p}$.

Note that the two roots of the equation $\lambda^2 - \frac{d^2-2p}{p}\lambda + 1 = 0$ are

$$\lambda_1 = \frac{d^2 - 2p + 3dbi\sqrt{3}}{2p} \quad \text{and} \quad \lambda_2 = \frac{d^2 - 2p - 3dbi\sqrt{3}}{2p}.$$

So from Equation (14) and its initial values we may immediately deduce the general term

$$U_k(p,\psi) = \left(\frac{d^2 - 2p + 3dbi\sqrt{3}}{2p}\right)^k + \left(\frac{d^2 - 2p - 3dbi\sqrt{3}}{2p}\right)^k,$$

where $i^2 = -1$. Now Theorem 1 has been finished.

Similarly, from Lemma 4 and the method of proving Theorem 1 we can easily obtain Theorem 2. Now, we have completed all the proofs of our Theorems.

4. Conclusions

The main results of this paper are Theorem 1 and 2. They give a new second-order linear recurrence formula for Equation (1) with the third-order character ψ mod p. Therefore, we can calculate the exact value of Equation (1). Note that $\left| \tau\left(\overline{\psi}\right) / \tau\left(\psi\right) \right| = 1$, so $\tau\left(\overline{\psi}\right) / \tau\left(\psi\right)$ is a unit root, thus, the results in this paper profoundly reveal the distributional properties of two different Gauss sums quotients on the unit circle.

For the other mod p characters, for example, the fifth-order character χ mod p with $p \equiv 1$ mod 5, we naturally ask whether there exists a similar formula as presented in our theorems. This is still an open problem. It will be the content of our future investigations.

Acknowledgments: The author wish to express her gratitude to the editors and the reviewers for their helpful comments.

References

1. Chen, Z.Y.; Zhang, W.P. On the fourth-order linear recurrence formula related to classical Gauss sums. *Open Math.* **2017**, *15*, 1251–1255.
2. Apostol, T.M. *Introduction to Analytic Number Theory*; Springer: New York, NY, USA, 1976.
3. Zhang, W.P.; Hu, J.Y. The number of solutions of the diagonal cubic congruence equation mod *p*. *Math. Rep.* **2018**, *20*, 73–80.
4. Chen, L.; Hu, J.Y. A linear recurrence formula involving cubic Gauss sums and Kloosterman sums. *Acta Math. Sin.* **2018**, *61*, 67–72.
5. Li, X.X.; Hu, J.Y. The hybrid power mean quartic Gauss sums and Kloosterman sums. *Open Math.* **2017**, *15*, 151–156.
6. Zhang, H.; Zhang, W.P. The fourth power mean of two-term exponential sums and its application. *Math. Rep.* **2017**, *19*, 75–83.
7. Zhang, W.P.; Liu, H.N. On the general Gauss sums and their fourth power mean. *Osaka J. Math.* **2005**, *42*, 189–199.
8. Berndt, B.C.; Evans, R.J. The determination of Gauss sums. *Bull. Am. Math. Soc.* **1981**, *5*, 107–128. [CrossRef]
9. Berndt, B.C.; Evans, R.J. Sums of Gauss, Jacobi, and Jacobsthal. *J. Number Theory* **1979**, *11*, 349–389. [CrossRef]
10. Hua, L.K. *Introduction to Number Theory*; Science Press: Beijing, China, 1979.
11. Kim, T. Power series and asymptotic series associated with the *q* analog of the two-variable *p*-adic *L*-function. *Russ. J. Math. Phys.* **2005**, *12*, 186–196.
12. Kim, H.S.; Kim, T. On certain values of *p*-adic *q*-*L*-function. *Rep. Fac. Sci. Eng. Saga Univ. Math.* **1995**, *23*, 1–2.
13. Chae, H.; Kim, D.S. L function of some exponential sums of finite classical groups. *Math. Ann.* **2003**, *326*, 479–487. [CrossRef]

6

Some Identities of Fully Degenerate Bernoulli Polynomials Associated with Degenerate Bernstein Polynomials

Jeong Gon Lee [1], Wonjoo Kim [2,*] and Lee-Chae Jang [3]

[1] Division of Applied Mathematics, Nanoscale Science and Technology Institute, Wonkwang University, Iksan 54538, Korea; jukolee@wku.ac.kr
[2] Department of Applied Mathematics, Kyunghee University, Yongin 17104, Korea
[3] Graduate School of Education, Konkuk University, Seoul 05029, Korea; lcjang@konkuk.ac.kr
* Correspondence: wjookim@khu.ac.kr

Abstract: In this paper, we investigate some properties and identities for fully degenerate Bernoulli polynomials in connection with degenerate Bernstein polynomials by means of bosonic p-adic integrals on $\mathbb{Z}_p$ and generating functions. Furthermore, we study two variable degenerate Bernstein polynomials and the degenerate Bernstein operators.

Keywords: degenerate Bernoulli polynomials; degenerate Bernstein operators

1. Introduction

Let p be a fixed prime number. Throughout this paper, $\mathbb{Z}$, $\mathbb{Z}_p$, $\mathbb{Q}_p$ and $\mathbb{C}_p$, will denote the ring of rational integers, the ring of p-adic integers, the field of p-adic rational numbers and the completion of algebraic closure of $\mathbb{Q}_p$, respectively. The p-adic norm $|q|_p$ is normalized as $|p|_p = \frac{1}{p}$.

For $\lambda, t \in \mathbb{C}_p$ with $|\lambda t|_p < p^{-\frac{1}{p-1}}$ and $|t|_p < 1$, the degenerate Bernoulli polynomials are defined by the generating function to be

$$\frac{t}{(1+\lambda t)^{\frac{1}{\lambda}} - 1}(1+\lambda t)^{\frac{x}{\lambda}} = \sum_{n=0}^{\infty} \beta_n(x|\lambda)\frac{t^n}{n!}, \tag{1}$$

(See [1–3]). When $x = 0$, $\beta_n(\lambda) = \beta_n(0|\lambda)$ are called the degenerate Bernoulli numbers. Note that $\lim_{\lambda \to 0} \beta_n(x|\lambda) = B_n(x)$, where $B_n(x)$ are the ordinary Bernoulli polynomials defined by

$$\frac{t}{e^t - 1}e^{xt} = \sum_{n=0}^{\infty} B_n(x)\frac{t^n}{n!}, \tag{2}$$

and $B_n = B_n(0)$ are called the Bernoulli numbers. The degenerate exponential function is defined by

$$e_\lambda^x(t) = (1+\lambda t)^{\frac{x}{\lambda}} = \sum_{n=0}^{\infty} (x)_{n,\lambda}\frac{t^n}{n!}, \tag{3}$$

where $(x)_{0,\lambda} = 1$, $(x)_{n,\lambda} = x(x-\lambda)(x-2\lambda)\cdots(x-(n-1)\lambda)$, for $n \geq 1$. From (1), we get

$$\beta_n(x|\lambda) = \sum_{l=0}^{n} \binom{n}{l}\beta_l(\lambda)(x)_{n-l,\lambda}. \tag{4}$$

Recently, Kim-Kim introduced the degenerate Bernstein polynomials given by

$$\frac{(x)_{k,\lambda}}{k!} t^k (1 + \lambda t)^{\frac{1-x}{\lambda}} = \sum_{n=k}^{\infty} B_{k,n}(x|\lambda) \frac{t^n}{n!}, \tag{5}$$

(See [4–6]). Thus, by (5), we note that

$$B_{k,n}(x|\lambda) = \begin{cases} \binom{n}{k}(x)_{k,\lambda}(1-x)_{n-k,\lambda}, & \text{if } n \geq k, \\ 0, & \text{if } n < k. \end{cases} \tag{6}$$

where n, k are non-negative integers. Let $UD(\mathbb{Z}_p)$ be the space of uniformly differentiable functions on $\mathbb{Z}_p$. For $f \in UD(\mathbb{Z}_p)$, the degenerate Bernstein operator of order n is given by

$$\begin{aligned} \mathbb{B}_{n,\lambda}(f|\lambda) &= \sum_{k=0}^{\infty} f\left(\frac{k}{n}\right) \binom{n}{k} (x)_{k,\lambda}(1-x)_{n-k,\lambda} \\ &= \sum_{k=0}^{\infty} f\left(\frac{k}{n}\right) B_{k,n}(x|\lambda), \end{aligned} \tag{7}$$

(See [4–6]). The bosonic p-adic integral on $\mathbb{Z}_p$ is defined by Volkenborn as

$$\int_{\mathbb{Z}_p} f(x) d\mu_0(x) = \lim_{N \to \infty} \frac{1}{p^N} \sum_{x=0}^{p^N-1} f(x), \tag{8}$$

(see [7]). By (8), we get

$$\int_{\mathbb{Z}_p} f(x+1) d\mu_0(x) - \int_{\mathbb{Z}_p} f(x) d\mu_0(x) = f'(0), \tag{9}$$

where $\frac{d}{dx} f(x)\big|_{x=0} = f'(0)$.

From (8), Kim-Seo [8] proposed fully degenerate Bernoulli polynomials which are reformulated in terms of bosonic p-adic integral on $\mathbb{Z}_p$ as

$$\int_{\mathbb{Z}_p} (1 + \lambda t)^{\frac{x+y}{\lambda}} d\mu_0(y) = \frac{\frac{1}{\lambda} \log(1 + \lambda t)}{(1 + \lambda t)^{\frac{1}{\lambda}} - 1} (1 + \lambda t)^{\frac{x}{\lambda}} = \sum_{n=0}^{\infty} B_n(x|\lambda) \frac{t^n}{n!}, \tag{10}$$

and for $x = 0$, $B_n(\lambda) = B_n(0|\lambda)$ are called fully degenerate Bernoulli numbers.

Note that the fully degenerate Bernoulli polynomial was named Daehee polynomials with α-parameter in [9]. On the other hand,

$$\int_{\mathbb{Z}_p} (1 + \lambda t)^{\frac{x+y}{\lambda}} d\mu_0(y) = \sum_{n=0}^{\infty} \int_{\mathbb{Z}_p} (x+y)_{n,\lambda} d\mu_0(y) \frac{t^n}{n!}. \tag{11}$$

By (10) and (11), we get

$$\int_{\mathbb{Z}_p} (x+y)_{n,\lambda} d\mu_0(y) = B_n(x|\lambda), \quad (n \geq 0). \tag{12}$$

Recall that the Daehee polynomials are defined by the generating function to be

$$\frac{\log(1+t)}{t} (1+t)^x = \sum_{n=0}^{\infty} D_n(x) \frac{t^n}{n!}, \tag{13}$$

and for $x = 0$, $D_n = D_n(0)$ are called the Daehee numbers (see [10,11]).

Also, the higher order Daehee polynomials are defined by the generating function to be

$$\left(\frac{\log(1+t)}{t}\right)^k (1+t)^x = \sum_{n=0}^{\infty} D_n^{(k)}(x)\frac{t^n}{n!}, \tag{14}$$

and for $x = 0$, $D_n^{(k)} = D_n^{(k)}(0)$ are called the higher order Daehee numbers. From (10), we observe

$$\begin{aligned}
\frac{\frac{1}{\lambda}\log(1+\lambda t)}{(1+\lambda t)^{\frac{1}{\lambda}}-1}(1+\lambda t)^{\frac{x}{\lambda}} &= \frac{t}{(1+\lambda t)^{\frac{1}{\lambda}}-1}(1+\lambda t)^{\frac{x}{\lambda}}\frac{\log(1+\lambda t)}{\lambda t}\\
&= \left(\sum_{m=0}^{\infty}\beta_m(x|\lambda)\frac{t^m}{m!}\right)\left(\sum_{l=0}^{\infty}D_l\frac{(\lambda t)^l}{l!}\right)\\
&= \sum_{n=0}^{\infty}\left(\sum_{m=0}^{n}\binom{n}{m}\beta_m(x|\lambda)D_{n-m}\lambda^{n-m}\right)\frac{t^n}{n!}.
\end{aligned} \tag{15}$$

By (10) and (14), we get

$$B_n(x|\lambda) = \sum_{m=0}^{n}\binom{n}{m}\beta_m(x|\lambda)D_{n-m}\lambda^{n-m}, \quad (n\geq 0). \tag{16}$$

From (3) and (10), we observe that

$$\begin{aligned}
\sum_{n=0}^{\infty}B_n(x|\lambda)\frac{t^n}{n!} &= \frac{\frac{1}{\lambda}\log(1+\lambda t)}{(1+\lambda t)^{\frac{1}{\lambda}}-1}(1+\lambda t)^{\frac{x}{\lambda}}\\
&= \left(\sum_{m=0}^{\infty}B_m(\lambda)\frac{t^m}{m!}\right)\left(\sum_{l=0}^{\infty}(x)_{l,\lambda}\frac{t^l}{l!}\right)\\
&= \sum_{n=0}^{\infty}\left(\sum_{m=0}^{n}\binom{n}{m}B_m(\lambda)(x)_{n-m,\lambda}\right)\frac{t^n}{n!}.
\end{aligned} \tag{17}$$

By (17), we get

$$B_n(x|\lambda) = \sum_{m=0}^{n}\binom{n}{m}B_m(\lambda)(x)_{n-m,\lambda}, \quad (n\geq 0). \tag{18}$$

From (1) and (3), we note that

$$\begin{aligned}
t &= \left((1+\lambda t)^{\frac{1}{\lambda}}-1\right)\sum_{m=0}^{\infty}\beta_m(\lambda)\frac{t^m}{m!}\\
&= \left(\sum_{l=0}^{\infty}(1)_{l,\lambda}\frac{t^l}{l!}\right)\left(\sum_{m=0}^{\infty}\beta_m(\lambda)\frac{t^m}{m!}\right) - \sum_{m=0}^{\infty}\beta_m(\lambda)\frac{t^m}{m!}\\
&= \sum_{n=0}^{\infty}\left(\sum_{m=0}^{n}\binom{n}{m}(1)_{n-m,\lambda}\beta_m(\lambda)\right)\frac{t^n}{n!} - \sum_{n=0}^{\infty}\beta_n(\lambda)\frac{t^n}{n!}\\
&= \sum_{n=0}^{\infty}\left(\sum_{m=0}^{n}\binom{n}{m}(1)_{n-m,\lambda}\beta_m(\lambda) - \beta_n(\lambda)\right)\frac{t^n}{n!}.
\end{aligned} \tag{19}$$

Comparing the cofficients on both sides of (19), we get

$$\sum_{m=0}^{n}\binom{n}{m}(1)_{n-m,\lambda}\beta_m(\lambda) - \beta_n(\lambda) = \delta_{1,n}, \quad (n\geq 0), \tag{20}$$

where $\delta_{k,n}$ is the Kronecker's symbol.

By (4) and (20), we have

$$\beta_n(1|\lambda) - \beta_n(\lambda) = \delta_{1,n}. \tag{21}$$

The generating function of fully degenerate Bernoulli polynomials introduced in (5) can be expressed as bosonic p-adic integral but the generating function of degenerate Bernoulli polynomials introduced in (1) is not expressed as a bosonic p-adic integral. This is why we considered the fully degenerate Bernoulli polynomials, and the motivation of this paper is to investigate some identities of them associated with degenerate Bernstein polynomials.

In this paper, we consider the fully degenerate Bernoulli polynomials and investigate some properties and identities for these polynomials in connection with degenerate Bernstein polynomials by means of bosonic p-adic integrals on $\mathbb{Z}_p$ and generating functions. Furthermore, we study two variable degenerate Bernstein polynomials and the degenerate Bernstein operators.

2. Fully Degenerate Bernoulli and Bernstein Polynomials

From (10), we observe that

$$
\begin{aligned}
\sum_{n=0}^{\infty} B_n(1-x|\lambda)\frac{t^n}{n!} &= \frac{\frac{1}{\lambda}\log(1+\lambda t)}{(1+\lambda t)^{\frac{1}{\lambda}} - 1}(1+\lambda t)^{\frac{1-x}{\lambda}} \\
&= \frac{\left(-\frac{1}{\lambda}\right)(1+(-\lambda)(-t))}{(1+(-\lambda)(-t))^{-\frac{1}{\lambda}} - 1}(1+(-\lambda)(-t))^{-\frac{x}{\lambda}} \\
&= \sum_{n=0}^{\infty} B_n(x|-\lambda)(-1)^n\frac{t^n}{n!}.
\end{aligned}
\tag{22}
$$

From (22), we obtain the following Lemma.

Lemma 1. *For $n \in \mathbb{N} \cup \{0\}$, we have*

$$B_n(1-x|\lambda) = (-1)^n B_n(x|-\lambda). \tag{23}$$

From (16) and (21), we get

$$
\begin{aligned}
B_n(1|\lambda) - B_n(\lambda) &= \sum_{m=0}^{n}\binom{n}{m}\left(\beta_m(1|\lambda) - \beta_m(\lambda)\right)D_{n-m}\lambda^{n-m} \\
&= \sum_{m=0}^{n}\binom{n}{m}\delta_{1,m}D_{n-m}\lambda^{n-m}, \quad (n \geq 0).
\end{aligned}
\tag{24}
$$

From (1), we observe that

$$
\begin{aligned}
\sum_{n=0}^{\infty}\beta_n(x+1|\lambda)\frac{t^n}{n!} &= (1+\lambda t)^{\frac{1}{\lambda}}\left(\sum_{m=0}^{\infty}\beta_m(x|\lambda)\frac{t^m}{m!}\right) \\
&= \left(\sum_{l=0}^{\infty}(1)_{l,\lambda}\frac{t^l}{l!}\right)\left(\sum_{m=0}^{\infty}\beta_m(x|\lambda)\frac{t^m}{m!}\right) \\
&= \sum_{n=0}^{\infty}\left(\sum_{m=0}^{n}\binom{n}{m}(1)_{n-m,\lambda}\beta_m(x|\lambda)\right)\frac{t^n}{n!}.
\end{aligned}
\tag{25}
$$

By (25), we get

$$\beta_n(x+1|\lambda) = \sum_{m=0}^{n}\binom{n}{m}(1)_{n-m,\lambda}\beta_m(x|\lambda). \tag{26}$$

By (26), with $x = 1$, we have

$$\beta_n(2|\lambda) = \sum_{m=0}^{n} \binom{n}{m} \beta_m(1|\lambda)(1)_{n-m,\lambda}$$

$$= (1)_{n,\lambda}\beta_0(1|\lambda) + n(1)_{n-1,\lambda}\beta_1(1|\lambda) + \sum_{m=2}^{n} \binom{n}{m} \beta_m(1|\lambda)(1)_{n-m,\lambda}$$

$$= (1)_{n,\lambda} + n(1)_{n-1,\lambda}(\beta_1(\lambda) - 1) + \sum_{m=2}^{n} \binom{n}{m} \beta_m(\lambda)(1)_{n-m,\lambda} \qquad (27)$$

$$= (1)_{n,\lambda} + n(1)_{n-1,\lambda}\beta_1(\lambda) - n(1)_{n-1,\lambda} + \sum_{m=2}^{n} \binom{n}{m} \beta_m(\lambda)(1)_{n-m,\lambda}$$

$$= -n(1)_{n-1,\lambda} + \sum_{m=0}^{n} \binom{n}{m} \beta_m(\lambda)(1)_{n-m,\lambda}$$

$$= -n(1)_{n-1,\lambda} + \beta_n(1|\lambda).$$

Therefore, by (27), we obtain the following theorem.

Theorem 1. *For $n \in \mathbb{N}$, we have*

$$\beta_n(2|\lambda) = -n(1)_{n-1,\lambda} + \beta_n(1|\lambda). \qquad (28)$$

Note that

$$(1 - x)_{n,\lambda} = (-1)^n (x - 1)_{n,-\lambda}, \quad (n \geq 0). \qquad (29)$$

Therefore by (12), (23), and (29), we get

$$\int_{\mathbb{Z}_p} (1 - x)_{n,\lambda} d\mu_0(x) = (-1)^n \int_{\mathbb{Z}_p} (x - 1)_{n,-\lambda} d\mu_0(x) = \int_{\mathbb{Z}_p} (x + 2)_{n,\lambda} d\mu_0(x). \qquad (30)$$

Therefore, by (30) and Theorem 1, we obtain the following theorem.

Theorem 2. *For $n \in \mathbb{N}$, we have*

$$\int_{\mathbb{Z}_p} (1 - x)_{n,\lambda} d\mu_0(x) = \int_{\mathbb{Z}_p} (x + 2)_{n,\lambda} d\mu_0(x) = n(1)_{n-1,\lambda}(\lambda - 1)B_1(\lambda) + \int_{\mathbb{Z}_p} (x)_{n,\lambda} d\mu_0(x). \qquad (31)$$

Corollary 1. *For $n \in \mathbb{N}$, we have*

$$(-1)^n B_n(-1| - \lambda) = (1)_{n-1,\lambda}(1 - nB_1(\lambda)) + B_n(\lambda) = B_n(2|\lambda). \qquad (32)$$

By (17), we get

$$B_n(1 - x|\lambda) = \sum_{m=0}^{n} \binom{n}{m} B_m(\lambda)(1 - x)_{n-m,\lambda}$$

$$= \sum_{m=0}^{n} \binom{n}{m} (x)_{m,\lambda}(1 - x)_{n-m,\lambda} \frac{B_m(\lambda)}{(x)_{m,\lambda}} \qquad (33)$$

$$= \sum_{m=0}^{n} B_{m,n}(x|\lambda) B_m(\lambda) \frac{1}{(x)_{m,\lambda}}.$$

In [8], we note that

$$\frac{1}{(x)_{m,\lambda}} = \frac{1}{x(x-\lambda)(x-2\lambda)\cdots(x-(m-1)\lambda)}$$
$$= \sum_{k=0}^{m-1} \frac{(-1)^k}{(m-1)!}\binom{m-1}{k}\frac{(-\lambda)^{1-m}}{x-k\lambda}, \quad (m \in \mathbb{N}). \tag{34}$$

By (33) and (34) we get

$$B_n(1-x|\lambda) = \sum_{m=0}^n B_{m,n}(x|\lambda)B_m(\lambda)\frac{1}{(x)_{m,\lambda}}$$
$$= (1-x)_{n,\lambda} + \sum_{m=1}^n B_{m,n}(x|\lambda)B_m(\lambda)\frac{1}{(x)_{m,\lambda}} \tag{35}$$
$$= (1-x)_{n,\lambda} + \sum_{m=1}^n B_{m,n}(x|\lambda)B_m(\lambda)\frac{(-\lambda)^{1-m}}{(m-1)!}\sum_{k=0}^{m-1}(-1)^k\binom{m-1}{k}\frac{1}{x-k\lambda}.$$

Therefore, by (35), we obtain the following theorem.

Theorem 3. *For $n \in \mathbb{N} \cup \{0\}$, we have*

$$B_n(1-x|\lambda) = (1-x)_{n,\lambda} + \sum_{m=1}^n B_{m,n}(x|\lambda)B_m(\lambda)\frac{(-\lambda)^{1-m}}{(m-1)!}\sum_{k=0}^{m-1}(-1)^k\binom{m-1}{k}\frac{1}{x-k\lambda}. \tag{36}$$

Corollary 2. *For $n \in \mathbb{N} \cup \{0\}$, we have*

$$B_n(2|\lambda) = (2)_{n,\lambda} + \sum_{m=1}^n B_{m,n}(-1|\lambda)B_m(\lambda)\frac{(-\lambda)^{1-m}}{(m-1)!}\sum_{k=0}^{m-1}(-1)^{k+1}\binom{m-1}{k}\frac{1}{1+k\lambda}. \tag{37}$$

For $k \in \mathbb{N}$, the higher-order fully degenerate Bernoulli polynomials are given by the generating function

$$\left(\frac{\frac{1}{\lambda}\log(1+\lambda t)}{(1+\lambda t)^{\frac{1}{\lambda}}-1}\right)^k(1+\lambda t)^{\frac{x}{\lambda}} = \sum_{n=0}^\infty B_n^{(k)}(x|\lambda)\frac{t^n}{n!}, \tag{38}$$

(See [8,12,13]). When $x = 0$, $B_n^{(k)}(\lambda) = B_n^{(k)}(x|0)$ are called the higher-order fully degenerate Bernoulli numbers. From (5) and (38), we note that

$$\left(\frac{\log(1+\lambda t)}{\lambda t}\right)^k\sum_{n=k}^\infty B_{k,n}(x|\lambda)\frac{t^n}{n!} = (x)_{k,\lambda}t^k(1+\lambda t)^{\frac{1-x}{\lambda}}\left(\frac{\log(1+\lambda t)}{\lambda t}\right)^k\frac{1}{k!}$$
$$= \frac{\left((1+\lambda t)^{\frac{1}{\lambda}}-1\right)^k}{\left((1+\lambda t)^{\frac{1}{\lambda}}-1\right)^k}(x)_{k,\lambda}\left(\frac{1}{\lambda}\log(1+\lambda t)\right)^k(1+\lambda t)^{\frac{1-x}{\lambda}}\frac{1}{k!}$$
$$= (x)_{k,\lambda}\sum_{m=0}^k\binom{k}{m}(-1)^{m-k}(1+\lambda t)^{\frac{m}{\lambda}}\left(\frac{\frac{1}{\lambda}\log(1+\lambda t)}{(1+\lambda t)^{\frac{1}{\lambda}}-1}\right)^k(1+\lambda t)^{\frac{1-x}{\lambda}}\frac{1}{k!} \tag{39}$$
$$= (x)_{k,\lambda}\sum_{m=0}^k\binom{k}{m}(-1)^{m-k}\left(\frac{\frac{1}{\lambda}\log(1+\lambda t)}{(1+\lambda t)^{\frac{1}{\lambda}}-1}\right)^k(1+\lambda t)^{\frac{1-x+m}{\lambda}}\frac{1}{k!}$$
$$= (x)_{k,\lambda}\sum_{m=0}^k\binom{k}{m}(-1)^{m-k}\sum_{n=0}^\infty B_n(1-x+m|\lambda)\frac{t^n}{n!}\frac{1}{k!}$$
$$= \sum_{n=0}^\infty\left((x)_{k,\lambda}\sum_{m=0}^k\binom{k}{m}(-1)^{m-k}B_n(1-x+m|\lambda)\frac{1}{k!}\right)\frac{t^n}{n!},$$

and hence, we get

$$
\left(\frac{\log(1+\lambda t)}{\lambda t}\right)^k \sum_{m=k}^{\infty} B_{k,m}(x|\lambda)\frac{t^m}{m!} = \left(\sum_{l=0}^{\infty} D_l^{(k)}\lambda^l \frac{t^l}{l!}\right)\left(\sum_{m=k}^{\infty} B_{k,m}(x|\lambda)\frac{t^m}{m!}\right)
$$
$$
= \sum_{n=k}^{\infty}\left(\sum_{l=0}^{n} D_l^{(k)}\lambda^l B_{k,n-l}(x|\lambda)\right)\frac{t^n}{n!}.
$$
(40)

Therefore, by (39) and (40), we obtain the following theorem.

Theorem 4. *For $k, n \in \mathbb{N}$, we have*

$$
\frac{1}{k!}(x)_{n,\lambda}\sum_{m=0}^{k}\binom{k}{m}(-1)^{m-k}B_n(1-x+m|\lambda) = \begin{cases} \sum_{l=0}^{n} D_l^{(k)}\lambda^l B_{k,n-l}(x|\lambda), & \text{if } n \geq k, \\ 0, & \text{if } n < k. \end{cases}
$$
(41)

Let $f \in UD(\mathbb{Z}_p)$. For $x_1, x_2 \in \mathbb{Z}_p$, we consider the degenerate Bernstein operator of order n given by

$$
\mathbb{B}_{n,\lambda}(f|x_1,x_2) = \sum_{k=0}^{n} f\left(\frac{k}{n}\right)\binom{n}{k}(x_1)_{k,\lambda}(1-x_2)_{n-k,\lambda} = \sum_{k=0}^{n} f\left(\frac{k}{n}\right) B_{k,n}(x_1,x_2|\lambda),
$$
(42)

where $B_{n,k}(x_1,x_2|\lambda)$ are called two variable degenerate Bernstein polynomials of degree n as followings (see, [2–6,9,14–27]):

$$
B_{k,n}(x_1,x_2|\lambda) = \binom{n}{k}(x_1)_{k,\lambda}(1-x_2)_{n-k,\lambda}, \quad (n \geq 0).
$$
(43)

The authors [3] obtained the following:

$$
\sum_{k=0}^{\infty} B_{k,n}(x_1,x_2|\lambda)\frac{t^n}{n!} = \frac{(x_1)_{k,\lambda}}{k!}t^k e_\lambda^{1-x_2}(t).
$$
(44)

The authors [8] obtained the following:

$$
B_{k,n}(x_1,x_2|\lambda) = \binom{n}{k}(1-(1-x_1))_{n-(n-k),\lambda}(1-x_2)_{n-k,\lambda}
$$
$$
= B_{n-k,n}(1-x_2,1-x_1|\lambda),
$$
(45)

and

$$
B_{k,n}(x_1,x_2|\lambda) = (1-x_2-(n-k-1)\lambda)B_{k,n-1}(x_1,x_2|\lambda)
$$
$$
+ (x_1-(k-1)\lambda)B_{k-1,n-1}(x_1,x_2|\lambda).
$$
(46)

From (42), we note that $x_1, x_2 \in \mathbb{Z}_p$, if $f(x) = 1$, then we have

$$
\mathbb{B}_{n,\lambda}(1|x_1,x_2) = \sum_{k=0}^{n} B_{k,n}(x_1,x_2|\lambda)
$$
$$
= \sum_{k=0}^{n}\binom{n}{k}(x_1)_{k,\lambda}(1-x_2)_{n-k,\lambda}
$$
$$
= (1+x_1-x_2)_{n,\lambda},
$$
(47)

and if $f(t) = t$, then we have

$$
\mathbb{B}_{n,\lambda}(t|x_1,x_2) = (x_1)_{1,\lambda}(x_1+1-\lambda-x_2)_{n-1,\lambda},
$$
(48)

and if $f(t) = t^2$, then we have

$$\mathbb{B}_{n,\lambda}(t^2|x_1, x_2) = \frac{1}{n}(x_1)_{1,\lambda}(x_1 + 1 - \lambda - x_2)_{n-1,\lambda} + \frac{n-1}{n}(x_1)_{2,\lambda}(1 + x_2 - 2\lambda - x_2)_{n-2,\lambda}. \quad (49)$$

The authors [3] obtained the following:

$$(x)_{1,\lambda} = \frac{1}{(x_1 + 1 - \lambda - x_2)_{n-1,\lambda}} \mathbb{B}_n(t|x_1, x_2), \quad (50)$$

and

$$(x)_{2,\lambda} = \frac{1}{(x_1 + 1 - 2\lambda - x_2)_{n-2,\lambda}} \sum_{k=2}^{n} \frac{\binom{k}{2}}{\binom{n}{2}} B_{k,n}(x_1, x_2|\lambda), \quad (51)$$

and

$$(x)_{i,\lambda} = \frac{1}{(1 + x_1 - x_2 - i\lambda)_{n-i,\lambda}} \sum_{k=i}^{n} \frac{\binom{k}{i}}{\binom{n}{i}} B_{k,n}(x_1, x_2|\lambda). \quad (52)$$

Taking double bosonic p-adic integral on $\mathbb{Z}_p$, we get the following equation:

$$\int_{\mathbb{Z}_p} \int_{\mathbb{Z}_p} B_{k,n}(x_1, x_2|\lambda) d\mu_0(x_1) d\mu_0(x_2) = \binom{n}{k} \int_{\mathbb{Z}_p}(x_1)_{k,\lambda} d\mu_0(x_1) \int_{\mathbb{Z}_p}(1 - x_2)_{n-k,\lambda} d\mu_0(x_2). \quad (53)$$

Therefore, by (53) and Theorem 2, we obtain the following theorem.

Theorem 5. *For $n, k \in \mathbb{N} \cup \{0\}$, we have*

$$\int_{\mathbb{Z}_p} \int_{\mathbb{Z}_p} B_{k,n}(x_1, x_2|\lambda) d\mu_0(x_1) d\mu_0(x_2)$$
$$= \begin{cases} \binom{n}{k} B_n(\lambda) \left((1)_{n-1,\lambda} n(\lambda - 1) B_n(\lambda) + B_{n-k}(\lambda)\right), & \text{if } n > k, \\ B_n(\lambda), & \text{if } n = k. \end{cases} \quad (54)$$

We get from the symmetric properties of two variable degenerate Bernstein polynomials that for $n, k \in \mathbb{N}$ with $n > k$,

$$\int_{\mathbb{Z}_p} \int_{\mathbb{Z}_p} B_{k,n}(x_1, x_2|\lambda) d\mu_0(x_1) d\mu_0(x_2)$$
$$= \sum_{m=0}^{k} \binom{n}{k} \binom{k}{m} (-1)^{k+m} (1)_{m,\lambda}$$
$$\times \int_{\mathbb{Z}_p} \int_{\mathbb{Z}_p} (1 - x_1)_{k-m,-\lambda} (1 - x_2)_{n-k,\lambda} d\mu_0(x_1) d\mu_0(x_2)$$
$$= \binom{n}{k} \int_{\mathbb{Z}_p} (1 - x_2)_{n-k} d\mu_0(x_2) \sum_{m=0}^{k} \binom{n}{k} \binom{k}{m} (-1)^{k+m} (1)_{m,\lambda} d\mu_0(x_2) \quad (55)$$
$$\times \left\{ (1)_{k-m,-\lambda}(k-m)(-\lambda - 1) B_1(-\lambda) + \int_{\mathbb{Z}_p} (x_1)_{k-m,-\lambda} d\mu_0(x_1) \right\}$$
$$= \binom{n}{k} B_{n-k,\lambda}(2) \sum_{m=0}^{k} \binom{n}{k} \binom{k}{m} (-1)^{k+m} (1)_{m,\lambda}$$
$$\times \left\{ (1)_{k-m,-\lambda}(k-m)(-\lambda - 1) B_1(-\lambda) + B_{k-m,-\lambda}(2) \right\}$$

Therefore, by Theorem 5, we obtain the following theorem.

Theorem 6. *For* $n, k \in \mathbb{N} \cup \{0\}$, *we have the following identities:*

1. *If* $n > k$, *then we have*

$$B_n \left((1)_{n-1,\lambda} n(\lambda - 1) B_1(\lambda) + B_{n-k}(\lambda) \right)$$

$$= B_{n-k,\lambda}(2) \sum_{m=0}^{k} \binom{n}{k} \binom{k}{m} (-1)^{k+m} (1)_{m,\lambda} \tag{56}$$

$$\times \left((1)_{k-m,-\lambda}(k-m)(-\lambda - 1) B_1(-\lambda) + B_{k-m,-\lambda}(2) \right).$$

2. *If* $n = k$, *then we have*

$$B_k(\lambda) = \sum_{m=0}^{k} \binom{k}{n} (1)^{k+m} (1)_{k,\lambda} \left((1)_{k-m,-\lambda}(k-m)(-\lambda - 1) B_1(-\lambda) + B_{k-m,-\lambda}(2) \right). \tag{57}$$

3. Remark

Let us assume that the probability of success in an experiment is p. We wondered if we could say the probability of success in the 9th trial is still p after failing eight times in a ten trial experiment, because there is a psychological burden to be successful. It seems plausible that the probability is less than p. The degenerate Bernstein polynomial $B_n(x|\lambda)$ is used in the probability of success. Thus, we give examples in our results as follows:

Example 1. *Let* $n = 2$, *we have*

$$B_2(2|\lambda) = 2(1)_{1,\lambda}(\lambda - 1) B_1(\lambda) + B_2(\lambda)$$

$$= 2(\lambda - 1) \left(-\frac{1}{2} \right) + \frac{\lambda}{2} + \frac{1}{6}$$

$$= -\frac{\lambda}{2} + \frac{7}{6}.$$

Example 2. *Let* $n = 1$, *we have*

$$B_1(1 - x|\lambda) = (1 - x)_{1,\lambda} + \sum_{m=1}^{1} B_{m,1}(x|\lambda) B_m(\lambda) \frac{(-1)^{1-m}}{(m-1)!} \sum_{k=0}^{m-1} (-1)^k \binom{m-1}{k} \frac{1}{x - k\lambda}$$

$$= (1 - x)_{1,\lambda} + B_{1,1}(x|\lambda) B_1(\lambda) \frac{1}{x}$$

$$= -x + \frac{1}{2}.$$

Example 3. *Let* $n = 1, k = 2$, *we have*

$$(x)_{1,\lambda} \sum_{m=0}^{2} \binom{2}{m} (-1)^{m-2} B_1(1 - x + m|\lambda) = x \left(B_1(1 - x|\lambda) - 2 B_1(2 - x|\lambda) + B_1(3 - x|\lambda) \right)$$

$$= -x \left(\left(-x + \frac{1}{2} \right) - 2 \left(-x + \frac{3}{2} \right) + \left(-x + \frac{5}{2} \right) \right)$$

$$= 0.$$

4. Conclusions

In this paper, we studied the fully degenerate Bernoulli polynomials associated with degenerate Bernstein polynomials. In Section 1, Equations (12), (18), (20) and (21) are some properties of them. In Section 2, Theorems 1–3 are results of identities for fully degenerate Bernoulli polynomials in connection with degenerate Bernstein polynomials by means of bosonic p-adic integrals on $\mathbb{Z}_p$

and generating functions. Theorems 4–6 are results of higher-order fully Bernoulli polynomials in connection with two variable degenerate Bernstein polynomials by means of bosonic p-adic integrals on $\mathbb{Z}_p$ and generating functions.

Author Contributions: Conceptualization, W.K. and L.-C.J.; Data curation, L.-C.J.; Formal analysis, L.-C.J.; Funding acquisition, J.G.L.; Investigation, J.G.L., W.K. and L.-C.J.; Methodology, W.K. and L.-C.J.; Project administration, L.-C.J.; Resources, L.-C.J.; Supervision, L.-C.J.; Visualization, L.-C.J.; Writing—original draft, W.K. and L.-C.J.; Writing—review & editing, J.G.L. and L.-C.J.

References

1. Kim, T. Barnes' type multiple degenerate Bernoulli and Euler polynomials. *Appl. Math. Comput.* **2015**, *258*, 556–564. [CrossRef]
2. Kim, D.S.; Kim, T. Identities for degenerate Bernoulli polynomials and Korobov polynomials of the first kind. *Sci. China Math.* **2019**, *62*, 999–1028. [CrossRef]
3. Kim, D.S.; Kim, H.Y.; Kim, D.J.; Kim, T. Identities of Symmetry for Type 2 Bernoulli and Euler Polynomials. *Symmetry* **2019**, *11*, 613. [CrossRef]
4. Kim, D.S.; Kim, T. Degenerate Bernstein polynomials. *Rev. R. Acad. Cienc. Exactas Fís. Nat. Ser. A Mater.* **2018**, 1–8. [CrossRef]
5. Kim, D.S.; Kim, T. Correction to: Degenerate Bernstein polynomials. *Rev. R. Acad. Cienc. Exactas Fís. Nat. Ser. A Mater.* **2019**, 1–2. [CrossRef]
6. Kim, D.S.; Kim, T. Some Identities on Degenerate Bernstein and Degenerate Euler Polynomials. *Mathematics* **2019**, *7*, 47. . [CrossRef]
7. Kim, T. *q*-Volkenborn integration. *Russ. J. Math. Phys.* **2002**, *9*, 288–299.
8. Kim, D.S.; Kim, T.; Mansour, T.; Seo, J.-J. Fully degenerate poly-Bernoulli polynomials with a *q* parameter. *Filomat* **2016**, *30*, 1029–1035. [CrossRef]
9. Kim, D.S.; Kim, T. A Note on polyexponential and unipoly functions. *Russ. J. Math. Phys.* **2019**, *26*, 40–49. [CrossRef]
10. Kim, T.; Kim, D.S. Extended Stirling numbers of the first kind associated with Daehee numbers and polynomials. *Rev. R. Acad. Cienc. Exactas Fis. Nat. Ser. A Mater.* **2019**, *113*, 1159–1171. [CrossRef]
11. Pyo, S.-S.; Kim, T.; Rim, S.-H. Degenerate Daehee Numbers of the Third Kind. *Mathematics* **2018**, *6*, 239. [CrossRef]
12. Kim, D.S.; Kim, T.; Seo, J.-J. Fully degenerate poly-Bernoulli numbers and polynomials. *Open Math.* **2016**, *14*, 545–556. [CrossRef]
13. Kim, T.; Kwon, H.I.; Mansour, T.; Rim, S.-H. Symmetric identities for the fully degenerate Bernoulli polynomials and degenerate Euler polynomials under symmetric group of degree n. *Util. Math.* **2017**, *103*, 61–72.
14. Araci, S.; Acikgoz, M. A note on the Frobenius-Euler numbers and polynomials associated with Bernstein polynomials. *Adv. Stud. Contemp. Math. (Kyungshang)* **2012**, *22*, 399–406.
15. Bayad, A.; Kim, T. Identities involving values of Bernstein, *q*-Bernoulli, and *q*-Euler polynomials. *Russ. J. Math. Phys.* **2011**, *18*, 133–143. [CrossRef]
16. Choi, J. A note on *p*-adic integrals associated with Bernstein and *q*–Bernstein polynomials'. *Adv. Stud. Contemp. Math. (Kyungshang)* **2011**, *21*, 133–138.
17. Kim, D.S.; Kim, T.; Jang, G.W.; Kwon, J. A note on degenerate Bernstein polynomials. *J. Inequal. Appl.* **2019**, *2019*, 129. [CrossRef]
18. Kim, T.; Choi, J.; Kim, Y.H.; Ryoo, C.S. On the fermionic *p*-adic integral representation of Bernstein polynomials associated with Euler numbers and polynomials. *J. Inequal. Appl.* **2010**, *1*, 864247. [CrossRef]
19. Kim, T.; Choi, J.; Kim, Y.H. On the *k*-dimensional generalization of *q*-Bernstein polynomials. *Proc. Jangjeon Math. Soc.* **2011**, *14*, 199–207.
20. Kim, T. A note on *q*-Bernstein polynomials. *Russ. J. Math. Phys.* **2011**, *18*, 73–82. [CrossRef]
21. Kim, T.; Kim, Y.H.; Bayad, A. A study on the *p*-adic *q*-integral representation on Z p associated with the weighted *q*-Bernstein and *q*-Bernoulli polynomials. *J. Inequal. Appl.* **2011**, 513821. [CrossRef]

22. Kim, T.; Lee, B.; Lee, S.-H.; Rim, S.-H. A note on q-Bernstein polynomials associated with p-adic integral on $\mathbb{Z}_p$. *J. Comput. Anal. Appl.* **2013**, *15*, 584–592.

23. Kim, W.J.; Kim, D.S.; Kim, H.Y.; Kim, T. Some identities of degenerate Euler polynomials associated with Degenerate Bernstein polynomials. *arXiv* **2019**, arXiv:1904.08592.

24. Kurt, V. Some relation between the Bernstein polynomials and second kind Bernoulli polynomials. *Adv. Stud. Contemp. Math. (Kyungshang)* **2013**, *23*, 43–48.

25. Ostrovska, S. On the q–Bernstein polynomials. *Adv. Stud. Contemp. Math. (Kyungshang)* **2005**, *11*, 193–204.

26. Park, J.-W.; Pak, H.K.; Rim, S.-H.; Kim, T.; Lee, S.-H. A note on the q-Bernoulli numbers and q–Bernstein polynomials. *J. Comput. Anal. Appl.* **2013**, *15*, 722–729.

27. Siddiqui, M.A.; Agrawal, R.R.; Gupta, N. On a class of modified new Bernstein operators. *Adv. Stud. Contemp. Math. (Kyungshang)* **2014**, *24*, 97–107.

7

Fibonacci and Lucas Numbers of the Form $2^a + 3^b + 5^c + 7^d$

Yunyun Qu [1,2], Jiwen Zeng [1,*] and Yongfeng Cao [3]

[1] School of Mathematical Sciences, Xiamen University, Xiamen 361005, China; qucloud@163.com
[2] School of Mathematical Sciences, Guizhou Normal University, Guiyang 550001, China
[3] School of Big Data and Computer Science, Guizhou Normal University, Guiyang 550001, China; cyfeis@whu.edu.cn
[*] Correspondence: jwzeng@xmu.edu.cn

Abstract: In this paper, we find all Fibonacci and Lucas numbers written in the form $2^a + 3^b + 5^c + 7^d$, in non-negative integers a, b, c, d, with $0 \leq \max\{a, b, c\} \leq d$.

Keywords: Fibonacci; Lucas; linear form in logarithms; continued fraction; reduction method

MSC: 11B39; 11J86; 11D61

1. Introduction

Let $\{F_n\}_{n \geq 0}$ be the Fibonacci sequence which is a second-order linear recursive sequence given by $F_{n+2} = F_{n+1} + F_n$, its initial values are $F_0 = 0$ and $F_1 = 1$, and its companion Lucas sequence $\{L_n\}_{n \geq 0}$ follows the same recursive pattern as the Fibonacci numbers, but with initial values $L_0 = 2$ and $L_1 = 1$. Fibonacci and Lucas numbers are very famous because they have amazing features (consult [1–3]). The problem of looking for a specific form of second-order recursive sequence has a very rich history. Bugeaud, Mignotte and Siksek [4] showed that $0, 1, 8, 144$ and $1, 4$ are the only Fibonacci and Lucas numbers, respectively, of the form y^t with $t > 1$ (perfect power). Other related papers searched for Fibonacci numbers of forms such as $px^2 + 1, px^3 + 1$ [5], $k^2 + k + 2$ [6], $p^a \pm p^b + 1$ [7]. In 1989, Luo [8] solved Vern Hoggatt's conjecture and proved that the only triangle numbers in the Fibonacci sequence $\{F_n\}$ are $1, 3, 21, 55$. In 1991, Luo [9] found all triangular numbers in the Lucas sequence $\{L_n\}$. In [10], Eric F. Bravo and Jhon J. Bravo found all positive integer solutions of the Diophantine equation $F_n + F_m + F_l = 2^a$ in non-negative integers n, m, l, and a with $n \geq m \geq l$. In [11], Normenyo, Luca and Togbé determined all base-10 repdigits that are expressible as sums of four Fibonacci or Lucas numbers. In [12], Marques and Togbé searched for Fibonacci numbers of the form $2^a + 3^b + 5^c$ which are sum of three perfect powers of some prescribed distinct bases.

In this paper, we are interested in Fibonacci numbers and Lucas numbers which are sum of four perfect powers of several prescribed distinct bases. The number of perfect powers involved in the Diophantine equation solved by the literature [12] is one less than the perfect powers involved in the equation solved by us and the amount of computation in the literature [12] is relatively small. More precisely, our results are the following.

Theorem 1. *The solutions of the Diophantine equation*

$$F_n = 2^a + 3^b + 5^c + 7^d \tag{1}$$

in non-negative integers n, a, b, c, d with $0 \leq \max\{a, b, c\} \leq d$ are $(n, a, b, c, d) \in \{(7, 1, 1, 0, 1), (10, 1, 1, 0, 2), (10, 2, 0, 0, 2), (14, 1, 3, 1, 3), (14, 3, 0, 2, 3)\}$.

Theorem 2. *The solutions of the Diophantine equation*

$$L_n = 2^a + 3^b + 5^c + 7^d \tag{2}$$

in non-negative integers n, a, b, c, d with $0 \leq \max\{a, b, c\} \leq d$ are $(n, a, b, c, d) \in \{(3, 0, 0, 0, 0), (5, 1, 0, 0, 1), (9, 0, 0, 2, 2)\}$.

2. Preliminaries

Before proceeding further, we recall some facts and tools which will be used later.

First, we recall the Binet's formulae for Fibonacci and Lucas sequences:

$$F_n = \frac{\gamma^n - \mu^n}{\gamma - \mu}$$

and

$$L_n = \gamma^n + \mu^n,$$

where $\gamma = \frac{1+\sqrt{5}}{2}$ and $\mu = \frac{1-\sqrt{5}}{2}$ are the roots of $F_n's$ characteristic polynomial $x^2 - x - 1 = 0$. For all positive integers n, the inequalities

$$\gamma^{n-2} \leq F_n \leq \gamma^{n-1}, \quad \gamma^{n-1} \leq L_n \leq 2\gamma^n \tag{3}$$

hold.

In order to prove our theorem, one tool used is a Baker type lower bound for a linear form in logarithms of algebraic numbers, and such a bound was given by the following result of Matveev (see [13]).

Lemma 1. *Let $\gamma_1, \gamma_2, \cdots, \gamma_t$ be real algebraic numbers and let $b_1, \cdots, b_t$ be non-zero rational integers. Let D be the degree of the number field $\mathbb{Q}(\gamma_1, \gamma_2, \cdots, \gamma_t)$ over $\mathbb{Q}$ and let A_j be a real number satisfying*

$$A_j \geq \max\{Dh(\gamma_j), |\log\gamma_j|, 0.16\}$$

for $j = 1, \cdots, t$. Assume that

$$B \geq \max\{|b_1|, \cdots, |b_t|\}.$$

If $\gamma_1^{b_1} \cdots \gamma_t^{b_t} \neq 1$, then

$$|\gamma_1^{b_1} \cdots \gamma_t^{b_t} - 1| \geq \exp(-1.4 \times 30^{t+3} \times t^{4.5} \times D^2(1 + \log D)(1 + \log B)A_1 \cdots A_t).$$

As usual, in the above statement, the logarithmic height of an $s-$degree algebraic number γ is defined as

$$h(\gamma) = \frac{1}{s}\left(\log|a| + \sum_{j=1}^{s} \log\max\{1, |\gamma^{(j)}|\}\right),$$

where a is the leading coefficient of the minimal polynomial of γ (over $\mathbb{Z}$) and $\gamma^{(j)}, 1 \leq j \leq s$ are the conjugates of γ (over $\mathbb{Q}$).

After finding an upper bound on n which is in general too large, the next step is to reduce it. For that, we need a variant of the famous Baker–Davenport lemma which was developed by Dujella and Pethő [14]. For a real number x, we use $\|x\| = \min\{|x - n| : n \in \mathbb{Z}\}$ for the distance from x to the nearest integer.

Lemma 2. *(see [10]) Let M be a positive integer, let $\frac{p}{q}$ be a convergent of the continued fraction of the irrational number α such that $q > 6M$, and let A, B, τ be some real numbers with $A > 0$ and $B > 1$.*

Let $\epsilon := \|\tau q\| - M\|\alpha q\|$, where $\|\cdot\|$ denotes the distance from the nearest integer. If $\epsilon > 0$, then no solution to the inequality

$$0 < |u\alpha - v + \tau| < AB^{-\omega}$$

exists in positive integers u, v, and ω with $u \leq M$ and $w \geq \frac{\log(Aq/\epsilon)}{\log B}$.

Next, we are ready to handle the proofs of our results.

3. Proof of Theorem 1

3.1. Bounding n

By combining the Binet formula together with (1), we get

$$\frac{\gamma^n}{\sqrt{5}} - 7^d = 2^a + 3^b + 5^c + \frac{\mu^n}{\sqrt{5}} > 0, \tag{4}$$

because $|\mu| < 1$ while $2^a \geq 1$. Thus,

$$\frac{\gamma^n 7^{-d}}{\sqrt{5}} - 1 = \frac{2^a}{7^d} + \frac{3^b}{7^d} + \frac{5^c}{7^d} + \frac{\mu^n}{7^d\sqrt{5}} > 0 \tag{5}$$

yields

$$\left| \frac{\gamma^n 7^{-d}}{\sqrt{5}} - 1 \right| < \frac{4}{7^{0.1d}}. \tag{6}$$

From the first inequality of (3), we obtain the estimate $\gamma^{n-2} < 4 \times 7^d$ and $7^d < \gamma^{n-1}$, which implies that $0.24n - 1.9 < d < 0.25(n - 1)$; also, this yields $d < n$.

We are in a situation where we can apply Matveev's result Lemma 1 to the left side of (6). The left expression of (6) is nonzero, since, if this expression is zero, it means that $\gamma^{2n} = 7^{2d} \times 5 \in \mathbb{Z}$, so $\gamma^{2n} \in \mathbb{Z}$ for some positive integer n, which is false. We take $t := 3$, $\gamma_1 := \gamma$, $\gamma_2 := 7$, $\gamma_3 := \sqrt{5}$ and $b_1 := n, b_2 := -d, b_3 := -1$. Then we have $D = [\mathbb{Q}(\sqrt{5}) : \mathbb{Q}] = 2$. Note that $h(\gamma_1) = \frac{1}{2}\log\gamma, h(\gamma_2) = \log 7$ and $h(\gamma_3) = \log\sqrt{5}$. Thus, we can take $A_1 := 0.5$, $A_2 := 3.9$ and $A_3 := 1.7$. Note that $\max\{|b_1|, |b_2|, |b_3|\} = \max\{n, d, 1\} = n$. We are in position to apply Matveev's result Lemma 1. This lemma together with a straightforward calculation gives

$$\left| \frac{\gamma^n 7^{-d}}{\sqrt{5}} - 1 \right| > \exp(-C(1 + \log n)), \tag{7}$$

where $C = 3.22 \times 10^{12}$. Thus, from (6), (7) and $d > 0.24n - 1.9$, taking logarithms in the inequalities (6), (7) and comparing the resulting inequalities, we get

$$0.046n - 1.8 < 3.22 \times 10^{12} \times (1 + \log n),$$

giving $n < 2.56 \times 10^{15}$. We summarize the conclusions of this section as follows.

Lemma 3. *If (n, a, b, c, d) is a solution in positive integers to Equation (1) with $0 \leq \max\{a, b, c\} \leq d$, then*

$$d < n < 2.56 \times 10^{15}.$$

3.2. Reducing the Bound on n

We use Lemma 2 several times to reduce the bound for n. We return to (6). Put

$$\Lambda_F := n\log\gamma - d\log 7 - \log\sqrt{5}.$$

Then (5), (6) implies that

$$0 < \Lambda_F < e^{\Lambda_F} - 1 < \frac{4}{7^{0.1d}}. \tag{8}$$

Dividing across by log7, we get

$$0 < |n\frac{\log\gamma}{\log 7} - d - \frac{\log\sqrt{5}}{\log 7}| < \frac{2.1}{7^{0.1d}}. \tag{9}$$

We are now ready to apply Lemma 2 with the obvious parameters,

$$\alpha := \frac{\log\gamma}{\log 7}, v := d, \tau := -\frac{\sqrt{5}}{\log 7}, A := 2.1, B := 1.2.$$

It is easy to see that α is irrational. In fact, we assume that $\alpha = \frac{p}{q}$, where $p, q \in \mathbb{Z}^+$ and $\gcd(p,q) = 1$. Then $\gamma^q = 7^p$, hence $\overline{\gamma}^q = 7^p$, where $\overline{\gamma}$ is the conjugate of γ. Thus, we can get $\gamma^q\overline{\gamma}^q = 7^{2p}$; hence, $(-1)^q = 7^{2p}$ which is an absurdity. We can take $M := 2.56 \times 10^{15}$. Let $\frac{p_k}{q_k}$ be the kth convergent of the continued fraction of α. By applying Lemma 2 and performing the calculations with $q_{39} > 6M$ and $\epsilon = \|\tau q_{39}\| - M\|\alpha q_{39}\| = 0.42904\cdots$, we get that if (n,a,b,c,d) is a solution in positive integers of Equation (1), then $d < 225$, which implies that

$$n < \frac{226.9}{0.24} = 945.417 < 946.$$

Then we can take $M := 946$. By applying Lemma 2 again and performing the calculations with $q_8 > 6M$ and $\epsilon = \|\tau q_8\| - M\|\alpha q_8\| = 0.07417\cdots$, we get that if (n,a,b,c,d) is a solution in positive integers of Equation (1), then $d < 73$, which implies that

$$n < 313.$$

Finally, we apply a program written in Mathematica to determine the solutions to (1) in the range $0 \le \max\{a,b,c\} \le d < 73$ and $n < 313$. Quickly, the program returns the following solutions: $(n,a,b,c,d) \in \{(7,1,1,0,1), (10,1,1,0,2), (10,2,0,0,2), (14,1,3,1,3), (14,3,0,2,3)\}$. This proof has been completed.

4. Proof of Theorem 2

4.1. Bounding n

By combining Binet formula together with (2), we get

$$\gamma^n - 7^d = 2^a + 3^b + 5^c - \mu^n > 0, \tag{10}$$

because $|\mu| < 1$ while $2^a \ge 1$. Thus,

$$\gamma^n 7^{-d} - 1 = \frac{2^a}{7^d} + \frac{3^b}{7^d} + \frac{5^c}{7^d} - \mu^n 7^{-d} > 0 \tag{11}$$

yields

$$|\gamma^n 7^{-d} - 1| < \frac{4}{7^{0.1d}}. \tag{12}$$

From the second inequality of (3) and (2), we obtain the estimate $\gamma^{n-1} < 4 \times 7^d$ and $7^d < 2 \times \gamma^n$, which implies that $4.04d - 1.45 < n < 4.05d + 3.89$; also, this yields $d \le n$.

We are also in a situation where we can apply Matveev's result Lemma 1 to the left side of (12). The left expression of (12) is nonzero, since, if this expression is zero, it means that $\gamma^n = 7^d \in \mathbb{Z}$, so $\gamma^n \in \mathbb{Z}$ for some positive integer n, which is false. We take $t := 2, \gamma_1 := \gamma, \gamma_2 := 7$ and $b_1 := n, b_2 := -d$.

Then we have $D = [\mathbb{Q}(\sqrt{5}) : \mathbb{Q}] = 2$. Note that $h(\gamma_1) = \frac{1}{2}\log\gamma, h(\gamma_2) = \log7$. Thus, we can take $A_1 := 0.5, A_2 := 3.9$. Note that $B = \max\{|b_1|, |b_2|\} = \max\{n, d\} = n$. We are in position to apply Matveev's result Lemma 1. This lemma together with a straightforward calculation gives

$$| \gamma^n 7^{-d} - 1 | > \exp(-C(1 + \log n)), \tag{13}$$

where $C = 1.02 \times 10^{10}$. Thus, from (12), (13) and $d > \frac{n-3.89}{4.05}$, taking logarithms in the inequalities (12), (13) and comparing the resulting inequalities, we get

$$0.1(n-1)\log\gamma - 1.1 \times \log4 < C \times (1 + \log n),$$

giving $n < 6.47 \times 10^{12}$. The conclusions of this section are as follows.

Lemma 4. *If (n, a, b, c, d) is a solution in positive integers to Equation (2) with $0 \leq \max\{a, b, c\} \leq d$, then*

$$d \leq n < 6.47 \times 10^{12}.$$

4.2. Reducing the Bound on n

We use the extremality property of continued fraction to reduce the bound for n. We return to (12) and put

$$\Lambda_L := n\log\gamma - d\log7.$$

Then (11), (12) implies that

$$0 < \Lambda_L < e^{\Lambda_L} - 1 < \frac{4}{7^{0.1d}}. \tag{14}$$

Dividing by $\log7$, we get

$$0 < n\frac{\log\gamma}{\log7} - d < \frac{2.1}{1.2^d}. \tag{15}$$

Let $[a_0, a_1, a_2, a_3, a_4, \cdots ,] = [0, 4, 22, 1, 5, 1, 1, 17, \cdots]$ be the continued fraction of $\frac{\log\gamma}{\log7}$, and let $\frac{p_k}{q_k}$ be its kth convergent. Recall that $n < 6.47 \times 10^{12}$ by Lemma 4. A quick inspection using Mathematica reveals that $q_{19} < 1.662 \times 10^{12} < q_{20}$. Furthermore, $a_M := \max\{a_i : i = 0, 1, \cdots, 27\} = a_{14} = 35$. So, in accordance with the extremality property of continued fraction, we obtain that

$$|n\frac{\log\gamma}{\log7} - d| > \frac{1}{(a_M + 2)n} = \frac{1}{37n}. \tag{16}$$

By comparing estimates (15) and (16), we get right away that

$$\frac{1}{37n} < \frac{2.1}{1.2^d}.$$

This leads to

$$d < \frac{\log(2.1 \times 37n)}{\log1.2} < 186,$$

which implies that

$$n < 757.$$

This can lead to

$$d < \frac{\log(2.1 \times 37n)}{\log1.2} < 61,$$

which implies that

$$n < 251.$$

Finally, we use a program written in Mathematica to find the solutions to (2) in the range $0 \leq \max\{a,b,c\} \leq d < 61$ and $n < 251$. Quickly, the program returns the following solutions: $(n,a,b,c,d) \in \{(3,0,0,0,0),(5,1,0,0,1),(9,0,0,2,2)\}$. This completes the proof.

5. Conclusions

In this paper, we find all the solutions of the Diophantine equation (1) by using a Baker type lower bound for a nonzero linear form in logarithms of algebraic numbers and the Lemma 2 from Diophantine approximation to reduce the upper bounds on the variables of the equation. For the Diophantine equation (2), we solve the equation by using the lower bound for a nonzero linear form in logarithms of algebraic numbers and the extremality properties of continued fraction to reduce the upper bounds on the variables of the equation.

6. Future Developments

We remark that we can further take advantage of our method to prove that there are only finitely many solutions (and all of them are effectively computable) for the Diophantine equation $F_n = -2^a - 3^b - 5^c + 7^d, L_n = -2^a - 3^b - 5^c + 7^d$ in non-negative integers n,a,b,c,d with $0 \leq \max\{a,b,c\} \leq d$. We leave this as a problem for other researchers.

Author Contributions: All authors contributed equally to this work. All authors read and approved the final manuscript.

Acknowledgments: The authors would like to express their sincere gratitude to the referees for their valuable comments which have significantly improved the presentation of this paper.

References

1. Koshy, T. *Fibonacci and Lucas Numbers with Applications*; Pure and Applied Mathematics (New York); Wiley-Interscience: New York, NY, USA, 2001.
2. Kim, T.; Kim, D.S.; Dolgy, D.V.; Park, J.-W. Sums of finite products of Chebyshev polynomials of the second kind and of Fibonacci polynomials. *J. Inequal. Appl.* **2018**, *2018*, 148. [CrossRef] [PubMed]
3. Kim, T.; Dolgy, D.V.; Kim, D.S.; Seo, J.J. Convolved Fibonacci numbers and their applications. *ARS Comb.* **2017**, *135*, 119–131.
4. Bugeaud, Y.; Mignotte, M.; Siksek, S. Classical and modular approaches to exponential Diophantine equations. I. Fibonacci and Lucas perfect powers. *Ann. Math.* **2006**, *163*, 969–1018. [CrossRef]
5. Robbins, N. Fibonacci numbers of the forms $pX^2 + 1, pX^3 + 1$, where p is prime. In Proceedings of the Applications of Fibonacci Numbers, San Jose, CA, USA, 13–16 August 1986; Kluwer Academic Publishers: Dordrecht, The Netherlands, 1988; pp. 77–88.
6. Luca, F. Fibonacci numbers of the form $k^2 + k + 2$. In Proceedings of the Applications of Fibonacci Numbers, Rochester, NY, USA, 22–26 June 1998; Kluwer Academic Publishers: Dordrecht, The Netherlands, 1999; Volume 8, pp. 241–249.
7. Luca, F.; Szalay, L. Fibonacci numbers of the form $p^a \pm p^b + 1$. *Fibonacci Quart.* **2007**, *45*, 98–103.
8. Luo, M. On triangular Fibonacci numbers. *Fibonacci Quart.* **1989**, *27*, 98–108.
9. Luo, M. On triangular Lucas numbers. In *Applications of Fibonacci Numbers*; Springer: Dordrecht, The Netherlands, 1991; pp. 231–240.
10. Bravo, E.F.; Bravo, J.J. Powers of two as sums of three Fibonacci numbers. *Lith. Math. J.* **2015**, *55*, 301–311. [CrossRef]
11. Normenyo, B.V.; Luca, F.; Togbé, A. Repdigits as Sums of Four Fibonacci or Lucas Numbers. *J. Integer Seq.* **2018**, *21*, 1–30.
12. Marques, D.; Togbé, A. Fibonacci and Lucas numbers of the form $2^a + 3^b + 5^c$. *Proc. Jpn. Acad. Ser. A Math. Sci.* **2013**, *89*, 47–50. [CrossRef]

13. Matveev, E.M. An explicit lower bound for a homogeneous rational linear form in logarithms of algebraic numbers, II. *Izv. Ross. Akad. Nauk Ser. Mat.* **2000**, *64*, 125–180. *English Translation in Izv. Math.* **2000**, *64*, 1217–1269. [CrossRef]

14. Dujella, A.; Pethő, A. A generalization of a theorem of Baker and Davenport. *Q. J. Math.* **1998**, *49*, 291–306. [CrossRef]

8

Some Identities on the Poly-Genocchi Polynomials and Numbers

Dmitry V. Dolgy [1] **and Lee-Chae Jang** [2,*]

[1]   Kwangwoon Glocal Education Center, Kwangwoon University, Seoul 139-701, Korea; d_dol@mail.ru
[2]   Graduate School of Education, Konkuk University, Seoul 143-701, Korea
*   Correspondence: Lcjang@konkuk.ac.kr

Abstract: Recently, Kim-Kim (2019) introduced polyexponential and unipoly functions. By using these functions, they defined type 2 poly-Bernoulli and type 2 unipoly-Bernoulli polynomials and obtained some interesting properties of them. Motivated by the latter, in this paper, we construct the poly-Genocchi polynomials and derive various properties of them. Furthermore, we define unipoly Genocchi polynomials attached to an arithmetic function and investigate some identities of them.

Keywords: polylogarithm functions; poly-Genocchi polynomials; unipoly functions; unipoly Genocchi polynomials

MSC: 11B83; 11S80

1. Introduction

The study of the generalized versions of Bernoulli and Euler polynomials and numbers was carried out in [1,2]. In recent years, various special polynomials and numbers regained the interest of mathematicians and quite a few results have been discovered. They include the Stirling numbers of the first and the second kind, central factorial numbers of the second kind, Bernoulli numbers of the second kind, Bernstein polynomials, Bell numbers and polynomials, central Bell numbers and polynomials, degenerate complete Bell polynomials and numbers, Cauchy numbers, and others (see [3–8] and the references therein). We mention that the study of a generalized version of the special polynomials and numbers can be done also for the transcendental functions like hypergeometric ones. For this, we let the reader refer to the papers [3,5,6,8,9]. The poly-Bernoulli numbers are defined by means of the polylogarithm functions and represent the usual Bernoulli numbers (more precisely, the values of Bernoulli polynomials at 1) when $k = 1$. At the same time, the degenerate poly-Bernoulli polynomials are defined by using the polyexponential functions (see [8]) and they are reduced to the degenerate Bernoulli polynomials if $k = 1$. The polyexponential functions were first studied by Hardy [10] and reconsidered by Kim [6,9,11,12] in view of an inverse to the polylogarithm functions which were studied by Zagier [13], Lewin [14], and Jaonquière [15]. In 1997, Kaneko [16] introduced poly-Bernoulli numbers which are defined by the polylogaritm function.

Recently, Kim-Kim introduced polyexponential and unipoly functions [9]. By using these functions, they defined type 2 poly-Bernoulli and type 2 unipoly-Bernoulli polynomials and obtained several interesting properties of them.

In this paper, we consider poly-Genocchi polynomials which are derived from polyexponential functions. Similarly motivated, in the final section, we define unipoly Genocchi polynomials attached to an arithmetic function and investigate some identities for them. In addition, we give explicit expressions and identities involving those polynomials.

It is well known, the Bernoulli polynomials of order α are defined by their generating function as follows (see [1–3,17,18]):

$$\left(\frac{t}{e^t - 1}\right)^\alpha e^{xt} = \sum_{n=0}^{\infty} B_n^{(\alpha)}(x)\frac{t^n}{n!}, \qquad . \tag{1}$$

We note that for $\alpha = 1$, $B_n(x) = B_n^{(1)}(x)$ are the ordinary Bernoulli polynomials. When $x = 0$, $B_n^\alpha = B_n^\alpha(0)$ are called the Bernoulli numbers of order α. The Genocchi polynomials $G_n(x)$ are defined by (see [19–24]).

$$\frac{2t}{e^t + 1}e^{xt} = \sum_{n=0}^{\infty} G_n(x)\frac{t^n}{n!}, \tag{2}$$

When $x = 0$, $G_n = G_n(0)$ are called the Genocchi numbers.

As is well-known, the Euler polynomials are defined by the generating function to be (see [1,4]).

$$\frac{2}{e^t + 1}e^{xt} = \sum_{n=0}^{\infty} E_n(x)\frac{t^n}{n!}, \tag{3}$$

For $n \geq 0$, the Stirling numbers of the first kind are defined by (see [5,7,25]),

$$(x)_n = \sum_{l=0}^{n} S_1(n, l)x^l, \tag{4}$$

where $(x)_0 = 1$, $(x)_n = x(x-1)\ldots(x-n+1)$, $(n \geq 1)$. From (4), it is easy to see that

$$\frac{1}{k!}(\log(1+t))^k = \sum_{n=k}^{\infty} S_1(n, k)\frac{t^n}{n!}. \tag{5}$$

In the inverse expression to (4), for $n \geq 0$, the Stirling numbers of the second kind are defined by

$$x^n = \sum_{l=0}^{n} S_2(n, l)(x)_l. \tag{6}$$

From (6), it is easy to see that

$$\frac{1}{k!}(e^t - 1)^k = \sum_{n=k}^{\infty} S_2(n, k)\frac{t^n}{n!}. \tag{7}$$

2. The Poly-Genocchi Polynomials

For $k \in \mathbb{Z}$, by (2) and (14), we define the poly-Genocchi polynomials which are given by

$$\frac{2e_k(\log(1+t))}{e^t + 1}e^{xt} = \sum_{n=0}^{\infty} G_n^{(k)}(x)\frac{t^n}{n!}. \tag{8}$$

When $x = 0$, $G_n^{(k)} = G_n^{(k)}(0)$ are called the poly-Genocchi numbers. From (8), we see that

$$G_n^{(1)}(x) = G_n(x), \quad (n \in \mathbb{N} \cup \{0\}) \tag{9}$$

are the ordinary Genocchi polynomials. From (2), (4) and (8) , we observe that

$$\sum_{n=0}^{\infty} G_n^{(k)} \frac{t^n}{n!}$$

$$= \frac{2e_k(\log(1+t))}{e^t + 1}$$

$$= \frac{2}{e^t + 1} \sum_{m=1}^{\infty} \frac{(\log(1+t))^m}{(m-1)!m^k}$$

$$= \frac{2}{e^t + 1} \sum_{m=0}^{\infty} \frac{(\log(1+t))^{m+1}}{m!(m+1)^k}$$

$$= \frac{2}{e^t - 1} \sum_{m=0}^{\infty} \frac{1}{(m+1)^{k-1}} \sum_{l=m+1}^{\infty} S_1(l, m+1) \frac{t^l}{l!}$$

$$= \frac{2t}{e^t + 1} \sum_{m=0}^{\infty} \frac{1}{(m+1)^{k-1}} \sum_{l=m}^{\infty} \frac{S_1(l+1, m+1)}{l+1} \frac{t^l}{l!}$$

$$= \left(\sum_{j=0}^{\infty} G_j \frac{t^j}{j!} \right) \sum_{l=0}^{\infty} \left(\sum_{m=0}^{l} \frac{1}{(m+1)^{k-1}} \frac{S_1(l+1, m+1)}{l+1} \right) \frac{t^l}{l!}$$

$$= \sum_{n=0}^{\infty} \left(\sum_{l=0}^{n} \sum_{m=0}^{l} \binom{n}{l} \frac{1}{(m+1)^{k-1}} \frac{S_1(l+1, m+1)}{l+1} G_{n-l} \right) \frac{t^n}{n!}. \qquad (10)$$

Therefore, by (10), we obtain the following theorem.

Theorem 1. *For $k \in \mathbb{Z}$ and $n \in \mathbb{N} \cup \{0\}$, we have*

$$G_n^{(k)} = \sum_{l=0}^{n} \sum_{m=0}^{l} \binom{n}{l} \frac{1}{(m+1)^{k-1}} \frac{S_1(l+1, m+1)}{l+1} G_{n-l}. \qquad (11)$$

Corollary 1. *For $n \in \mathbb{N} \cup \{0\}$, we have*

$$G_n^{(1)} = G_n = \sum_{l=0}^{n} \sum_{m=0}^{l} \binom{n}{l} \frac{S_1(l+1, m+1)}{l+1} G_{n-l}. \qquad (12)$$

Moreover,

$$\sum_{l=1}^{n} \sum_{m=0}^{l} \binom{n}{l} \frac{S_1(l+1, m+1)}{l+1} G_{n-l} = 0, \quad (n \in \mathbb{N}). \qquad (13)$$

Kim-Kim ([9]) defined the polyexponential function by (see [6,9–12,26]).

$$e_k(x) = \sum_{n=1}^{\infty} \frac{x^n}{(n-1)!n^k}, \qquad (14)$$

In [18], it is well known that for $k \geq 2$,

$$\frac{d}{dx} e_k(x) = \frac{1}{x} e_{k-1}(x). \qquad (15)$$

Thus, by (15), for $k \geq 2$, we get

$$e_k(x) = \int_0^x \frac{1}{t_1} \underbrace{\int_0^{t_1} \frac{1}{t_1} \cdots \int_0^{t_{k-2}}}_{(k-2)times} \frac{1}{t_{k-1}} (e^{t_{k-1}} - 1) d_{k-1} t dt_{k-1} \cdots dt_1. \qquad (16)$$

From (16), we obtain the following equation.

$$\sum_{n=0}^{\infty} G_n^{(k)} \frac{x^n}{n!}$$

$$= \frac{2}{e^x + 1} e_k(\log(1+x))$$

$$= \frac{2}{e^x + 1} \int_0^x \frac{1}{(1+t)\log(1+t)} e_{k-1}(\log(1+t)) dt$$

$$= \frac{2}{e^x + 1} \int_0^x \frac{1}{(1+t_1)\log(1+t_1)}$$

$$\underbrace{\int_0^{t_1} \frac{1}{(1+t_2)\log(1+t_2)} \cdots \int_0^{t_{k-2}}}_{(k-2)\,times} \frac{t_{k-1}}{(1+t_{k-1})\log(1+t_{k-1})} dt_{k-1} dt_{k-2} \cdots dt_1, (k \geq 2). \quad (17)$$

Let us take $k = 2$. Then, by (2) and (16), we get

$$\sum_{n=0}^{\infty} G_n^{(2)} \frac{x^n}{n!} = \frac{2}{e^x + 1} \int_0^x \frac{t}{(1+t)\log(1+t)} dt$$

$$= \frac{2}{e^x + 1} \sum_{l=0}^{\infty} \frac{B_l^{(l)}}{l!} \int_0^x t^l dt$$

$$= \frac{2}{e^x + 1} \sum_{l=0}^{\infty} \frac{B_l^{(l)}}{l+1} \frac{x^{l+1}}{l!}$$

$$= \frac{2x}{e^x + 1} \sum_{l=0}^{\infty} \frac{B_l^{(l)}}{l+1} \frac{x^l}{l!}$$

$$= \left(\sum_{m=0}^{\infty} G_m \frac{x^m}{m!} \right) \left(\sum_{l=0}^{\infty} \frac{B_l^{(l)}}{l+1} \frac{x^l}{l!} \right)$$

$$= \sum_{n=0}^{\infty} \left(\sum_{l=0}^{n} \binom{n}{l} \frac{B_l^{(l)}}{l+1} G_{n-l} \right) \frac{x^n}{n!}. \quad (18)$$

Therefore, by (18), we obtain the following theorem.

Theorem 2. *Let* $n \in \mathbb{N} \cup \{0\}$, *we have*

$$G_n^{(2)} = \sum_{l=0}^{n} \binom{n}{l} \frac{B_l^{(l)}}{l+1} G_{n-l}. \quad (19)$$

From (3) and (16), we also get

$$
\begin{aligned}
\sum_{n=0}^{\infty} G_n^{(2)} \frac{x^n}{n!} &= \frac{2}{e^x+1} \int_0^x \frac{t}{(1+t)\log(1+t)}\,dt \\
&= \frac{2}{e^x+1} \sum_{l=0}^{\infty} B_l^{(l)} \frac{x^{l+1}}{(l+1)!} \\
&= \frac{2}{e^x+1} \sum_{l=1}^{\infty} B_{l-1}^{(l-1)} \frac{x^l}{l!} \\
&= \left(\sum_{m=0}^{\infty} E_m \frac{x^m}{m!} \right) \left(\sum_{l=1}^{\infty} B_{l-1}^{(l-1)} \frac{x^l}{l!} \right) \\
&= \sum_{n=1}^{\infty} \left(\sum_{l=1}^{n} \binom{n}{l} B_{l-1}^{(l-1)} E_{n-l} \right) \frac{x^n}{n!}.
\end{aligned}
\tag{20}
$$

Therefore, by (20), we obtain the following theorem.

Theorem 3. *Let $n \geq 1$, we have*

$$
G_n^{(2)} = \sum_{l=1}^{n} \binom{n}{l} B_{l-1}^{(l-1)} E_{n-l}.
\tag{21}
$$

From (8), we observe that

$$
\begin{aligned}
\sum_{n=0}^{\infty} G_n^{(k)}(x) \frac{t^n}{n!} &= \frac{2 e_k(\log(1+t))}{e^t+1} e^{xt} \\
&= \left(\sum_{l=0}^{\infty} G_l^{(k)} \frac{t^l}{l!} \right) \left(\sum_{m=0}^{\infty} x^m \frac{t^m}{m!} \right) \\
&= \sum_{n=0}^{\infty} \left(\sum_{l=0}^{n} \binom{n}{l} G_l^{(k)} x^{n-l} \right) \frac{t^n}{n!} \\
&= \sum_{n=0}^{\infty} \left(\sum_{l=0}^{n} \binom{n}{l} G_{n-l}^{(k)} x^l \right) \frac{t^n}{n!}.
\end{aligned}
\tag{22}
$$

From (22), we obtain the following theorem.

Theorem 4. *Let $n \in \mathbb{N}$, we have*

$$
G_n^{(k)}(x) = \sum_{l=0}^{n} \binom{n}{l} G_{n-l}^{(k)} x^l.
\tag{23}
$$

From (23), we observe that

$$
\begin{aligned}
\frac{d}{dx}G_n^{(k)}(x) &= \sum_{l=1}^{n}\binom{n}{l}G_{n-l}^{(k)}lx^{l-1} \\
&= \sum_{l=0}^{n-1}\binom{n}{l+1}G_{n-l-1}^{(k)}(l+1)x^{l} \\
&= \sum_{l=0}^{n-1}\frac{n!}{(l+1)!(n-l-1)!}G_{n-1-l}^{(k)}(l+1)x^{l} \\
&= n\sum_{l=0}^{n-1}\frac{(n-1)!}{l!(n-1-l)!}G_{n-1-l}^{(k)}x^{l} \\
&= nG_{n-1}^{(k)}(x).
\end{aligned}
\tag{24}
$$

From (24), we obtain the following theorem.

Theorem 5. *Let $n \in \mathbb{N} \cup \{0\}$ and $k \in \mathbb{Z}$, we have*

$$
\frac{d}{dx}G_n^{(k)}(x) = nG_{n-1}^{(k)}(x).
\tag{25}
$$

3. The Unipoly Genocchi Polynomials and Numbers

Let p be any arithmetic function which is real or complex valued function defined on the set of positive integers $\mathbb{N}$. Then, Kim-Kim ([9]) defined the unipoly function attached to polynomials by

$$
u_k(x|p) = \sum_{n=1}^{\infty}\frac{p(n)x^n}{n^k}, \quad (k \in \mathbb{Z}).
\tag{26}
$$

It is well known that

$$
u_k(x|1) = \sum_{n=1}^{\infty}\frac{x^n}{n^k} = Li_k(x)
\tag{27}
$$

is the ordinary polylogarithm function, and for $k \geq 2$,

$$
\frac{d}{dx}u_k(x|p) = \frac{1}{x}u_{k-1}(x|p),
\tag{28}
$$

and

$$
u_k(x|p) = \int_0^x \frac{1}{t}\underbrace{\int_0^t \frac{1}{t}\cdots\int_0^t \frac{1}{t}}_{(k-2)\,times}u_1(t|p)dtdt\cdots dt
\tag{29}
$$

By using (26), we define the unipoly Genocchi polynomials as follows:

$$
\frac{2}{e^t+1}u_k(\log(1+t)|p)e^{xt} = \sum_{n=0}^{\infty}G_{n,p}^{(k)}(x)\frac{t^n}{n!}.
\tag{30}
$$

Let us take $p(n) = \frac{1}{(n-1)!}$. Then we have

$$
\begin{aligned}
\sum_{n=0}^{\infty} G_{n,p}^{(k)}(x) \frac{t^n}{n!} &= \frac{2}{e^t+1} u_k \left(\log(1+t) \,\middle|\, \frac{1}{(n-1)!} \right) e^{xt} \\
&= \frac{2}{e^t+1} \sum_{m=1}^{\infty} \frac{(\log(1+t))^m}{m^k (m-1)!} e^{xt} \\
&= \frac{2 e_k(\log(1+t))}{e^t+1} e^{xt} \\
&= \sum_{n=0}^{\infty} G_n^{(k)}(x) \frac{t^n}{n!}.
\end{aligned}
\tag{31}
$$

Thus, by (31), we have the following theorem.

Theorem 6. *If we take $p(n) = \frac{1}{(n-1)!}$ for $n \in \mathbb{N} \cup \{0\}$ and $k \in \mathbb{Z}$, then we have*

$$
G_{n,p}^{(k)}(x) = G_n^{(k)}(x).
\tag{32}
$$

From (4) and (30) with $x = 0$, we have

$$
\begin{aligned}
&\sum_{n=0}^{\infty} G_{n,p}^{(k)} \frac{t^n}{n!} \\
&= \frac{2}{e^t+1} \sum_{m=1}^{\infty} \frac{p(m)}{m^k} (\log(1+t))^m \\
&= \frac{2}{e^t+1} \sum_{m=0}^{\infty} \frac{p(m+1)(m+1)!}{(m+1)^k} \sum_{l=m+1}^{\infty} S_1(l, m+1) \frac{t^l}{l!} \\
&= \frac{2}{e^t+1} \sum_{m=0}^{\infty} \frac{p(m+1)(m+1)!}{(m+1)^k} \sum_{l=m}^{\infty} S_1(l+1, m+1) \frac{t^{l+1}}{(l+1)!} \\
&= \frac{2t}{e^t+1} \sum_{m=0}^{\infty} \frac{p(m+1)(m+1)!}{(m+1)^k} \sum_{l=m}^{\infty} S_1(l+1, m+1) \frac{t^l}{(l+1)!} \\
&= \left(\sum_{j=0}^{\infty} G_j \frac{t^j}{j!} \right) \sum_{l=0}^{\infty} \left(\sum_{m=0}^{l} \frac{p(m+1)(m+1)!}{(m+1)^k} \frac{S_1(l+1, m+1)}{l+1} \right) \frac{t^l}{l!} \\
&= \sum_{n=0}^{\infty} \left(\sum_{l=0}^{n} \sum_{m=0}^{l} \binom{n}{l} \frac{p(m+1)(m+1)!}{(m+1)^k} \frac{S_1(l+1, m+1)}{l+1} G_{n-l} \right) \frac{t^n}{n!}.
\end{aligned}
\tag{33}
$$

Therefore, by comparing the coefficients on both sides of (33), we obtain the following theorem.

Remark 1. *Let $n \in \mathbb{N}$ and $k \in \mathbb{Z}$. Then, we have*

$$
G_{n,p}^{(k)} = \sum_{l=0}^{n} \sum_{m=0}^{l} \binom{n}{l} \frac{p(m+1)(m+1)!}{(m+1)^k} \frac{S_1(l+1, m+1)}{l+1} G_{n-l}.
\tag{34}
$$

In particular,

$$
G_{n, \frac{1}{(n-1)!}}^{(k)} = \sum_{l=0}^{n} \sum_{m=0}^{l} \binom{n}{l} \frac{G_{n-l}}{(m+1)^{k-1}} \frac{S_1(l+1, m+1)}{l+1}
\tag{35}
$$

arrives at (11).

From (30), we easily obtain the following theorem.

Theorem 7. *Let* $n \in \mathbb{N} \cup \{0\}$ *and* $k \in \mathbb{Z}$. *Then, we have*

$$G_{n,p}^{(k)}(x) = \sum_{l=0}^{n} \binom{n}{l} G_{n-l,p}^{(k)} x^l. \tag{36}$$

From (36), we easily obtain the following theorem.

Theorem 8. *Let* $n \in \mathbb{N} \cup \{0\}$ *and* $k \in \mathbb{Z}$. *Then, we have*

$$\frac{d}{dx} G_{n,p}^{(k)}(x) = n G_{n-1,p}^{(k)}(x). \tag{37}$$

Finally, by (4) and (30), we observe that

$$\sum_{n=0}^{\infty} G_{n,p}^{(k)} \frac{t^n}{n!}$$

$$= \frac{2}{e^t + 1} \sum_{m=1}^{\infty} \frac{p(m)}{m^k} \frac{m!}{m!} (\log(1+t))^m$$

$$= \frac{2}{e^t + 1} \sum_{m=1}^{\infty} \frac{p(m+1)}{(m+1)^k} \frac{(m+1)!}{(m+1)!} (\log(1+t))^{m+1}$$

$$= \sum_{j=0}^{\infty} E_j \frac{t^j}{j!} \sum_{m=0}^{\infty} \frac{p(m+1)(m+1)!}{(m+1)^k} \sum_{l=m+1}^{\infty} S_1(l, m+1) \frac{t^l}{l!}$$

$$= \sum_{j=0}^{\infty} E_j \frac{t^j}{j!} \sum_{l=0}^{\infty} \sum_{m=0}^{l} \frac{p(m+1)(m+1)!}{(m+1)^k} \sum_{l=m}^{\infty} S_1(l+1, m+1) \frac{t^{l+1}}{(l+1)!}$$

$$= \sum_{n=0}^{\infty} \left(\sum_{l=0}^{n} \sum_{m=0}^{l} \binom{n}{l} \frac{p(m+1)(m+1)!}{(m+1)^k} \frac{S_1(l+1, m+1)}{l+1} E_{n-l} \right) \frac{t^n}{n!}. \tag{38}$$

From (37), we obtain the following theorem.

Theorem 9. *Let* $n \in \mathbb{N}$ *and* $k \in \mathbb{Z}$, *we have*

$$G_{n,p}^{(k)} = \sum_{l=0}^{n} \sum_{m=0}^{l} \binom{n}{l} \frac{p(m+1)(m+1)!}{(m+1)^k} \frac{S_1(l+1, m+1)}{l+1} E_{n-l}. \tag{39}$$

4. Conclusions

In 2019, Kim-Kim considered the polyexponential functions and poly-Bernoulli polynomials. In the same view as these functions and polynomials, we defined the poly-Genocchi polynomials (Equation (8)) and obtained some identities (Theorem 1 and Corollary 1). In particular, we observed explicit poly-Genocchi numbers for $k = 2$ (Theorems 2, 3 and 4). Furthermore, by using the unipoly functions, we defined the unipoly Genocchi polynomials (Equation (30)) and obtained some their properties (Theorems 6 and 7). Finally, we obtained the derivative of the unipoly Genocchi polynomials (Theorem 8) and gave the identity indicating the relationship of unipoly Genocchi polynomials and Euler polynomials (Theorem 9). It is recommended that our readers look at references [27–31] if they want to know the applications related to this paper.

Author Contributions: L.-C.J. and D.V.D. conceived the framework and structured the whole paper; D.V.D. and L.-C.J. checked the results of the paper and completed the revision of the article. All authors have read and agreed to the published version of the manuscript.

References

1. Bayad, A.; Kim, T. Identities for the Bernoulli, the Euler and the Genocchi numbers and polynomials. *Adv. Stud. Contemp. Math. (Kyungshang)* **2010**, *20*, 247–253.
2. Bayad, A.; Chikhi, J. Non linear recurrences for Apostol-Bernoulli-Euler numbers of higher order. *Adv. Stud. Contemp. Math. (Kyungshang)* **2012**, *22*, 1–6.
3. Kim, T.; Kim, D.S.; Lee, H.; Kwon, J. Degenerate binomial coefficients and degenerate hypergeometric functions. *Adv. Differ. Equ.* **2020**, *2020*, 115. [CrossRef]
4. Kim, T. Some identities for the Bernoulli, the Euler and the Genocchi numbers and polynomials. *Adv. Stud. Contemp. Math. (Kyungshang)* **2010**, *20*, 23–28.
5. Kim, T.; Kim, D.S. Some identities of extended degenerate *r*-central Bell polynomials arising from umbral calculus. *Rev. R. Acad. Cienc. Exactas Fis. Nat. Ser. A Mat. RASAM* **2020**, *114*, 1. [CrossRef]
6. Kim, T.; Kim, D.S. Degenerate polyexponential functions and degenerate Bell polynomials. *J. Math. Anal. Appl.* **2020**, *487*, 124017. [CrossRef]
7. Kim, T.; Kim, D.S. A note on central Bell numbers and polynomilas. *Russ. J. Math. Phys.* **2020**, *27*, 76–81.
8. Kim, T.; Kim, D.S.; Kim, H.Y.; Jang, L.C. Degenerate poly-Bernoulli numbers and polynomials. *Informatica* **2020**, *31*, 1–7.
9. Kim, D.S.; Kim, T. A note on polyexponential and unipoly functions. *Russ. J. Math. Phys.* **2019**, *26*, 40–49. [CrossRef]
10. Hardy, G.H. On a class a functions. *Proc. Lond. Math. Soc.* **1905**, *3*, 441–460. [CrossRef]
11. Kim, T.; Kim, D.S.; Kwon, J.K.; Lee, H.S. Degenerate polyexponential functions and type 2 degenerate poly-Bernoulli numbers and polynomials. *Adv. Differ. Equ.* **2020**, *2020*, 168. [CrossRef]
12. Kim, T.; Kim, D.S.; Kwon, J.K.; Kim, H.Y. A note on degenerate Genocchi and poly-Genocchi numbers and polynomials. *J. Inequal. Appl.* **2020**, *2020*, 110. [CrossRef]
13. Zagier, D. The Bloch-Wigner-Ramakrishnan polylogarithm function. *Math. Ann.* **1990**, *286*, 613–624. [CrossRef]
14. Lewin, L. *Polylogarithms and Associated Functions*; With a foreword by A. J. Van der Poorten; North-Holland Publishing Co.: New York, NY, USA; Amsterdam, The Netherlands, 1981; p. xvii+359.
15. Jonquière, A. Note sur la serie $\sum_{n=1}^{\infty} \frac{x^n}{n^s}$. *Bull. Soc. Math. France* **1889**, *17*, 142–152.
16. Kaneko, M. Poly-Bernoulli numbers. *J. Theor. Nombres Bordeaux* **1997**, *9*, 221–228. [CrossRef]
17. Dolgy, D.V.; Kim, T.; Kwon, H.-I.; Seo, J.J. Symmetric identities of degenerate *q*-Bernoulli polynomials under symmetric group S_3. *Proc. Jangjeon Math. Soc.* **2016**, *19*, 1–9.
18. Gaboury, S.; Tremblay, R.; Fugere, B.-J. Some explicit formulas for certain new classes of Bernoulli, Euler and Genocchi polynomials. *Proc. Jangjeon Math. Soc.* **2014**, *17*, 115–123.
19. Cangul, I.N.; Kurt, V.; Ozden, H.; Simsek, Y. On the higher-order $w-q$-Genocchi numbers. *Adv. Stud. Contemp. Math. (Kyungshang)* **2009**, *19*, 39–57.
20. Duran, U.; Acikgoz, M.; Araci, S. Symmetric identities involving weighted *q*-Genocchi polynomials under S_4. *Proc. Jangjeon Math. Soc.* **2015**, *18*, 445–465.
21. Jang, L.C.; Ryoo, C.S.; Lee, J.G.; Kwon, H.I. On the *k*-th degeneration of the Genocchi polynomials. *J. Comput. Anal. Appl.* **2017**, *22*, 1343–1349.
22. Jang, L.C. A study on the distribution of twisted *q*-Genocchi polynomials. *Adv. Stud. Contemp. Math. (Kyungshang)* **2009**, *18*, 181–189.
23. Kim, D.S.; Kim, T. A note on a new type of degenerate Bernoulli numbers. *Russ. J. Math. Phys.* **2020**, *27*, 227–235. [CrossRef]
24. Kurt, B.; Simsek, Y. On the Hermite based Genocchi polynomials. *Adv. Stud. Contemp. Math. (Kyungshang)* **2013**, *23*, 13–17.
25. Kwon, J.; Kim, T.; Kim, D.S.; Kim, H.Y. Some identities for degenerate complete and inomplete *r*-Bell polynomials. *J. Inequal. Appl.* **2020**, 23. [CrossRef]
26. Roman, S. *The Umbral Calculus, Pure and Applied Mathematics, 111*; Academic Press, Inc.: Harcourt Brace Jovanovich: New York, NY, USA, 1984; p. x+193.
27. Jang, L.-C.; Kim, D.S.; Kim, T.; Lee, H. *p*-Adic integral on $\mathbb{Z}_p$ associated with degenerate Bernoulli polynomials of the second kind. *Adv. Differ. Equ.* **2020**, *2020*, 278. [CrossRef]

28. kim, T.; Kim, D.S.; Jang, L.-C.; Lee, H. Jindalrae and Gaenari numbers and polynomials in connection with Jindalrae–Stirling numbers. *Adv. Differ. Equ.* **2020**, *2020*, 245. [CrossRef]

29. Kim, T.; Kim, D.S. Some relations of two type 2 polynomials and discrete harmonic numbers and polynomials. *Symmetry* **2020**, *12*, 905. [CrossRef]

30. Kwon, J.; Jang, L.-C. A note on the type 2 poly-Apostol-Bernoulli polynomials. *Adv. Stud. Contemp. Math. (Kyungshang)* **2020**, *30*, 253–262.

31. Kim, D.S.; Kim, T.; Kwon, J.; Lee, H. A note on λ-Bernoulli numbers of the second kind. *Adv. Stud. Contemp. Math. (Kyungshang)* **2020**, *30*, 187–195.

9

A New Class of Hermite-Apostol type Frobenius-Euler Polynomials and its Applications

Serkan Araci [1,*], **Mumtaz Riyasat** [2], **Shahid Ahmad Wani** [2] and **Subuhi Khan** [2]

[1] Department of Economics, Faculty of Economics, Administrative and Social Sciences, Hasan Kalyoncu University, Gaziantep TR-27410, Turkey
[2] Department of Mathematics, Faculty of Science, Aligarh Muslim University, Aligarh 202 002, India; mumtazrst@gmail.com (M.R.); shahidwani177@gmail.com (S.A.W.); subuhi2006@gmail.com (S.K.)
* Correspondence: serkan.araci@hku.edu.tr.

Abstract: The article is written with the objectives to introduce a multi-variable hybrid class, namely the Hermite–Apostol-type Frobenius–Euler polynomials, and to characterize their properties via different generating function techniques. Several explicit relations involving Hurwitz–Lerch Zeta functions and some summation formulae related to these polynomials are derived. Further, we establish certain symmetry identities involving generalized power sums and Hurwitz–Lerch Zeta functions. An operational view for these polynomials is presented, and corresponding applications are given. The illustrative special cases are also mentioned along with their generating equations.

Keywords: Apostol-type Frobenius–Euler polynomials; three-variable Hermite polynomials; symmetric identities; explicit relations; operational connection

MSC: 11B68; 05A10; 11B65

1. Introduction and Preliminaries

The multi-variable forms of the special polynomials of mathematical physics help in deriving several useful identities and in introducing new families of special polynomials. We know that the generalized Hermite polynomials are important to deal with quantum mechanical and optical beam transport problems [1] (also see [2,3]). The generating equation for the three-variable Hermite polynomials (3VHP) $H_n(x, y, z)$ [4] is given by:

$$e^{xt+yt^2+zt^3} = \sum_{n=0}^{\infty} H_n(x,y,z) \frac{t^n}{n!}, \tag{1}$$

which for $z = 0$ reduce to the two-variable Hermite–Kampé de Fériet polynomials (2VHKdFP) $H_n(x,y)$ [5] and for $z = 0$, $x = 2x$ and $y = -1$ become the classical Hermite polynomials $H_n(x)$ [6].

For $u \in \mathbb{C}$, $u \neq 1$, the generating equation for the Apostol-type Frobenius–Euler polynomials (ATFEP) $\mathfrak{F}_n^{(\alpha)}(x; u; \lambda)$, of order $\alpha \in \mathbb{C}$, is given by [7]:

$$\left(\frac{1-u}{\lambda e^t - u} \right)^{\alpha} e^{xt} = \sum_{n=0}^{\infty} \mathfrak{F}_n^{(\alpha)}(x; u; \lambda) \frac{t^n}{n!}, \tag{2}$$

which for $x = 0$ gives the Apostol-type Frobenius–Euler numbers (ATFEN) $\mathfrak{F}_n^{(\alpha)}(u; \lambda)$, of order α such that:

$$\left(\frac{1-u}{\lambda e^t - u} \right)^{\alpha} = \sum_{n=0}^{\infty} \mathfrak{F}_n^{(\alpha)}(u; \lambda) \frac{t^n}{n!}. \tag{3}$$

For $u = -1$, the ATFEP reduce to the Apostol–Euler polynomials $\mathfrak{E}_n^{(\alpha)}(x; \lambda)$ [8], which for $\lambda = 1$, become the Euler polynomials $E_n^{(\alpha)}(x)$ [9]. Furthermore, the ATFEP for $\lambda = 1$ becomes the Frobenius–Euler polynomials $\mathfrak{F}_n^{(\alpha)}(x; u)$ [10].

The generating equations for the special polynomials are important from different view points and help in finding connection formulas, recursive relations and difference equations and in solving enumeration problems in combinatorics and encoding their solutions.

We intended to introduce a new hybrid class, namely the class of three-variable Hermite–Apostol-type Frobenius–Euler polynomials (3VHATFEP).

Upon replacing the powers x^n by the polynomials $H_n(x, y, z)$ for $(n = 0, 1, 2, \ldots)$ in Equation (2) and upon the use of Equation (1), we have:

For $u, \lambda \in \mathbb{C}$, $u \neq 1$, the three-variable Hermite–Apostol-type Frobenius–Euler polynomials $_H\mathcal{F}_n^{(\alpha)}(x, y, z; u; \lambda)$, of order $\alpha \in \mathbb{C}$, are defined by the following generating function:

$$\left(\frac{1-u}{\lambda e^t - u} \right)^\alpha e^{xt + yt^2 + zt^3} = \sum_{n=0}^{\infty} {}_H\mathfrak{F}_n^{(\alpha)}(x, y, z; u; \lambda) \frac{t^n}{n!}, \tag{4}$$

which for $\lambda = 1$ becomes the three-variable Hermite–Frobenius–Euler polynomials $_H\mathfrak{F}_n^{(\alpha)}(x, y, z; u)$, of order α, which again for $\alpha = 1$, give the three-variable Hermite-Frobenius–Euler polynomials $_H\mathfrak{F}_n(x, y, z; u)$.

Again, the 3VHATFEP for $u = -1$ give the three-variable Hermite–Apostol–Euler polynomials $_H\mathfrak{E}_n^{(\alpha)}(x, y, z; \lambda)$ of order α, which for $\lambda = 1$ reduce to the three-variable Hermite–Euler polynomials $_HE_n^{(\alpha)}(x, y, z)$.

The 3VHATFEP are also defined as the discrete Apostol-type Frobenius–Euler convolution of the 3VHP given by:

$$_H\mathfrak{F}_n^{(\alpha)}(x, y, z; u; \lambda) = n! \sum_{k=0}^{n} \sum_{r=0}^{[k/3]} \frac{\mathfrak{F}_{n-k}^{(\alpha)}(u; \lambda) z^r H_{k-3r}(x, y)}{(n-k)! r! (k - 3r)!}, \tag{5}$$

where $H_n(x, y)$ are the 2VHKdFP.

Next, we deduce certain special cases related to the 3VHATFEP family. Some of these cases are known in the literature. These polynomials are given in Table 1 below.

In this article, the 3VHATFEP are introduced, and certain properties including the explicit relations, summation formulae and symmetric identities for these polynomials are proven using different generating function methods. Some applications for the aforementioned hybrid class of polynomials are given.

Table 1. Special polynomials related to the $_H\mathfrak{F}_n^{(\alpha)}(x,y,z;u;\lambda)$ family.

S.No.	Cases	Name of Polynomial	Generating Function
I.	$z=0$	2-variable Hermite–Apostol-type Frobenius–Euler polynomials of order α	$\left(\dfrac{1-u}{\lambda e^t-u}\right)^\alpha e^{xt+yt^2} = \sum_{n=0}^\infty {}_H\mathfrak{F}_n^{(\alpha)}(x,y;u;\lambda)\dfrac{t^n}{n!}$
	$z=0,\lambda=1$	2-variable Hermite–Frobenius–Euler polynomials of order α	$\left(\dfrac{1-u}{e^t-u}\right)^\alpha e^{xt+yt^2} = \sum_{n=0}^\infty {}_H\mathfrak{F}_n^{(\alpha)}(x,y;u)\dfrac{t^n}{n!}$
	$z=0,\lambda=\alpha=1$	2-variable Hermite–Frobenius–Euler polynomials	$\left(\dfrac{1-u}{e^t-u}\right) e^{xt+yt^2} = \sum_{n=0}^\infty {}_H\mathfrak{F}_n(x,y;u)\dfrac{t^n}{n!}$
II.	$x=2x,$ $y=-1; z=0$	Hermite–Apostol-type Frobenius–Euler polynomials of order α	$\left(\dfrac{1-u}{\lambda e^t-u}\right)^\alpha e^{2xt-t^2} = \sum_{n=0}^\infty {}_H\mathfrak{F}_n^{(\alpha)}(x;u;\lambda)\dfrac{t^n}{n!}$
	$x=2x, y=-1,$ $z=0; \lambda=1$	Hermite–Frobenius–Euler polynomials of order α	$\left(\dfrac{1-u}{e^t-u}\right)^\alpha e^{2xt-t^2} = \sum_{n=0}^\infty {}_H\mathfrak{F}_n^{(\alpha)}(x;u)\dfrac{t^n}{n!}$
	$x=2x, y=-1,$ $z=0; \alpha=\lambda=1$	Hermite–Frobenius–Euler polynomials	$\left(\dfrac{1-u}{e^t-u}\right) e^{2xt-t^2} = \sum_{n=0}^\infty {}_H\mathfrak{F}_n(x;u)\dfrac{t^n}{n!}$
III.	$u=-1$	3-variable Hermite–Apostol–Euler polynomials of order α [11]	$\left(\dfrac{2}{\lambda e^t+1}\right)^\alpha e^{xt+yt^2+zt^3} = \sum_{n=0}^\infty {}_H\mathfrak{E}_n^{(\alpha)}(x,y,z;\lambda)\dfrac{t^n}{n!}$
	$u=-1,$ $\lambda=1$	3-variable Hermite–Euler polynomials of order α [11]	$\left(\dfrac{2}{e^t+1}\right)^\alpha e^{xt+yt^2+zt^3} = \sum_{n=0}^\infty {}_H E_n^{(\alpha)}(x,y,z)\dfrac{t^n}{n!}$
	$u=-1,$ $\lambda=\alpha=1$	3-variable Hermite–Euler polynomials [11]	$\left(\dfrac{2}{e^t+1}\right) e^{xt+yt^2+zt^3} = \sum_{n=0}^\infty {}_H E_n(x,y,z)\dfrac{t^n}{n!}$
IV.	$u=-1, z=0$	2-variable Hermite–Apostol–Euler polynomials of order α [11]	$\left(\dfrac{2}{\lambda e^t+1}\right)^\alpha e^{xt+yt^2} = \sum_{n=0}^\infty {}_H\mathfrak{E}_n^{(\alpha)}(x,y;\lambda)\dfrac{t^n}{n!}$
	$u=-1, \lambda=1;$ $z=0$	2-variable Hermite–Euler polynomials of order α [11]	$\left(\dfrac{2}{e^t+1}\right)^\alpha e^{xt+yt^2} = \sum_{n=0}^\infty {}_H E_n^{(\alpha)}(x,y)\dfrac{t^n}{n!}$
	$u=-1, \lambda=\alpha=1;$ $z=0$	2-variable Hermite–Euler polynomials [11]	$\left(\dfrac{2}{e^t+1}\right) e^{xt+yt^2} = \sum_{n=0}^\infty {}_H E_n(x,y)\dfrac{t^n}{n!}$
V.	$u=-1, x=2x,$ $y=-1; z=0$	Hermite–Apostol–Euler polynomials of order α [12]	$\left(\dfrac{2}{\lambda e^t+1}\right)^\alpha e^{2xt-t^2} = \sum_{n=0}^\infty {}_H\mathfrak{E}_n^{(\alpha)}(x;\lambda)\dfrac{t^n}{n!}$
	$u=-1, \lambda=1;$ $x=2x, y=-1; z=0$	Hermite–Euler polynomials of order α [12]	$\left(\dfrac{2}{e^t+1}\right)^\alpha e^{2xt-t^2} = \sum_{n=0}^\infty {}_H E_n^{(\alpha)}(x)\dfrac{t^n}{n!}$
	$u=-1, \lambda=\alpha=1;$ $x=2x, y=-1; z=0$	Hermite–Euler polynomials [12]	$\left(\dfrac{2}{e^t+1}\right) e^{2xt-t^2} = \sum_{n=0}^\infty {}_H E_n(x)\dfrac{t^n}{n!}$

2. Relations

To derive some relations for the 3VHATFEP, the following results are proven:

Theorem 1. *Let $\alpha, \beta \in \mathbb{Z}$, then we have the following relation for the 3VHATFEP of order α:*

$$_H\mathfrak{F}_n^{(\alpha\pm\beta)}(x,y,z;u;\lambda) = \sum_{k=0}^{n} \binom{n}{k} \mathfrak{F}_k^{(\alpha)}(u;\lambda)\,_H\mathfrak{F}_{n-k}^{(\pm\beta)}(x,y,z;u;\lambda). \tag{6}$$

Proof. We write the generating Function (4) in the following form:

$$\sum_{n=0}^{\infty} {_H\mathfrak{F}_n^{(\alpha\pm\beta)}(x,y,z;u;\lambda)} \frac{t^n}{n!} = \left(\frac{1-u}{\lambda e^t - u}\right)^{(\alpha\pm\beta)} e^{xt+yt^2+zt^3}, \tag{7}$$

for which, upon using Equations (3) and (4) and then after simplification, we get Equation (6). □

Corollary 1. *For $\alpha, \beta \in \mathbb{Z}$, the following relation for the 3VHAEP of order α holds true:*

$$_H\mathfrak{E}_n^{(\alpha\pm\beta)}(x,y,z;\lambda) = \sum_{k=0}^{n} \binom{n}{k} \mathfrak{E}_k^{(\alpha)}(\lambda)\,_H\mathfrak{E}_{n-k}^{(\pm\beta)}(x,y,z;\lambda), \tag{8}$$

$\mathfrak{E}_k^{(\alpha)}(\lambda)$ *means Apostol–Euler numbers of order α.*

Theorem 2. *The following recurrence relation for the 3VHATFEP holds true:*

$$_H\mathfrak{F}_{n+1}(x,y,z;u;\lambda) = x\,_H\mathfrak{F}_n(x,y,z;u;\lambda) + 2yn\,_H\mathfrak{F}_{n-1}(x,y,z;u;\lambda) + 3zn(n-1)$$
$$_H\mathfrak{F}_{n-2}(x,y,z;u;\lambda) - \frac{\lambda}{1-u}\sum_{k=0}^{n}\binom{n}{k}{_H\mathfrak{F}_{n-k}(x,y,z;u;\lambda)}\,_H\mathfrak{F}_k(1,0,0;u;\lambda). \tag{9}$$

Proof. Taking $\alpha = 1$ and then taking the derivative with respect to t in Equation (4), we find:

$$\sum_{n=0}^{\infty} {_H\mathfrak{F}_{n+1}(x,y,z;u;\lambda)} \frac{t^n}{n!} = \left(\frac{1-u}{\lambda e^t - u}\right) e^{xt+yt^2+zt^3}(x+2yt+3zt^2) - \frac{(1-u)\lambda e^t}{(\lambda e^t - u)^2} e^{xt+yt^2+zt^3}, \tag{10}$$

from which, upon using Equation (4) (for $\alpha = 1$) and after simplifying the resultant equation, it follows that:

$$\sum_{n=0}^{\infty} {_H\mathfrak{F}_{n+1}(x,y,z;u;\lambda)} \frac{t^n}{n!} = x\sum_{n=0}^{\infty} {_H\mathfrak{F}_n(x,y,z;u;\lambda)} \frac{t^n}{n!} + 2y\sum_{n=0}^{\infty} {_H\mathfrak{F}_n(x,y,z;u;\lambda)} \frac{t^{n+1}}{n!} + 3z$$
$$\sum_{n=0}^{\infty} {_H\mathfrak{F}_n(x,y,z;u;\lambda)} \frac{t^{n+2}}{n!} - \frac{\lambda}{1-u}\sum_{n=0}^{\infty} {_H\mathfrak{F}_n(x,y,z;u;\lambda)} \frac{t^n}{n!}\sum_{k=0}^{\infty} {_H\mathfrak{F}_k(1,0,0;u;\lambda)} \frac{t^k}{k!}. \tag{11}$$

Replacing $n \to n-1, n-2$ and $n-k$ consecutively in the second, third and last term of the above equation on the r.h.s., it follows that:

$$\sum_{n=0}^{\infty} {_H\mathfrak{F}_{n+1}(x,y,z;u;\lambda)} \frac{t^n}{n!} = x\sum_{n=0}^{\infty} {_H\mathfrak{F}_n(x,y,z;u;\lambda)} \frac{t^n}{n!} + 2y\sum_{n=0}^{\infty} {_H\mathfrak{F}_{n-1}(x,y,z;u;\lambda)} \frac{t^n}{(n-1)!} + 3z$$
$$\sum_{n=0}^{\infty} {_H\mathfrak{F}_{n-2}(x,y,z;u;\lambda)} \frac{t^n}{(n-2)!} - \frac{\lambda}{1-u}\sum_{n=0}^{\infty}\sum_{k=0}^{\infty} {_H\mathfrak{F}_{n-k}(x,y,z;u;\lambda)}\,_H\mathfrak{F}_k(1,0,0;u;\lambda) \frac{t^n}{k!(n-k)!},$$

which, upon comparing the coefficients of like powers of $t^n/n!$ on both sides, gives the recurrence Relation (9). □

Corollary 2. *The following recurrence relation for the 3VHAEP holds true:*

$$_H\mathfrak{E}_{n+1}(x,y,z;\lambda) = x\,_H\mathfrak{E}_n(x,y,z;\lambda) + 2yn\,_H\mathfrak{E}_{n-1}(x,y,z;\lambda) + 3zn(n-1)_H\mathfrak{E}_{n-2}(x,y,z;\lambda)$$
$$-\frac{\lambda}{2}\sum_{k=0}^{n}\binom{n}{k}{}_H\mathfrak{E}_{n-k}(x,y,z;\lambda)_H\mathfrak{E}_k(1,0,0;\lambda). \tag{12}$$

Theorem 3. *For $\gamma > 0$, the following relation for the 3VHATFEP of order α holds true:*

$$(1-u)^\gamma\,_H\mathfrak{F}_n^{(\alpha-\gamma)}(x,y,z;u;\lambda) = \sum_{k=0}^{n}\binom{n}{k}{}_H\mathfrak{F}_{n-k}^{(\alpha)}(x,y,z;u;\lambda)\sum_{p=0}^{\gamma}\binom{\gamma}{p}\lambda^p\,p^k(-u)^{\gamma-p}. \tag{13}$$

Proof. We write the generating Function (4) in the following form:

$$\sum_{n=0}^{\infty}{}_H\mathfrak{F}_n^{(\alpha-\gamma)}(x,y,z;u;\lambda)\frac{t^n}{n!} = \left(\frac{1-u}{\lambda e^t - u}\right)^\alpha e^{xt+yt^2+zt^3}(\lambda e^t - u)^\gamma(1-u)^{-\gamma}, \tag{14}$$

which, upon simplifying and again using Equation (4), gives:

$$\sum_{n=0}^{\infty}{}_H\mathfrak{F}_n^{(\alpha-\gamma)}(x,y,z;u;\lambda)\frac{t^n}{n!} = (1-u)^{-\gamma}\sum_{n=0}^{\infty}{}_H\mathfrak{F}_n^{(\alpha)}(x,y,z;u;\lambda)\frac{t^n}{n!}\sum_{k=0}^{\infty}\sum_{p=0}^{\gamma}\binom{\gamma}{p}\lambda^p\,p^k(-u)^{\gamma-p}\frac{t^k}{k!}. \tag{15}$$

Now, simplifying and then comparing the coefficients of the same powers of t in the resultant equation yield Assertion (13). $\square$

Corollary 3. *For $\gamma > 0$, the following relation for the 3VHAEP of order α holds true:*

$$2^\gamma\,_H\mathfrak{E}_n^{(\alpha-\gamma)}(x,y,z;\lambda) = \sum_{k=0}^{n}\binom{n}{k}{}_H\mathfrak{E}_{n-k}^{(\alpha)}(x,y,z;\lambda)\sum_{p=0}^{\gamma}\binom{\gamma}{p}\lambda^p\,p^k. \tag{16}$$

Theorem 4. *For $u, \alpha \in \mathbb{C}$, $u \neq 1$, there is the following relationship between the 3VHATFEP of order α and the generalized Hurwitz–Lerch Zeta function (GHLZF) $\Phi_\mu(z,s,a)$:*

$$_H\mathfrak{F}_n^{(\alpha)}(x,y,z;u;\lambda) = \left(\frac{u-1}{u}\right)^\alpha\sum_{l=0}^{n}\binom{n}{l}\Phi_\alpha\left(\frac{\lambda}{u},1-n,x\right)H_l(0,y,z). \tag{17}$$

Proof. We write the generating Function (4) in the following form:

$$\sum_{n=0}^{\infty}{}_H\mathfrak{F}_n^{(\alpha)}(x,y,z;u;\lambda)\frac{t^n}{n!} = (1-u)^\alpha(\lambda e^t - u)^{-\alpha}e^{xt+yt^2+zt^3}, \tag{18}$$

which, upon simplification, becomes:

$$\sum_{n=0}^{\infty}{}_H\mathfrak{F}_n^{(\alpha)}(x,y,z;u;\lambda)\frac{t^n}{n!} = (1-u)^\alpha(-u)^{-\alpha}\sum_{n=0}^{\infty}\sum_{k=0}^{\infty}\frac{(\alpha)_k}{k!}\left(\frac{\lambda}{u}\right)^k\frac{(k+x)^n t^n}{n!}e^{yt^2+zt^3}. \tag{19}$$

Using Equation (1) and the following formula for the GHLZF $\Phi_\mu(z,s,a)$ [13]:

$$\Phi_\mu(z,s,a) = \sum_{n=0}^{\infty}\frac{(\mu)_n}{n!}\frac{z^n}{(n+a)^s}, \tag{20}$$

and after simplifying the resultant equation yield Relation (17). $\square$

Corollary 4. *There is the following relationship between the 3VHAEP of order α and generalized Hurwitz–Lerch Zeta function $\Phi_\mu(z, s, a)$:*

$$_H\mathfrak{E}_n^{(\alpha)}(x, y, z; \lambda) = 2^\alpha \sum_{l=0}^{n} \binom{n}{l} \Phi_\alpha\left(-\lambda, l - n, x\right) H_l(0, y, z). \tag{21}$$

Theorem 5. *Let α and γ be nonnegative integers. There is the following relationship between the numbers $S(n, k, \lambda)$ and the 3VHATFEP of order α:*

$$\alpha! \sum_{l=0}^{n} \binom{n}{l} {}_H\mathfrak{F}_{n-l}^{(\alpha)}(x, y, z; u; \lambda) S\left(l, \alpha, \frac{\lambda}{u}\right) = \left(\frac{1-u}{u}\right)^\alpha H_n(x, y, z), \tag{22}$$

$$_H\mathfrak{F}_n^{(\alpha-\gamma)}(x, y, z; u; \lambda) = \gamma! \left(\frac{u}{1-u}\right)^\gamma \sum_{l=0}^{n} \binom{n}{l} {}_H\mathfrak{F}_{n-l}^{(\alpha)}(x, y, z; u; \lambda) S\left(l, \gamma, \frac{\lambda}{u}\right). \tag{23}$$

Proof. The generating Equation (4) can be formulated as:

$$\sum_{n=0}^{\infty} {}_H\mathcal{F}_n^{(\alpha)}(x, y, z; u; \lambda) \frac{t^n}{n!} = (1-u)^\alpha \frac{1}{(\lambda e^t - u)^\alpha} e^{xt + yt^2 + zt^3}, \tag{24}$$

which, upon rearranging the terms using Equation (1) and the following expansion:

$$\frac{(\lambda e^t - 1)^k}{k!} = \sum_{n=0}^{\infty} S(n, k, \lambda) \frac{t^n}{n!}. \tag{25}$$

becomes:

$$\alpha! \sum_{n=0}^{\infty} {}_H\mathfrak{F}_n^{(\alpha)}(x, y, z; u; \lambda) \frac{t^n}{n!} \sum_{l=0}^{\infty} S\left(l, \alpha, \frac{\lambda}{u}\right) \frac{t^l}{l!} = \left(\frac{1-u}{u}\right)^\alpha \sum_{n=0}^{\infty} H_n(x, y, z) \frac{t^n}{n!}. \tag{26}$$

which, upon rearranging the summation and then simplifying the resultant equation, yields Relation (22).

Again, we consider the following arrangement of the generating Function (4):

$$\sum_{n=0}^{\infty} {}_H\mathfrak{F}_n^{(\alpha-\gamma)}(x, y, z; u; \lambda) \frac{t^n}{n!} = \left(\frac{1-u}{\lambda e^t - u}\right)^\alpha e^{xt + yt^2 + zt^3} \left(\frac{u}{1-u}\right)^\gamma \gamma! \frac{(\frac{\lambda}{u} e^t - 1)^\gamma}{\gamma!}, \tag{27}$$

which, upon the use of Equations (4) and (25), applying the Cauchy product rule and then canceling the same powers of t in resultant the equation, yields Relation (23). □

Corollary 5. *There is the following relationship between the numbers $S(n, k, \lambda)$ and the 3VHAEP of order α:*

$$\alpha! \sum_{l=0}^{n} \binom{n}{l} {}_H\mathfrak{E}_{n-l}^{(\alpha)}(x, y, z; \lambda) S\left(l, \alpha, -\lambda\right) = (-2)^\alpha H_n(x, y, z).$$

$$_H\mathfrak{E}_n^{(\alpha-\gamma)}(x, y, z; \lambda) = \gamma! \left(\frac{-1}{2}\right)^\gamma \sum_{l=0}^{n} \binom{n}{l} {}_H\mathfrak{E}_{n-l}^{(\alpha)}(x, y, z; \lambda) S\left(l, \gamma, -\lambda\right). \tag{28}$$

In the next section, we derive some summation formulae for the 3VHATFEP.

3. Summation Formulae

In order to prove the summation formulae for the 3VHATFEP $_H\mathfrak{F}_n^{(\alpha)}(x, y, z; u; \lambda)$, we have the following theorems:

Theorem 6. *The following implicit summation formula for the 3VHATFEP of order α holds true:*

$$_H\mathfrak{F}_n^{(\alpha)}(x+w,y,z;u;\lambda) = \sum_{k=0}^{n} \binom{n}{k} {}_H\mathfrak{F}_k^{(\alpha)}(x,y,z;u;\lambda)w^{n-k}. \tag{29}$$

Proof. Substituting $x \to x+w$ in (4), then making use of Equation (4) and with the series expansion of e^{wt} in the resultant equation, we have:

$$\sum_{n=0}^{\infty} {}_H\mathfrak{F}_n^{(\alpha)}(x+w,y,z;u;\lambda)\frac{t^n}{n!} = \sum_{n=0}^{\infty}\sum_{k=0}^{\infty} {}_H\mathfrak{F}_k^{(\alpha)}(x,y,z;u;\lambda)w^n\frac{t^{n+k}}{n!k!}, \tag{30}$$

which, upon simplification, gives Assertion (29). $\quad\square$

Corollary 6. *For $w=1$ in Equation (29), we have:*

$$_H\mathfrak{F}_n^{(\alpha)}(x+1,y,z;u;\lambda) = \sum_{k=0}^{n} \binom{n}{k} {}_H\mathfrak{F}_k^{(\alpha)}(x,y,z;u;\lambda). \tag{31}$$

Theorem 7. *The following implicit summation formula for the 3VHATFEP of order α holds true:*

$$_H\mathfrak{F}_n^{(\alpha)}(x+v,y+w,z+r;u;\lambda) = \sum_{k=0}^{n} \binom{n}{k} {}_H\mathfrak{F}_{n-k}^{(\alpha)}(x,y,z;u;\lambda)\,H_k(v,w,r). \tag{32}$$

Proof. Replacing $x \to x+v$, $y \to y+w$ and $z \to z+r$ in the generating Function (4) and by the help of Equations (1) and (4), we find:

$$\sum_{n=0}^{\infty} {}_H\mathfrak{F}_n^{(\alpha)}(x+v,y+w,z+r;u;\lambda)\frac{t^n}{n!} = \sum_{n=0}^{\infty}\sum_{k=0}^{\infty} {}_H\mathfrak{F}_n^{(\alpha)}(x,y,z;\lambda;u)H_k(v,w,r)\frac{t^{n+k}}{n!k!}, \tag{33}$$

which, after simplification, gives Formula (32). $\quad\square$

Corollary 7. *For $r=0$ in Equation (32), we have:*

$$_H\mathfrak{F}_n^{(\alpha)}(x+v,y+w,z;u;\lambda) = \sum_{k=0}^{n} \binom{n}{k} {}_H\mathfrak{F}_{n-k}^{(\alpha)}(x,y,z;u;\lambda)\,H_k(v,w). \tag{34}$$

Theorem 8. *The following implicit summation formula for the 3VHATFEP of order α holds true:*

$$_H\mathfrak{F}_{n+k}^{(\alpha)}(p,y,z;u;\lambda) = \sum_{l,m=0}^{n,k} \binom{n}{l}\binom{k}{m}(p-x)^{l+m}\,{}_H\mathfrak{F}_{n+k-l-m}^{(\alpha)}(x,y,z;u;\lambda). \tag{35}$$

Proof. Reestablishing t by $t+v$ and after using the following rule:

$$\sum_{N=0}^{\infty} f(N)\frac{(x+y)^N}{N!} = \sum_{l,m=0}^{\infty} f(l+m)\frac{x^l\,y^m}{l!\,m!} \tag{36}$$

in Equation (4) and then simplifying the resultant equation, it follows that:

$$e^{-x(t+v)} \sum_{n,k=0}^{\infty} {}_H\mathfrak{F}_{n+k}^{(\alpha)}(x,y,z;\lambda;u)\frac{t^n\,v^k}{n!\,k!} = \left(\frac{1-u}{\lambda e^{t+v}-u}\right)^{\alpha} e^{y(t+v)^2+z(t+v)^3}. \tag{37}$$

Replacing x by p in the above equation, equating the resultant equation to the above equation and then expanding the exponential function give:

$$\sum_{n,k=0}^{\infty} {}_{H}\mathfrak{F}_{n+k}^{(\alpha)}(p,y,z;u;\lambda)\frac{t^n v^k}{n!\, k!} = \sum_{N=0}^{\infty}(p-x)^N \frac{(t+v)^N}{N!} \sum_{n,k=0}^{\infty} {}_{H}\mathfrak{F}_{n+k}^{(\alpha)}(x,y,z;u;\lambda)\frac{t^n v^k}{n!\, k!}. \tag{38}$$

Now, using Formula (36) in the above equation and then replacing $n \to n-l$ and $k \to k-m$ in the resultant equation, it follows that:

$$\sum_{n,k=0}^{\infty} {}_{H}\mathfrak{F}_{n+k}^{(\alpha)}(p,y,z;u;\lambda)\frac{t^n v^k}{n!\, k!} = \sum_{n,k=0}^{\infty} \sum_{l,m=0}^{n,k} \frac{(p-x)^{l+m}}{l!\, m!} {}_{H}\mathfrak{F}_{n+k-l-m}^{(\alpha)}(x,y,z;u;\lambda)\frac{t^n v^k}{(n-l)!\,(k-m)!}, \tag{39}$$

which gives Formula (35). $\square$

Corollary 8. *For $n = 0$ in Equation (35), we have:*

$$ {}_{H}\mathfrak{F}_{k}^{(\alpha)}(p,y,z;u;\lambda) = \sum_{m=0}^{k} \binom{k}{m}(p-x)^m {}_{H}\mathfrak{F}_{k-m}^{(\alpha)}(x,y,z;u;\lambda). \tag{40}$$

Corollary 9. *Replacing p by $p+x$ and taking $z = 0$ in Equation (35), we have:*

$$ {}_{H}\mathfrak{F}_{n+k}^{(\alpha)}(p+x,y;u;\lambda) = \sum_{l,m=0}^{n,k} \binom{n}{l}\binom{k}{m} p^{l+m} {}_{H}\mathfrak{F}_{n+k-l-m}^{(\alpha)}(x,y;u;\lambda). \tag{41}$$

Corollary 10. *Replacing p by $p+x$ and taking $y = 0$ $z = 0$ in Equation (35), we have:*

$$ {}_{H}\mathfrak{F}_{n+k}^{(\alpha)}(p+x;u;\lambda) = \sum_{l,m=0}^{n,k} \binom{n}{l}\binom{k}{m} p^{l+m} {}_{H}\mathfrak{F}_{n+k-l-m}^{(\alpha)}(x;u;\lambda). \tag{42}$$

Corollary 11. *For $p = 0$ in Equation (35), we have:*

$$ {}_{H}\mathfrak{F}_{n+k}^{(\alpha)}(y,z;u;\lambda) = \sum_{l,m=0}^{n,k} \binom{n}{l}\binom{k}{m} (-x)^{l+m} {}_{H}\mathfrak{F}_{n+k-l-m}^{(\alpha)}(x,y,z;u;\lambda). \tag{43}$$

Theorem 9. *The following relation for the 3VHATFEP of order α holds true:*

$$ {}_{H}\mathfrak{F}_{n}^{(\alpha)}(x,y,z;u;\lambda) = \sum_{k=0}^{\left[\frac{n}{3}\right]} \frac{n!}{(n-3k)!k!} {}_{H}\mathfrak{F}_{n-3k}^{(\alpha)}(x,y;u;\lambda)z^k. \tag{44}$$

Proof. Using the equation from Table 1(I), the expansion of e^{zt^3} in Equation (4) and then simplifying the resulting equation give:

$$\sum_{n=0}^{\infty} {}_{H}\mathfrak{F}_{n}^{(\alpha)}(x,y,z;u;\lambda)\frac{t^n}{n!} = \sum_{n=0}^{\infty}\left(\sum_{k=0}^{\left[\frac{n}{3}\right]}\frac{n!}{(n-3k)!k!}{}_{H}\mathfrak{F}_{n-3k}^{(\alpha)}(x,y;u;\lambda)z^k\right)\frac{t^n}{n!}. \tag{45}$$

After comparing the coefficients of same powers of $t^n/n!$ in the above equation, we are led to Relation (44). $\square$

Theorem 10. *The following relation for the 3VHATFEP of order α holds true:*

$$_H\mathfrak{F}_n^{(\alpha)}(x,y,z;u;\lambda) = \sum_{k=0}^{n} \sum_{s=0}^{[\frac{k}{3}]} \frac{n!}{(n-k)!(k-3s)!s!} \mathfrak{F}_{n-k}^{(\alpha)}(u;\lambda) H_{k-3s}(x,y)z^s. \tag{46}$$

Proof. Using Equations (3) and (1) (for $z = 0$), the expansion of e^{zt^3} in Equation (4) and after rearranging the terms, it follows that:

$$\sum_{n=0}^{\infty} {}_H\mathfrak{F}_n^{(\alpha)}(x,y,z;u;\lambda)\frac{t^n}{n!} = \sum_{n=0}^{\infty} \sum_{k=0}^{n} \binom{n}{k} \mathfrak{F}_{n-k}^{(\alpha)}(u;\lambda) \left(\sum_{s=0}^{[\frac{k}{3}]} \frac{k!}{(k-3s)!s!} H_{k-3s}(x,y)z^s \right) \frac{t^n}{n!}. \tag{47}$$

Upon canceling the coefficients of like powers of t in Equation (47), we get Assertion (46). $\square$

Theorem 11. *The following relation for the 3VHATFEP of order α holds true:*

$$_H\mathfrak{F}_n^{(\alpha)}(x,y,z;u;\lambda) = \sum_{s=0}^{[\frac{n}{3}]} \sum_{k=0}^{[\frac{n-3s}{2}]} \frac{n!}{s!(n-3s-2k)!k!} \mathfrak{F}_{n-3s-2k}^{(\alpha)}(x;u;\lambda)y^k z^s. \tag{48}$$

Proof. With the use of Equation (2), the expansions of e^{yt^2} and e^{zt^3} in Equation (4) and upon simplifying the resulting equation, we obtain:

$$\sum_{n=0}^{\infty} {}_H\mathfrak{F}_n^{(\alpha)}(x,y,z;u;\lambda)\frac{t^n}{n!} = \sum_{n=0}^{\infty} \left(\sum_{s=0}^{[\frac{n}{3}]} \sum_{k=0}^{[\frac{n-3s}{2}]} \frac{n!}{s!(n-3s-2k)!k!} \mathfrak{F}_{n-3s-2k}^{(\alpha)}(x;u;\lambda)y^k z^s \right) \frac{t^n}{n!} \tag{49}$$

Finally, upon equating the coefficients of the same powers of t in the above equation, Relation (48) is proven. $\square$

In the next section, we establish some symmetric identities for the 3VHATFEP.

4. Symmetric Identities

The identities for the generalized special functions are useful in electromagnetic processes, combinatorics, numerical analysis, etc. Several types of identities and relations related to Apostol-type polynomials and related polynomials are considered in [14–27]. This provides the motivation to explore symmetry identities for the 3VHATFEP. We recall the following:

For any $\gamma \in \mathbb{R}$ or $\mathbb{C}$, the generalized sum of integer powers $\mathcal{S}_k(p;\gamma)$ is given by:

$$\frac{\gamma^{p+1}e^{(p+1)t} - 1}{\gamma e^t - 1} = \sum_{k=0}^{\infty} \mathcal{S}_k(p;\gamma)\frac{t^k}{k!}, \tag{50}$$

which gives:

$$\mathcal{S}_k(p;\gamma) = \sum_{l=0}^{k} \gamma^l l^k.$$

For any $\gamma \in \mathbb{R}$ or $\mathbb{C}$, the multiple power sums $\mathcal{S}_k^{(l)}(m;\gamma)$ are given by:

$$\left(\frac{1-\gamma^m e^{mt}}{1-\gamma e^t} \right)^l = \frac{1}{\gamma^l} \sum_{n=0}^{\infty} \left\{ \sum_{p=0}^{n} \binom{n}{p} (-1)^{n-p} \mathcal{S}_k^{(l)}(m;\gamma) \right\} \frac{t^n}{n!}. \tag{51}$$

To prove the symmetry identities for the 3VHATFEP, we have the following theorems:

Theorem 12. *For all integers* $c, d > 0$ *and* $n \geq 0$, $\alpha \geq 1$, $\lambda, u \in \mathbb{C}$, *the following symmetry relation between the 3VHATFEP of order* α *and the generalized integer power sums holds true:*

$$\sum_{k=0}^{n} \binom{n}{k} c^{n-k} {}_H\mathfrak{F}_{n-k}^{(\alpha)}(dx, d^2y, d^3z; \lambda; u) \sum_{l=0}^{k} \binom{k}{l} d^k u^{c-1} \mathcal{S}_l(c-1; \tfrac{\lambda}{u}) {}_H\mathfrak{F}_{k-l}^{(\alpha-1)}(cX, c^2Y, c^3Z; \lambda; u)$$
$$= \sum_{k=0}^{n} \binom{n}{k} d^{n-k} u^{d-1} {}_H\mathfrak{F}_{n-k}^{(\alpha)}(cx, c^2y, c^3z; \lambda; u) \sum_{l=0}^{k} \binom{k}{l} c^k \mathcal{S}_l(d-1; \tfrac{\lambda}{u}) {}_H\mathfrak{F}_{k-l}^{(\alpha-1)}(dX, d^2Y, d^3Z; \lambda; u). \tag{52}$$

Proof. Let

$$G(t) := \frac{(1-u)^{2\alpha-1} e^{cdxt+y(cdt)^2+z(cdt)^3} \left(\lambda^c e^{cdt} - u^c\right) e^{cdXt+Y(cdt)^2+Z(cdt)^3}}{(\lambda e^{ct} - u)^\alpha \; (\lambda e^{dt} - u)^\alpha}, \tag{53}$$

which, upon rearranging the powers and then using Equations (4) and (50) in the resultant equation, yields:

$$G(t) = \left(\sum_{n=0}^{\infty} {}_H\mathfrak{F}_n^{(\alpha)}(dx, d^2y, d^3z; \lambda; u) \frac{(ct)^n}{n!} \right) \left(u^{c-1} \sum_{l=0}^{\infty} \mathcal{S}_l(c-1; \tfrac{\lambda}{u}) \frac{(dt)^l}{l!} \right)$$
$$\times \left(\sum_{k=0}^{\infty} {}_H\mathfrak{F}_k^{(\alpha-1)}(cX, c^2Y, c^3Z; \lambda; u) \frac{(dt)^k}{k!} \right). \tag{54}$$

Upon applying the Cauchy product rule in the above equation, we get:

$$G(t) = \sum_{n=0}^{\infty} \left(\sum_{k=0}^{n} \binom{n}{k} c^{n-k} d^k u^{c-1} {}_H\mathfrak{F}_{n-k}^{(\alpha)}(dx, d^2y, d^3z; \lambda; u) \sum_{l=0}^{k} \binom{k}{l} \mathcal{S}_l(c-1; \tfrac{\lambda}{u}) \right.$$
$$\left. \times {}_H\mathfrak{F}_{k-l}^{(\alpha-1)}(cX, c^2Y, c^3Z; \lambda; u) \right) \frac{t^n}{n!}. \tag{55}$$

In a similar manner, we obtain:

$$G(t) = \sum_{n=0}^{\infty} \left(\sum_{k=0}^{n} \binom{n}{k} d^{n-k} c^k u^{d-1} {}_H\mathfrak{F}_{n-k}^{(\alpha)}(cx, c^2y, c^3z; \lambda; u) \sum_{l=0}^{k} \binom{k}{l} \mathcal{S}_l(d-1; \tfrac{\lambda}{u}) \right.$$
$$\left. \times {}_H\mathfrak{F}_{k-l}^{(\alpha-1)}(dX, d^2Y, d^3Z; \lambda; u) \right) \frac{t^n}{n!}. \tag{56}$$

Equating the coefficients of the like powers of t in the r.h.s. of Expansions (55) and (56), we are led to Identity (52). $\square$

Theorem 13. *For each pair of positive integers* c, d *and for all integers* $n \geq 0$, $\alpha \geq 1$, $\lambda, u \in \mathbb{C}$, *the following symmetry identity for the 3VHATFEP of order* α *holds true:*

$$\sum_{k=0}^{n} \binom{n}{k} \sum_{i=0}^{c-1} \sum_{j=0}^{d-1} u^{c+d-2} (\tfrac{\lambda}{u})^{i+j} c^{n-k} d^k {}_H\mathfrak{F}_k^{(\alpha)}\left(cX + \tfrac{c}{d}j, c^2Y, c^3Z; \lambda; u\right) {}_H\mathfrak{F}_{n-k}^{(\alpha)}\left(dx + \tfrac{d}{c}i, d^2y, d^3z; \lambda; u\right)$$
$$= \sum_{k=0}^{n} \binom{n}{k} \sum_{i=0}^{d-1} \sum_{j=0}^{c-1} u^{c+d-2} (\tfrac{\lambda}{u})^{i+j} d^{n-k} c^k {}_H\mathfrak{F}_k^{(\alpha)}\left(dX + \tfrac{d}{c}j, d^2Y, d^3Z; \lambda; u\right) {}_H\mathfrak{F}_{n-k}^{(\alpha)}\left(cx + \tfrac{c}{d}i, c^2y, c^3z; \lambda; u\right). \tag{57}$$

Proof. Let

$$H(t) := \frac{(1-u)^{2\alpha} e^{cdxt+y(cdt)^2+z(cdt)^3} \left(\lambda^c e^{cdt} - u^c\right)\left(\lambda^d e^{cdt} - u^d\right) e^{cdXt+Y(cdt)^2+Z(cdt)^3}}{(\lambda e^{ct} - u)^{\alpha+1} (\lambda e^{dt} - u)^{\alpha+1}}, \tag{58}$$

from which, upon rearranging the powers and using the series expansions for $\left(\frac{\lambda^c e^{cdt} - u^c}{\lambda e^{dt} - u}\right)$ and $\left(\frac{\lambda^d e^{cdt} - u^d}{\lambda e^{ct} u}\right)$ in the resultant equation, it follows that:

$$H(t) = \left(\frac{1-u}{\lambda e^{ct} - u}\right)^{\alpha} e^{dx(ct) + d^2 y(ct)^2 + d^3 z(ct)^3} u^{c-1} \sum_{i=0}^{c-1} \left(\frac{\lambda}{u}\right)^{i} e^{dti}$$

$$\times \left(\frac{1-u}{\lambda e^{dt} - u}\right)^{\alpha} e^{cX(dt) + c^2 Y(dt)^2 + c^3 Z(dt)^3} u^{d-1} \sum_{j=0}^{d-1} \left(\frac{\lambda}{u}\right)^{j} e^{ctj}. \tag{59}$$

Now, by making use of Equation (4) and the application of the Cauchy product rule in the resultant equation, we have:

$$H(t) = \sum_{k=0}^{n} \binom{n}{k} \sum_{i=0}^{c-1} \sum_{j=0}^{d-1} u^{c+d-2} (\tfrac{\lambda}{u})^{i+j} c^{n-k} d^k {}_H\mathfrak{F}_k^{(\alpha)} \left(cX + \tfrac{c}{d}j, c^2 Y, c^3 Z; \lambda; u\right)$$
$${}_H\mathfrak{F}_{n-k}^{(\alpha)} \left(dx + \tfrac{d}{c}i, d^2 y, d^3 z; \lambda; u\right). \tag{60}$$

Following the same lines of proof as above gives another identity:

$$H(t) = \sum_{k=0}^{n} \binom{n}{k} \sum_{i=0}^{d-1} \sum_{j=0}^{c-1} u^{d+c-2} (\tfrac{\lambda}{u})^{i+j} d^{n-k} c^k {}_H\mathfrak{F}_k^{(\alpha)} \left(dX + \tfrac{d}{c}j, d^2 Y, d^3 Z; \lambda; u\right)$$
$${}_H\mathfrak{F}_{n-k}^{(\alpha)} \left(cx + \tfrac{c}{d}i, c^2 y, c^3 z; \lambda; u\right). \tag{61}$$

Comparing the coefficients of the same powers of t in the r.h.s. of Expressions (60) and (61) gives Identity (57). □

Theorem 14. *For each pair of positive integers c, d and for all integers $n \geq 0$, $\alpha \geq 1$, $\lambda, u \in \mathbb{C}$, the following symmetry identity for the 3VHATFEP holds true:*

$$\sum_{m=0}^{d-1} u^{d-1} (\tfrac{\lambda}{u})^m \sum_{l=0}^{n} \binom{n}{l} {}_H\mathfrak{F}_{n-l}\left(cx, c^2 y, c^3 z; \lambda; u\right) d^{n-l} (cm)^l$$
$$= \sum_{m=0}^{c-1} u^{c-1} (\tfrac{\lambda}{u})^m \sum_{l=0}^{n} \binom{n}{l} {}_H\mathfrak{F}_{n-l}\left(dx, d^2 y, d^3 z; \lambda; u\right) c^{n-l} (dm)^l. \tag{62}$$

Proof. Let

$$N(t) := \frac{(1-u)e^{cdxt + y(cdt)^2 + z(cdt)^3}(\lambda^d e^{cdt} - u^d)}{(\lambda e^{ct} - u)(\lambda e^{dt} - u)}. \tag{63}$$

Proceeding on the same lines of proof as in Theorem 13, we get Identity (62). Thus, we omit the proof. □

Theorem 15. *For each pair of positive integers c, d and for all integers $n \geq 0$, $\alpha \geq 1$, $\lambda, u \in \mathbb{C}$, the following symmetry relation between the 3VHATFEP and multiple power sums holds true:*

$$\sum_{l=0}^{n} \binom{n}{l} {}_H\mathfrak{F}_{n-l}(dx, d^2 y, d^3 z; \lambda; u) \, u^{d\alpha} \lambda^{-\alpha} \sum_{m=0}^{l} \binom{l}{m} \sum_{r=0}^{m} \binom{m}{r} (-\alpha)^{m-r} S_k^{(\alpha)}\left(d; \tfrac{\lambda}{u}\right)$$
$$\times {}_H\mathfrak{F}_{l-m}^{(\alpha+1)}(cX, c^2 Y, c^3 Z; \lambda; u) c^{n-l+m} d^{l-m}$$
$$= \sum_{l=0}^{n} \binom{n}{l} {}_H\mathfrak{F}_{n-l}(cx, c^2 y, c^3 z; \lambda; u) \, u^{c\alpha} \lambda^{-\alpha} \sum_{m=0}^{l} \binom{l}{m} \sum_{r=0}^{m} \binom{m}{r} (-\alpha)^{m-r} S_k^{(\alpha)}\left(c; \tfrac{\lambda}{u}\right)$$
$$\times {}_H\mathfrak{F}_{l-m}^{(\alpha+1)}(dX, d^2 Y, d^3 Z; \lambda; u) d^{n-l+m} c^{l-m}. \tag{64}$$

Proof. Let:

$$F(t) := \frac{(1-u)^{\alpha+2} \, e^{dx(ct)+d^2y(ct)^2+d^3z(ct)^3} \left(\lambda^d e^{dct} - u^d\right)^\alpha \, e^{cX(dt)+c^2Y(dt)^2+c^3Z(dt)^3}}{\left(\lambda e^{dt} - u\right)^{\alpha+1} \left(\lambda e^{ct} - u\right)^{\alpha+1}}, \tag{65}$$

which, upon rearranging the powers and use of Equations (4) and (51) in the resultant equation, yields:

$$F(t) := \sum_{n=0}^{\infty} {}_H\mathfrak{F}_n(dx, d^2y, d^3z; \lambda; u)c^n \frac{t^n}{n!} \, u^{d\alpha}\lambda^{-\alpha} \sum_{m=0}^{\infty}\sum_{r=0}^{m} \binom{m}{r} (-\alpha)^{m-r} S_k^{(\alpha)}\left(d; \frac{\lambda}{u}\right) c^m \frac{t^m}{m!}$$
$$\sum_{l=0}^{\infty} {}_H\mathfrak{F}_l^{(\alpha+1)}(cX, c^2Y, c^3Z; \lambda; u) d^l \frac{t^l}{l!}. \tag{66}$$

Now, appropriately applying the using Cauchy product rule in the above equation leads to:

$$F(t) := \sum_{n=0}^{\infty}\sum_{l=0}^{n} \binom{n}{l}{}_H\mathfrak{F}_{n-l}(dx, d^2y, d^3z; \lambda; u)c^{n-l} \, u^{d\alpha}\lambda^{-\alpha} \sum_{m=0}^{l} \binom{l}{m} \sum_{r=0}^{m} \binom{m}{r} (-\alpha)^{m-r} S_k^{(\alpha)}\left(d; \frac{\lambda}{u}\right)$$
$${}_H\mathfrak{F}_{l-m}^{(\alpha+1)}(cX, c^2y, c^3z; \lambda; u)c^m d^{l-m} \frac{t^n}{n!}. \tag{67}$$

Similarly, we can find:

$$F(t) := \sum_{n=0}^{\infty}\sum_{l=0}^{n} \binom{n}{l}{}_H\mathfrak{F}_{n-l}(cx, c^2y, c^3z; \lambda; u)d^{n-l} \, u^{c\alpha}\lambda^{-\alpha} \sum_{m=0}^{l} \binom{l}{m} \sum_{r=0}^{m} \binom{m}{r} (-\alpha)^{m-r} S_k^{(\alpha)}\left(c; \frac{\lambda}{u}\right)$$
$${}_H\mathfrak{F}_{l-m}^{(\alpha+1)}(dx, d^2y, d^3z; \lambda; u)d^m c^{l-m} \frac{t^n}{n!}. \tag{68}$$

Equating the coefficients of the like powers of $t^n/n!$ in the r.h.s. of Expansions (67) and (68) gives Identity (64). $\square$

Theorem 16. *For each pair of positive integers c, d and for all integers $n \geq 0$, $\alpha \geq 1$, $\lambda, u \in \mathbb{C}$, the following symmetry relation between the 3VHATFEP of order α and multiple power sums holds true:*

$$\sum_{m=0}^{n} \binom{n}{m}{}_H\mathfrak{F}_{n-m}^{(\alpha)}(dx, d^2y, d^3z; \lambda; u)c^{n-m} \, u^{c\alpha}\lambda^{-\alpha} \sum_{r=0}^{m} \binom{m}{r} (-\alpha)^{m-r} S_k^{(\alpha)}\left(c; \frac{\lambda}{u}\right)d^m$$
$$= \sum_{m=0}^{n} \binom{n}{m}{}_H\mathfrak{F}_{n-m}^{(\alpha)}(cx, c^2y, c^3z; \lambda; u)d^{n-m} \, u^{d\alpha}\lambda^{-\alpha} \sum_{r=0}^{m} \binom{m}{r} (-\alpha)^{m-r} S_k^{(\alpha)}\left(d; \frac{\lambda}{u}\right)c^m. \tag{69}$$

Proof. Let:

$$M(t) := \frac{(1-u)^\alpha \, e^{dx(ct)+d^2y(ct)^2+d^3z(ct)^3} \left(\lambda^c e^{cdt} - u^c\right)^\alpha}{\left(\lambda e^{dt} - u\right)^\alpha \left(\lambda e^{ct} - u\right)^\alpha}. \tag{70}$$

Proceeding on the same lines of proof as in Theorem 15, we get Identity (69). Thus, we omit the proof. $\square$

Theorem 17. *For each pair of positive integers c, d and for all integers $n \geq 0$, $\alpha \geq 1$, $\lambda, u \in \mathbb{C}$, the following symmetry relation between the 3VHATFEP of order α and the Hurwitz–Lerch Zeta function holds true:*

$$\left(\frac{1-u}{u}\right)^\alpha (-1)^\alpha \left(\sum_{p=0}^{n} \binom{n}{p} \sum_{s=0}^{n-p} \binom{n-p}{s} \Phi_\alpha\left(\frac{\lambda}{u}, s-n+p, cx\right) H_s(0, c^2y, c^3z)d^n \, u^c\lambda^{-1} \right.$$
$$\left. \sum_{r=0}^{p} \binom{r}{p} \sum_{q=0}^{p-r} \binom{p-r}{q} (-1)^{p-r-q} S_q\left(c, \frac{\lambda}{u}\right){}_H\mathfrak{F}_r^{(\alpha)}(dX, d^2Y, d^3Z; \lambda; u)c^r d^{p-r} \right)$$
$$= \left(\frac{1-u}{u}\right)^\alpha (-1)^\alpha \left(\sum_{p=0}^{n} \binom{n}{p} \sum_{s=0}^{n-p} \binom{n-p}{s} \Phi_\alpha\left(\frac{\lambda}{u}, s-n+p, cx\right) H_s(0, d^2y, d^3z)c^n \, u^d\lambda^{-1} \right.$$
$$\left. \sum_{r=0}^{p} \binom{r}{p} \sum_{q=0}^{p-r} \binom{p-r}{q} (-1)^{p-r-q} S_q\left(d, \frac{\lambda}{u}\right){}_H\mathfrak{F}_r^{(\alpha)}(cX, c^2Y, c^3Z; \lambda; u)d^r c^{p-r} \right). \tag{71}$$

Proof. Let:

$$P(t) := \frac{(1-u)^{2\alpha}\, e^{cx(dt)+c^2y(dt)^2+c^3z(dt)^3}\left(\lambda^c e^{cdt} - u^c\right) e^{dX(ct)+d^2Y(ct)^2+d^3Z(ct)^3}}{\left(\lambda e^{dt} - u\right)^{\alpha+1} \left(\lambda e^{ct} - u\right)^{\alpha}}, \tag{72}$$

which, upon rearranging the powers and after using Equations (4) and (51) (for $\alpha = 1$) and the following formula for the generalized binomial theorem:

$$(1+w)^{-\alpha} = \sum_{m=0}^{\infty} \binom{m+\alpha-1}{m}(-w)^m; \quad |w| < 1, \tag{73}$$

in the resultant equation becomes:

$$\begin{aligned}
P(t) := \quad &\left(\tfrac{1-u}{u}\right)^{\alpha}(-1)^{\alpha} \sum_{m=0}^{\infty} \binom{m+\alpha-1}{m}\left(\tfrac{\lambda}{u}\right)^m e^{mdt}\, e^{cx(dt)+c^2y(dt)^2+c^3z(dt)^3}\, u^c \lambda^{-1} \sum_{p=0}^{\infty}\sum_{q=0}^{p}\binom{p}{q}(-1)^{p-q} \\
&S_q\left(c,\tfrac{\lambda}{u}\right) d^p \tfrac{t^p}{p!} \sum_{r=0}^{\infty} {}_H\mathfrak{F}_r^{(\alpha)}(dX,d^2Y,d^3Z;\lambda;u)\tfrac{(ct)^r}{r!}.
\end{aligned} \tag{74}$$

Simplifying the above equation with the use of Equations (1) and (20) and then using the Cauchy product rule in the resultant equation, we get:

$$\begin{aligned}
P(t): \quad = &\left(\tfrac{1-u}{u}\right)^{\alpha}(-1)^{\alpha} \sum_{n=0}^{\infty}\Bigg(\sum_{p=0}^{n}\binom{n}{p}\sum_{s=0}^{n-p}\binom{n-p}{s}\Phi_{\alpha}\left(\tfrac{\lambda}{u},s-n+p,cx\right) H_s(0,c^2y,c^3z)d^n\, u^c\lambda^{-1} \\
&\sum_{r=0}^{p}\binom{r}{p}\sum_{q=0}^{p-r}\binom{p-r}{q}(-1)^{p-r-q}S_q\left(c,\tfrac{\lambda}{u}\right){}_H\mathfrak{F}_r^{(\alpha)}(dX,d^2Y,d^3Z;\lambda;u)c^r d^{p-r}\Bigg)\tfrac{t^n}{n!}.
\end{aligned} \tag{75}$$

In a similar manner, we have:

$$\begin{aligned}
P(t): \quad = &\left(\tfrac{1-u}{u}\right)^{\alpha}(-1)^{\alpha} \sum_{n=0}^{\infty}\Bigg(\sum_{p=0}^{n}\binom{n}{p}\sum_{s=0}^{n-p}\binom{n-p}{s}\Phi_{\alpha}\left(\tfrac{\lambda}{u},s-n+p,dx\right) H_s(0,d^2y,d^3z)c^n\, u^d\lambda^{-1} \\
&\sum_{r=0}^{p}\binom{r}{p}\sum_{q=0}^{p-r}\binom{p-r}{q}(-1)^{p-r-q}S_q\left(d,\tfrac{\lambda}{u}\right){}_H\mathfrak{F}_r^{(\alpha)}(cX,c^2Y,c^3Z;\lambda;u)d^r c^{p-r}\Bigg)\tfrac{t^n}{n!}.
\end{aligned} \tag{76}$$

Finally, canceling the coefficients of the same powers of t in the r.h.s. of Expansions (75) and (76), Identity (71) is proven. $\square$

Note: The results established above for the 3VHATFEP can be reduced to the illustrative special cases mentioned in Table 1 simply by substituting special values of the variables or parameters. Therefore, we omit them.

5. Operational Representation

The classical and Apostol-type Frobenius–Euler numbers and polynomials are the generalization of Euler numbers and polynomials, and these are associated with the Brouwer fixed-point theorem and vector fields [28].

From generating Equation (4), we find that the 3VHATFEP are the solutions of the following equations:

$$\frac{\partial}{\partial y}{}_H\mathfrak{F}_n^{(\alpha)}(x,y,z;u;\lambda) = \frac{\partial^2}{\partial x^2}{}_H\mathfrak{F}_n^{(\alpha)}(x,y,z;u;\lambda), \tag{77}$$

$$\frac{\partial}{\partial z}{}_H\mathfrak{F}_n^{(\alpha)}(x,y,z;u;\lambda) = \frac{\partial^3}{\partial x^3}{}_H\mathfrak{F}_n^{(\alpha)}(x,y,z;u;\lambda), \tag{78}$$

under the following initial condition:

$$_H\mathfrak{F}_n^{(\alpha)}(x,0,0;u;\lambda) = \mathfrak{F}_n^{(\alpha)}(x;u;\lambda). \tag{79}$$

Thus, in view of the above equation, we find that, for the 3VHATFEP, the following operational representation holds true:

$$_H\mathfrak{F}_n^{(\alpha)}(x,y,z;u;\lambda) = \exp\left(y\frac{\partial^2}{\partial x^2} + z\frac{\partial^3}{\partial x^3}\right)\{\mathfrak{F}_n^{(\alpha)}(x;u;\lambda)\}. \tag{80}$$

The operational formalism developed above can be used to obtain the corresponding identities for the 3VHATFEP and for their special cases. To give the applications of the operational representation (80), we apply the operation $\mathcal{O}$ given below:

$\mathcal{O}$: Operating $\exp\left(y\frac{\partial^2}{\partial x^2} + z\frac{\partial^3}{\partial x^3}\right)$ on both sides of a given result.

Consider the following identities for the FEP $\mathfrak{F}_n^{(\alpha)}(x;u)$ from [17]:

$$u\mathfrak{F}_n(x;u^{-1}) + \mathfrak{F}_n(x;u) = (1+u)\sum_{k=0}^{n}\binom{n}{k}\mathfrak{F}_{n-k}(u^{-1})\mathfrak{F}_k(x;u), \tag{81}$$

$$\frac{1}{n+1}\mathfrak{F}_k(x;u) + \mathfrak{F}_{n-k}(x;u) = \sum_{k=0}^{n-1}\frac{\binom{n}{k}}{n-k+1}\sum_{l=k}^{n}((-u)\mathfrak{F}_{l-k}(u)\mathfrak{F}_{n-l}(u) + 2u\mathfrak{F}_{n-k}(u)) \tag{82}$$
$$\mathfrak{F}_k(x;u)\,\mathfrak{F}_n(x;u),$$

$$\mathfrak{F}_n^{(\alpha)}(x;u) = \sum_{k=0}^{n}\binom{n}{k}\mathfrak{F}_{n-k}^{(\alpha-1)}(u)\mathfrak{F}_k(x;u) \qquad (n \in \mathbb{Z}_+), \tag{83}$$

$$\mathfrak{F}_n(x;u) = \frac{1}{(1-u)^{\alpha}}\sum_{k=0}^{n}\binom{n}{k}\left(\sum_{j=0}^{\alpha}\binom{\alpha}{j}(-u)^{\alpha-j}\mathfrak{F}_{n-k}(j;u)\right)\mathfrak{F}_k^{(\alpha)}(x;u) \qquad (n \in \mathbb{Z}_+), \tag{84}$$

which, upon using operation ($\mathcal{O}$) in both sides, yields the following identities for the polynomials $_H\mathfrak{F}_n^{(\alpha)}(x,y,z;u)$:

$$u_H\mathfrak{F}_n(x,y,z;u^{-1}) + {}_H\mathfrak{F}_n(x,y,z;u) = (1+u)\sum_{k=0}^{n}\binom{n}{k}\mathfrak{F}_{n-k}(u^{-1})_H\mathfrak{F}_k(x,y,z;u), \tag{85}$$

$$\frac{1}{n+1}{}_H\mathfrak{F}_k(x,y,z;u) + {}_H\mathfrak{F}_{n-k}(x,y,z;u) = \sum_{k=0}^{n-1}\frac{\binom{n}{k}}{n-k+1}\sum_{l=k}^{n}((-u)\mathfrak{F}_{l-k}(u)\mathfrak{F}_{n-l}(u) + 2u\mathfrak{F}_{n-k}(u)) \tag{86}$$
$$_H\mathfrak{F}_k(x,y,z;u)\,\mathfrak{F}_n(x;u),$$

$$_H\mathfrak{F}_n^{(\alpha)}(x,y,z;u) = \sum_{k=0}^{n}\binom{n}{k}\mathfrak{F}_{n-k}^{(\alpha-1)}(u)_H\mathfrak{F}_k(x,y,z;u) \qquad (n \in \mathbb{Z}_+), \tag{87}$$

$$_H\mathfrak{F}_n(x,y,z;u) = \frac{1}{(1-u)^{\alpha}}\sum_{k=0}^{n}\binom{n}{k}\left(\sum_{j=0}^{\alpha}\binom{\alpha}{j}(-u)^{\alpha-j}\mathfrak{F}_{n-k}(j;u)\right)_H\mathfrak{F}_k^{(\alpha)}(x,y,z;u) \quad (n \in \mathbb{Z}_+). \tag{88}$$

Thus, we find that the aforementioned polynomials, which include the polynomials as their special cases given in Table 1 along with the underlying operational formalism, offer a powerful tool for the investigation of the properties of a wide class of polynomials. Thus, the combination of Hermite and Frobenius–Euler polynomials yields such interesting results.

Further, motivated by the ATFEP $\mathfrak{F}_n^{(\alpha)}(x;u;\lambda)$, we introduce the Apostol type Frobenius–Genocchi polynomials $\mathfrak{H}_n^{(\alpha)}(x;u;\lambda)$ (ATFGP). For $u \in \mathbb{C}$, $u \neq 1$, the ATFGP of order $\alpha \in \mathbb{C}$ are defined by:

$$\left(\frac{(1-u)t}{\lambda e^t - u}\right)^{\alpha}e^{xt} = \sum_{n=0}^{\infty}\mathfrak{H}_n^{(\alpha)}(x;u;\lambda)\frac{t^n}{n!}, \tag{89}$$

which, for $\lambda = \alpha = 1$, reduce to the Frobenius–Genocchi polynomials $G_n^F(x;u)$ [29].

Using the previous approach, we introduce the three-variable Hermite–Apostol-type Frobenius–Genocchi polynomials (3VHATFGP) $_H\mathfrak{H}_n^{(\alpha)}(x,y,z;u;\lambda)$ of order $\alpha \in \mathbb{C}$ defined by:

$$\left(\frac{(1-u)t}{\lambda e^t - u}\right)^\alpha e^{xt+yt^2+zt^3} = \sum_{n=0}^{\infty} {_H\mathfrak{H}_n^{(\alpha)}}(x,y,z;u;\lambda)\frac{t^n}{n!}. \tag{90}$$

The special members related to the 3VHATFGP $_H\mathfrak{H}_n^{(\alpha)}(x,y,z;u;\lambda)$ can be obtained, and corresponding results for these polynomials and for their special cases can be obtained easily. Thus, we omit them.

6. Conclusions

In this paper, a multi-variable hybrid class of the Hermite–Apostol-type Frobenius–Euler polynomials is introduced and their properties are explored using various generating function methods. Several explicit and recurrence relations, summation formulae and symmetry identities are established for these hybrid polynomials. A brief view of the operational approach is also given for these polynomials. The operational representations combined with integral transforms may lead to other interesting results, which may be helpful to the theory of fractional calculus. Several techniques and methods are used in [30,31], which are applicable to the other fields of mathematics. The applicability of these techniques to the hybrid polynomial families can also be explored. These aspects will be undertaken in further investigation.

Author Contributions: All authors contributed equally

References

1.   Dattoli, G.; Lorenzutta, S.; Maino, G.; Torre, A.; Cesarano, C. Generalized Hermite polynomials and super-Gaussian forms. *J. Math. Anal. Appl.* **1996**, *203*, 597–609. [CrossRef]
2.   Cesarano, C. Operational methods and new identities for Hermite polynomials. *Math. Model. Nat. Phenom.* **2017**, *12*, 44–50. [CrossRef]
3.   Cesarano, C.; Fornaro, C.; Vázquez, L. A note on a special class of Hermite polynomials. *Int. J. Pure Appl. Math.* **2015**, *98*, 261–273. [CrossRef]
4.   Dattoli, G. Generalized polynomials operational identities and their applications. *J. Comput. Appl. Math.* **2000**, *118*, 111–123. [CrossRef]
5.   Appell, P.; de Fériet, J.K. *Fonctions Hypergéométriques et Hypersphériques: Polynômes d' Hermite*; Gauthier-Villars: Paris, France, 1926.
6.   Andrews, L.C. *Special Functions for Engineers and Applied Mathematicians*; Macmillan Publishing Company: New York, NY, USA, 1985.
7.   Özarslan, M.A. Unified Apostol-Bernoulli, Euler and Genocchi polynomials. *Comput. Math. Appl.* **2011**, *62*, 2452–2462. [CrossRef]
8.   Luo, Q.M. Apostol–Euler polynomials of higher order and the Gaussian hypergeometric function. *Taiwan. J. Math.* **2006**, *10*, 917–925. [CrossRef]
9.   Erdélyi, A.; Magnus, W.; Oberhettinger, F.; Tricomi, F.G. *Higher Transcendental Functions*; McGraw-Hill Book Company: New York, NY, USA; Toronto, ON, Canada; London, UK, 1955; Volume III.
10. Carlitz, L. Eulerian numbers and polynomials. *Math. Mag.* **1959**, *32*, 247–260. [CrossRef]
11. Khan, S.; Yasmin, G.; Khan, R.; Hassan, N.A.M. Hermite-based Appell polynomials: Properties and applications. *J. Math. Anal. Appl.* **2009**, *351*, 756–764. [CrossRef]
12. Khan, S.; Riyasat, M. A determinantal approach to Sheffer-Appell polynomials via monomiality principle. *J. Math. Anal. Appl.* **2015**, *421*, 806–829. [CrossRef]
13. Goyal, S.P.; Laddha, R.K. On the generalized Riemann zeta functions and the generalized Lambert transform. *Ganita Sandesh* **1997**, *11*, 99–108.

14. Jang, G.-W.; Kwon, H.-I.; Kim, T. A note on degenerate Apostol-Bernoulli numbers and polynomials. *Adv. Stud. Contemp. Math. (Kyungshang)* **2017**, *27*, 279–288.

15. Kim, T. An identity of the symmetry for the Frobenius–Euler polynomials associated with the fermionic *p*-adic invariant *q*-integrals on Z_p. *Rocky Mt. J. Math.* **2011**, *41*, 239–247. [CrossRef]

16. Kim, T. Identities involving Frobenius–Euler polynomials arising from non-linear differential equations. *J. Number Theory* **2012**, *132*, 2854–2865. [CrossRef]

17. Kim, D.S.; Kim, T. Some new identities of Frobenius–Euler numbers and polynomials. *J. Inequal. Appl.* **2012**, *307*, 1–10. [CrossRef]

18. Kim, D.S.; Kim, T. Higher-order Frobenius–Euler and poly-Bernoulli mixed-type polynomials. *Adv. Differ. Equ.* **2013**, *251*, 13. [CrossRef]

19. Kim, T.; Kim, D.S. Identities for degenerate Bernoulli polynomials and Korobov polynomials of the first kind. *Sci. China Math.* **2018**. [CrossRef]

20. Kim, T.; Kim, D.S. An identity of symmetry for the degenerate Frobenius–Euler polynomials. *Math. Slov.* **2018**, *68*, 239–243. [CrossRef]

21. Kim, T.; Kwon, H.-I.; Seo, J.J. Some identities of degenerate Frobenius–Euler polynomials and numbers. *Proc. Jangjeon Math. Soc.* **2016**, *19*, 157–163.

22. Kim, T.; Mansour, T. Umbral calculus associated with Frobenius-type Eulerian polynomials. *Russ. J. Math. Phys.* **2014**, *21*, 484–493. [CrossRef]

23. Kurt, V. Some symmetry identities for the Apostol-type polynomials related to multiple alternating sums. *Adv. Differ. Equ.* **2013**, *32*, 1–32. [CrossRef]

24. Bayad, A.; Kim, T. Identities for Apostol-type Frobenius–Euler polynomials resulting from the study of a nonlinear operator. *Russ. J. Math. Phys.* **2016**, *23*, 164–171. [CrossRef]

25. Duran, U.; Acikgoz, M.; Araci, S. Hermite based poly-Bernoulli polynomials with a *q*-parameter. *Adv. Stud. Contemp. Math. (Kyungshang)* **2018**, *28*, 285–296.

26. Yang, S.L. An identity of symmetry for the Bernoulli polynomials. *Discrete Math.* **2008**, *308*, 550–554. [CrossRef]

27. Zhang, Z.; Yang, H. Several identities for the generalized Apostol-Bernoulli polynomials. *Comput. Math. Appl.* **2008**, *56*, 2993–2999. [CrossRef]

28. Milnor, J.W. *Topology from the Differentiable View Point*; University of Virginia Press: Charlottesville, VA, USA, 1965.

29. Yilmaz, B.; Özarslan, M.A. Frobenius–Euler and Frobenius-Genocchi polynomials and their differential equations. *New Trends Math. Sci.* **2015**, *3*, 172–180.

30. Marin, M.; Florea, O. On temporal behaviour of solutions in thermoelasticity of porous micropolar bodies. *An. St. Univ. Ovidius Constanta-Ser. Math.* **2014**, *22*, 169–188. [CrossRef]

31. Marin, M. Weak solutions in elasticity of dipolar porous materials. *Math. Probl. Eng.* **2008**, *2008*, 158908. [CrossRef]

10

The Extended Minimax Disparity RIM Quantifier Problem

Dug Hun Hong

Department of Mathematics, Myongji University, Yongin Kyunggido 449-728, Korea; dhhong@mju.ac.kr

Abstract: An interesting regular increasing monotone (RIM) quantifier problem is investigated. Amin and Emrouznejad [Computers & Industrial Engineering 50(2006) 312–316] have introduced the extended minimax disparity OWA operator problem to determine the OWA operator weights. In this paper, we propose a corresponding continuous extension of an extended minimax disparity OWA model, which is the extended minimax disparity RIM quantifier problem, under the given orness level and prove it analytically.

Keywords: fuzzy sets; RIM quantifier; extended minimax disparity; OWA model; RIM quantifier problem

1. Introduction

One of the important topic in the theory of ordered weighted averaging (OWA) operators is the determination of the associated weights. Several authors have suggested a number of methods for obtaining associated weights in various areas such as decision making, approximate reasoning, expert systems, data mining, fuzzy systems and control [1–18]. Researchers can easily see most of OWA papers in the recent bibliography published in Emrouznejad and Marra [5]. Yager [16] proposed RIM quantifiers as a method for finding OWA weight vectors through fuzzy linguistic quantifiers. Liu [19] and Liu and Da [20] gave solutions to the maximum-entropy RIM quantifier model when the generating functions are differentiable. Liu and Lou [21] studied the equivalence of solutions to the minimax ratio and maximum-entropy RIM quantifier models, and the equivalence of solutions to the minimax disparity and minimum-variance RIM quantifier problems. Hong [22,23] gave the proof of the minimax ratio RIM quantifier problem and the minimax disparity RIM quantifier model when the generating functions are absolutely continuous. He also gave solutions to the maximum-entropy RIM quantifier model and the minimum-variance RIM quantifier model when the generating functions are Lebesgue integrable. Liu [24] proposed a general RIM quantifier determination model, proved it analytically using the optimal control method and investigated the solution equivalence to the minimax problem for the RIM quantifier. However, Hong [11] recently provided a modified model for the general RIM quantifier model and the correct formulation of Liu's result.

Amin and Emrouznejad [1] have introduced the following the extended minimax disparity OWA operator model to determine the OWA operator weights:

$$\text{Minimize} \quad \max_{i\in\{1,\cdots,n-1\},\, j\in\{i+1,\cdots,n\}} |w_i - w_j|$$

$$\text{subject to} \quad orness(W) = \sum_{i=1}^{n} \frac{n-i}{n-1} w_i = \alpha, \ 0 \le \alpha \le 1,$$

$$w_1 + \cdots + w_n = 1, 0 \le w_i, i = 1, \cdots, n.$$

In this paper, we propose a corresponding extended minimax disparity model for RIM quantifier determination under given orness level and prove it analytically. This paper is organized as follows: Section 2 presents the preliminaries and Section 3 reviews some models for the RIM quantifier problems and propose the extended minimax disparity model for the RIM quantifier problem. In Section 4, we prove the extended minimax disparity model problem mathematically for the case in which the generating functions are Lesbegue integrable functions.

2. Preliminaries

Yager [15] introduced a new aggregation technique based on the OWA operators. An OWA operator of dimension n is a function $F : R^n \rightarrow R$ that has an associated weighting vector $W = (w_1, \cdots, w_n)^T$ of having the properties $0 \leq w_i \leq 1$, $i = 1, \cdots, n$, $w_1 + \cdots + w_n = 1$, and such that

$$F(a_1, \cdots, a_n) = \sum_{i=1}^{n} w_i b_i,$$

where b_j is the jth largest element of the collection of the aggregated objects $\{a_1, \cdots, a_n\}$. In [15], Yager defined a measure of "orness" associated with the vector W of an OWA operator as

$$orness(W) = \sum_{i=1}^{n} \frac{n-i}{n-1} w_i,$$

and it characterizes the degree to which the aggregation is like an *or* operation.

The RIM quantifiers was introduced by Yager [16] as a method for obtaining the OWA weight vectors via fuzzy linguistic quantifiers. The RIM quantifiers can provide information aggregation procedures guided by a dimension independent description and verbally expressed concepts of the desired aggregation.

Definition 1 ([14]). *A fuzzy subset Q is called a **RIM quantifier** if $Q(0) = 0, Q(1) = 1$ and $Q(x) \geq Q(y)$ for $x > y$.*

The quantifier *for all* is represented by the fuzzy set

$$Q_*(r) = \begin{cases} 1, & x = 1, \\ 0, & x \neq 1. \end{cases}$$

The quantifier *there exist*, not none, is defined as

$$Q^*(r) = \begin{cases} 0, & x = 0, \\ 1, & x \neq 0. \end{cases}$$

Both of these are examples of RIM quantifier. To analyze the relationship between OWA and RIM quantifier, a generating function representation of RIM quantifier was proposed.

Definition 2. *For $f(t)$ on [0, 1] and a RIM quantifier $Q(x)$, $f(t)$ is called **generating function** of $Q(x)$, if it satisfies*

$$Q(x) = \int_0^x f(t)dt$$

where $f(t) \geq 0$ and $\int_0^1 f(t)dt = 1$.

If $Q(x)$ is an absolutely continuous function, then $f(x)$ is a Lesbegue integrable function; moreover, $f(x)$ is unique in the sense of "almost everywhere" in abbreviated form, *a.e.*

Yager extended the *orness* measure of OWA operator, and defined the *orness* of a RIM quantifier [16].

$$orness(Q) = \int_0^1 Q(x)dx = \int_0^1 (1-t)f(t)dt.$$

As the RIM quantifier can be seen as the continuous form of OWA operator with generating function, OWA optimization problem is extended to the RIM quantifier case.

The definitions of *essential supremum* and *essential infimum* [21] of f are as follows:

$$ess\ sup f = \inf\{t : |\{x \in [0,1] : f(x) > t\}| = 0\},$$

$$ess\ inf f = \sup\{t : |\{x \in [0,1] : f(x) < t\}| = 0\},$$

where $|E|$ is the Lebesgue measure of the Lebesgue measurable set E.

3. Models for the RIM Quantifier Problems

Fullér and Majlender [8] proposed the minimum variance model, which minimizes the variance of OWA operator weights under a given level of orness. Their method requires the proof of the following mathematical programming problem:

$$\text{Minimize} \quad D(W) = \frac{1}{n}\sum_{i=1}^{n-1}\left(w_i - \frac{1}{n}\right)^2$$

$$\text{subject to} \quad orness(W) = \sum_{i=1}^{n}\frac{n-i}{n-1}w_i = \alpha,\ 0 \le \alpha \le 1,$$

$$w_1 + \cdots + w_n = 1, 0 \le w_i, i = 1, \cdots, n.$$

Liu [19,24] extended the minimum variance problem for OWA operator to the RIM quantifier problem case:

$$\text{Minimize} \quad D_f = \int_0^1 f^2(r)dr - 1$$

$$\text{subject to} \quad \int_0^1 rf(r)dr = 1 - \alpha,\ 0 < \alpha < 1,$$

$$\int_0^1 f(r)dr = 1,\ f(r) \ge 0.$$

Wang and Parkan [13] proposed the minimax disparity problem as follows:

$$\text{Minimize} \quad \max_{i \in \{1,\cdots,n-1\}} |w_i - w_{i+1}|$$

$$\text{subject to} \quad orness(W) = \sum_{i=1}^{n}\frac{n-i}{n-1}w_i = \alpha,\ 0 \le \alpha \le 1,$$

$$w_1 + \cdots + w_n = 1, 0 \le w_i, i = 1, \cdots, n.$$

Similar to the minimax disparity OWA operator problem, Hong [11] proposed the minimax disparity RIM quantifier problem as follows:

$$\text{Minimize} \quad ess\ sup_{t \in [0,1]} |f'(t)|$$

$$\text{subject to} \quad \int_0^1 rf(r)dr = 1 - \alpha,\ 0 < \alpha < 1,$$

$$\int_0^1 f(r)dr = 1,\ \text{absolutely continuous } f(r) \ge 0.$$

Wang et al. [14] have introduced the following least squares deviation (LSD) method as an alternative approach to determine the OWA operator weights.

$$\text{Minimize} \quad \sum_{i=1}^{n-1} (w_i - w_{i-1})^2$$

$$\text{subject to} \quad orness(W) = \sum_{i=1}^{n} \frac{n-i}{n-1} w_i = \alpha, \ 0 \le \alpha \le 1,$$

$$w_1 + \cdots + w_n = 1, 0 \le w_i, i = 1, \cdots, n.$$

Hong [25] proposed the following corresponding least squares disparity RIM quantifier problem under a given orness level:

$$\text{Minimize} \quad D_f = \int_0^1 (f')^2(r) dr$$

$$\text{subject to} \quad \int_0^1 (1-r) f(r) dr = \alpha, \ 0 < \alpha < 1,$$

$$\int_0^1 f(r) dr = 1,$$

$$f(r) > 0.$$

Recently, Amin and Emrouznejad [1] proposed a problem of minimizing the maximum disparity of any distinct pairs of weights instead of adjacent weights. that is:

$$\text{Minimize} \quad \max_{i \in \{1,\cdots,n-1\}, j \in \{i+1,\cdots,n\}} |w_i - w_j| \qquad (1)$$

$$\text{subject to} \quad orness(W) = \sum_{i=1}^{n} \frac{n-i}{n-1} w_i = \alpha, \ 0 \le \alpha \le 1,$$

$$w_1 + \cdots + w_n = 1, 0 \le w_i, i = 1, \cdots, n.$$

We consider the following easy important fact.
Note

$$max_{i \in \{1,\cdots,n-1\}, j \in \{i+1,\cdots,n\}} |w_i - w_j| = max \ w_i - min \ w_i.$$

For this, first it is trivial that

$$max_{i \in \{1,\cdots,n-1\}, j \in \{i+1,\cdots,n\}} |w_i - w_j| \le max \ w_i - min \ w_i.$$

Next, suppose that $max \ w_i = w_{i_0}$, $min \ w_i = w_{j_0}$. If $i_0 < j_0$, then

$$\begin{aligned} max \ w_i - min \ w_i &= w_{i_0} - w_{j_0} \\ &= |w_{i_0} - w_{j_0}| \\ &\le max_{i \in \{1,\cdots,n-1\}, j \in \{i_0+1,\cdots,n\}} |w_i - w_j| \end{aligned}$$

If $i_0 > j_0$, then

$$\begin{aligned} max \ w_i - min \ w_i &= w_{i_0} - w_{j_0} \\ &= |w_{j_0} - w_{i_0}| \\ &\le max_{i \in \{1,\cdots,n-1\}, j \in \{j_0+1,\cdots,n\}} |w_i - w_j|. \end{aligned}$$

and hence the equality holds.

Then the corresponding extended minimax disparity model for RIM quantifier problem with given orness level can be proposed as follows:

$$\text{Minimize} \quad ess\ sup\ f - ess\ inf\ f \tag{2}$$

$$\text{subject to} \quad \int_0^1 r f(r)dr = 1 - \alpha, \ 0 < \alpha < 1,$$

$$\int_0^1 f(r)dr = 1, \ f(r) \geq 0.$$

4. Relation of Solutions between OWA Operator Model and RIM Quantifier Model

The following result is the solution of the extended minimax OWA operator problem given by Hong [26].

Theorem 1 ($n = 2k$:even). *An optimal weight for the constrained optimization problem (2) for a given level of* $\alpha = orness(W)$ *should satisfy the following equation:*

$$H(\alpha) = Minimize \left\{ \max_{i \in \{1, \cdots, n-1\},\ j \in \{i+1, \cdots, n\}} |w_i - w_j| \right\} = \left| \frac{(1 - 2\alpha)(n - 1)}{(n - m)m} \right|$$

$$w_1^* = w_2^* = \cdots = w_m^*, \ w_{k+1}^* = w_{k+2}^* = \cdots = w_n^*,$$

where

$$w_1^* = \frac{m - (1 - 2\alpha)(n - 1)}{nm}$$

and

$$w_{m+1}^* = \frac{n - m - (2\alpha - 1)(n - 1)}{n(n - m)}.$$

Here m satisfies the following:

$$m = \begin{cases} \lceil (1 - 2\alpha)(n - 1) \rceil, & if \quad 0 \leq \alpha \leq \frac{n-2}{4(n-1)}, \\ k, & if \quad \frac{n-2}{4(n-1)} \leq \alpha \leq \frac{3n-2}{4(n-1)}, \\ n - \lceil (2\alpha - 1)(n - 1) \rceil, & if \quad \frac{3n-2}{4(n-1)} \leq \alpha \leq 1. \end{cases}$$

where $\lceil x \rceil = m + 1 \iff m < x \leq m + 1$ *for any integer m.*

Can we get a hint about the solution of the extended minimax Rim quantifier problem? Here, we suggest an idea.

For a given associated weighting vector $W_n = (w_1, \cdots, w_n)$ of having the property $w_1 + \cdots + w_n = 1$, $0 \leq w_i \leq 1$, $i = 1, \cdots, n$, we define a generating function $f(t)$

$$f_{W_n}(x) = nw_i, \quad x \in \left[\frac{i}{n}, \frac{i+1}{n}\right), \quad i = 0, 1, \cdots, n - 1,$$

having the property $\int_0^1 f_W^n(x)dx = 1$ and let

$$f^*(x) = \lim_{n \to \infty} = f_{W_n}(x).$$

Can this function $f^*(x)$ be a solution of the corresponding extended minimax Rim quantifier problem? Maybe, yes! Let's try to follow this idea.

For given $W_n^* = (w_1^*, \cdots, w_n^*)$ from above Theorem 1, we have for $0 < \alpha \leq \frac{1}{4}$,

$$f_{W_n^*}(x) = \begin{cases} \frac{\lceil (1-2\alpha)(n-1)\rceil - (1-2\alpha)(n-1)}{\lceil (1-2\alpha)(n-1)\rceil}, & \text{if} \quad x \in \left[0, \frac{\lceil (1-2\alpha)(n-1)\rceil}{n}\right) \\ \frac{n - \lceil (1-2\alpha)(n-1)\rceil - (2\alpha-1)(n-1)}{n - \lceil (1-2\alpha)(n-1)\rceil}, & \text{if} \quad x \in \left[\frac{\lceil (1-2\alpha)(n-1)\rceil}{n}, 1\right] \end{cases}.$$

for $\frac{1}{4} \leq \alpha \leq \frac{3}{4}$,

$$f_{W_n^*}(x) = \begin{cases} \frac{n/2 - (1-2\alpha)(n-1)}{n/2}, & \text{if} \quad x \in \left[0, \frac{1}{2}\right) \\ \frac{n/2 - (2\alpha-1)(n-1)}{(n/2)}, & \text{if} \quad x \in \left[\frac{1}{2}, 1\right] \end{cases}.$$

for $3/4 \leq \alpha \leq 1$,

$$f_{W_n^*}(x) = \begin{cases} \frac{n - \lceil (2\alpha-1)(n-1)\rceil - (1-2\alpha)(n-1)}{n - \lceil (2\alpha-1)(n-1)\rceil}, & \text{if} \quad x \in \left[0, 1 - \frac{\lceil (1-2\alpha)(n-1)\rceil}{n}\right) \\ \frac{\lceil (2\alpha-1)(n-1)\rceil - (2\alpha-1)(n-1)}{\lceil (2\alpha-1)(n-1)\rceil}, & \text{if} \quad x \in \left[1 - \frac{\lceil (1-2\alpha)(n-1)\rceil}{n}, 1\right] \end{cases}.$$

Let $\lim_{n \to \infty} f_{W_n^*}(x) = f^*(x)$, then

1. for $0 < \alpha \leq \frac{1}{4}$,

$$f^*(r) = \begin{cases} 0, & \text{if} \quad r \in [0, 1-2\alpha), \\ \frac{1}{2\alpha}, & \text{if} \quad r \in [1-2\alpha, 1]. \end{cases}$$

2. for $\frac{1}{4} \leq \alpha \leq \frac{3}{4}$,

$$f^*(r) = \begin{cases} 4\alpha - 1, & \text{if} \quad r \in \left[0, \frac{1}{2}\right), \\ 3 - 4\alpha, & \text{if} \quad r \in \left[\frac{1}{2}, 1\right]. \end{cases}$$

3. for $\frac{3}{4} < \alpha \leq 1$,

$$f^*(r) = \begin{cases} \frac{1}{2(1-\alpha)}, & \text{if} \quad r \in [0, 2\alpha], \\ 0, & \text{elsewhere.} \end{cases}$$

In the following section, we will show that f^* can be the solution of the extended minimax RIM quantifier problem.

5. Proof of the Extended Minimax RIM Quantifier Problem

In this section, we prove the following main result.

Theorem 2. *The optimal solution for problem (2) for given orness level α is the weighting function f^* such that*

1. for $0 < \alpha \leq \frac{1}{4}$,

$$f^*(r) = \begin{cases} 0 \text{ a.e.,} & \text{if} \quad r \in [0, 1-2\alpha), \\ \frac{1}{2\alpha} \text{ a.e.,} & \text{if} \quad r \in [1-2\alpha, 1]. \end{cases}$$

2. for $\frac{1}{4} \leq \alpha \leq \frac{3}{4}$,

$$f^*(r) = \begin{cases} 4\alpha - 1 \text{ a.e.,} & \text{if} \quad r \in \left[0, \frac{1}{2}\right), \\ 3 - 4\alpha \text{ a.e.,} & \text{if} \quad r \in \left[\frac{1}{2}, 1\right]. \end{cases}$$

3.   *for $\frac{3}{4} < \alpha \leq 1$,*

$$f^*(r) = \begin{cases} \frac{1}{2(1-\alpha)} & a.e., & if \quad r \in [0, 2\alpha], \\ 0 & a.e., & elsewhere. \end{cases}$$

and

$$H(\alpha) = \text{Minimize} \ |ess\, sup f - ess\, inf f| = \begin{cases} \frac{1}{2\alpha} & if \quad 0 < \alpha \leq \frac{1}{4}, \\ 4|(1 - 2\alpha)| & if \quad \frac{1}{4} \leq \alpha \leq \frac{3}{4}, \\ \frac{1}{2\alpha} & if \quad \frac{3}{4} < \alpha \leq 1. \end{cases}$$

We need the following two lemma's to prove the main result. We denote $D_f(x) = \int_0^x f(t)dt$, $0 \leq x \leq 1$ and $E(f) = \int_0^1 rf(r)dr$.

The following result is known.

Lemma 1. $E(f) = \int_0^1 (1 - D_f(t))dt$.

Lemma 2. *Let ess inf $f = \beta_0 \geq 0$ and ess sup $f = \beta_1 > 0$ such that $\int_0^1 f(r)dr = 1$ and define a function f_0 as*

$$f_0(r) = \begin{cases} \beta_0 \ a.e., & if \quad r \in [0, c_0), \\ \beta_1 \ a.e., & if \quad r \in [c_0, 1]. \end{cases}$$

for some $c_0 \in (0, 1)$ such that $\int_0^1 f_0(r)dr = 1$. Then we have $E(f) \leq E(f_0)$ and the equality holds iff $f = f_0$ a.e.

Proof. The result follows immediately from Lemma 1 if we show that $D_{f_0}(x) \leq D_f(x), x \in [0, 1]$. It is clear that $D_{f_0}(x) \leq D_f(x), x \in [0, c_0]$. Suppose that there exists a point $t_0 \in (c_0, 1)$ such that $D_{f_0}(t_0) > D_f(t_0)$. Then

$$\int_{t_0}^1 \beta_1 dr = \int_{t_0}^1 f_0(r)dr = 1 - D_{f_0}(t_0) < 1 - D_f(t_0) = \int_{t_0}^1 f(r)dr$$

which implies *ess* $sup_{(t_0, 1)} f > \beta_1$. It is a contradiction. $\square$

Proof of Theorem 2. If $\alpha = \frac{1}{2}$, we clearly have the optimal solution is $f^*(r) = 1$ *a.e.* for $r \in [0, 1]$. Note that *ess inf $f^* < 1 < ess\, sup f^*$* for $\alpha \in \left(0, \frac{1}{2}\right)$. Without loss of generality, we can assume that $\alpha \in \left(0, \frac{1}{2}\right)$, since if a weighting function $f^*(r)$ is optimal to problem (2) for some given level of preference $\alpha \in \left(0, \frac{1}{2}\right]$, then $f^*(1-r)$ is optimal to the problem (2) for a given level of preference $1 - \alpha$. Indeed, since $D_f = D_{f^R}$, $\int_0^1 f(r)dr = \int_0^1 f^R(r)dr$ and $E(f^R) = 1 - E(f)$, where $f^R(r) = f(1 - r)$ hence for $\alpha > \frac{1}{2}$, we can consider problem (2) for the level of preference with index $1 - \alpha$, and then take the reverse of that optimal solution. We can easily check that the weighting functions, f^*, given above are feasible for problem (2). We show that f^* is the unique optimal solution for a given α. Let nonnegative function f satisfy $1 = \int_0^1 f(r)dr$ and $E(f) = \int_0^1 rf(r)dr = 1 - \alpha$. Let *ess inf $f = \beta_0$* and *ess sup $f = \beta_1$*.

Case (A): $\alpha \in \left(0, \frac{1}{4}\right]$.

We note that $ess\ inf\ f^* - ess\ inf\ f^* = \frac{1}{2\alpha}$. We will show that $\beta_1 - \beta_0 \geq \frac{1}{2\alpha}$. To show this, we define a function f_0 as

$$f_0(r) = \begin{cases} \beta_0 & \text{if} \quad r \in [0, x_0), \\ \beta_1 & \text{if} \quad r \in [x_0, 1], \end{cases}$$

for some $x_0 \in (0,1)$ such that $\int_0^1 f_0(r)dr = 1$. Then by Lemma 2, $E(f) \leq E(f_0)$. Suppose that $\beta_1 - \beta_0 < \frac{1}{2\alpha}$ and define another function f_0^* as

$$f_0^*(r) = \begin{cases} \beta_0 & \text{if} \quad r \in [0, x_0^*), \\ \beta_0 + \frac{1}{2\alpha} & \text{if} \quad r \in [x_0^*, 1], \end{cases}$$

for some $x_0^* \in (0,1)$ such that $\int_0^1 f_0^*(r)dr = 1$. Then $E(f_0) < E(f_0^*)$. We note that $1 = \beta_0 x_0^* + (1 - x_0^*)(\beta_0 + \frac{1}{2\alpha})$. Then

$$x_0^* = 2\alpha\beta_0 + 1 - 2\alpha. \tag{3}$$

We know that

$$\begin{aligned} E(f_0^*) &= \beta_0 \int_0^{x_0^*} x dx + \left(\beta_0 + \frac{1}{2\alpha}\right) \int_{x_0^*}^1 x dx \\ &= \frac{\beta_0}{2} + \frac{1}{4\alpha} - \frac{x_0^{*2}}{4\alpha} \end{aligned}$$

and

$$E(f^*) = \frac{1}{2\alpha} \int_{1-2\alpha}^1 x dx = 1 - \alpha.$$

And we have

$$\begin{aligned} E(f^*) - E(f_0^*) &= \frac{1}{2}\frac{x_0^{*2}}{2\alpha} - \frac{1}{2}\frac{(1-2\alpha)^2}{2\alpha} - \frac{\beta_0}{2} \\ &= \frac{1}{2}\left[\frac{1}{2\alpha}x_0^{*2} - \frac{(1-2\alpha)^2}{2\alpha} - \beta_0\right] \\ &= \frac{1}{2}\left[\frac{1}{2\alpha}(2\alpha\beta_0 + 1 - 2\alpha)^2 - \frac{(1-2\alpha)^2}{2\alpha} - \beta_0\right] \\ &= \frac{\beta_0}{2}[2\alpha\beta_0 + 2(1-2\alpha) - 1] \\ &\geq 0 \end{aligned}$$

where the third equality comes from (3) and the last inequality comes from the facts that $1 - 2\alpha \geq \frac{1}{2}$, $\beta_0 \geq 0$ and $\alpha > 0$. This proves $E(f) < E(f_0^*) \leq E(f^*) = 1 - \alpha$, which is a contradiction. Hence f^* is an optimal solution for the case of $\alpha \in \left(0, \frac{1}{4}\right]$.

Case (B): $\alpha \in \left(\frac{1}{4}, \frac{1}{2}\right)$.

We note that $ess\ inf\ f^* - ess\ inf\ f^* = 4(1 - 2\alpha)$. We will show that $\beta_1 - \beta_0 \geq 4(1 - 2\alpha)$. As in the Case (A), we define a function f_0 as

$$f_0(r) = \begin{cases} \beta_0 & \text{if} \quad r \in [0, x_0), \\ \beta_1 & \text{if} \quad r \in [x_0, 1], \end{cases}$$

for some $x_0 \in (0,1)$ such that $\int_0^1 f_0(r)dr = 1$. Then by lemma 2, $E(f) \leq E(f_0)$. Suppose that $\beta_1 - \beta_0 < \frac{1}{2\alpha}$ and define another function f_1^* as

$$
f_1^*(r) = \begin{cases} \beta_0 & \text{if} \quad r \in [0, x_1^*), \\ \beta_0 + 4(1 - 2\alpha) & \text{if} \quad r \in [x_1^*, 1], \end{cases}
$$

for some $x_1^* \in (0,1)$ such that $\int_0^1 f_1^*(r)dr = 1$. Then, since $x_0 < x_1^*$, by lemma 2 $E(f_0) < E(f_1^*)$. We note that $1 = \beta_0 x_1^* + (1 - x_1^*)(\beta_0 + 4(1 - 2\alpha))$. Then

$$
x_1^* = 1 + \frac{\beta_0 - 1}{4(1 - 2\alpha)}
$$

and

$$
x_1^{*2} = 1 + \frac{\beta_0 - 1}{2(1 - 2\alpha)} + \frac{(\beta_0 - 1)^2}{16(1 - 2\alpha)^2} \tag{4}
$$

We know that

$$
\begin{aligned}
E(f_0^*) &= \beta_0 \int_0^{x_1^*} x\,dx + (\beta_0 + 4(1 - 2\alpha)) \int_{x_1^*}^1 x\,dx \\
&= \frac{1}{2}[\beta_0 + 4(1 - 2\alpha)] - 2(1 - 2\alpha)x_1^{*2}
\end{aligned}
$$

and

$$
\begin{aligned}
E(f^*) &= (4\alpha - 1) \int_0^{\frac{1}{2}} x\,dx + (3 - 4\alpha) \int_{\frac{1}{2}}^1 x\,dx \\
&= 1 - \alpha.
\end{aligned}
$$

Then we have that

$$
\begin{aligned}
E(f^*) - E(f_1^*) &= 3\alpha - 1 - \frac{\beta_0}{2} + 2(1 - 2\alpha)x_1^{*2} \\
&= \frac{(\beta_0 - 1)^2}{8(1 - 2\alpha)} + \frac{\beta_0}{2} - \alpha \\
&= \frac{[\beta_0 - (4\alpha - 1)]^2}{8(1 - 2\alpha)} \\
&\geq 0
\end{aligned}
$$

where the second equality comes from (4) and hence $E(f) < E(f_1^*) \leq E(f^*) = 1 - \alpha$, which is a contradiction. This completes the proof. $\square$

6. Conclusions

Previous studies have suggested a number of methods for obtaining optimal solution of the RIM quantifier problem. This paper proposes the extended minimax disparity RIM quantifier problem under a given orness level. We completely prove it analytically.

References

1. Amin, G.R.; Emrouznejad, A. An extended minimax disparity to determine the OWA operator weights. *Comput. Ind. Eng.* **2006**, *50*, 312–316. [CrossRef]
2. Amin, G.R. Notes on priperties of the OWA weights determination model. *Comput. Ind. Eng.* **2007**, *52*, 533–538. [CrossRef]
3. Emrouznejad, A.; Amin, G.R. Improving minimax disparity model to determine the OWA operator weights. *Inf. Sci.* **2010**, *180*, 1477–1485. [CrossRef]
4. Emrouznejad, A. MP-OWA: The most preferred OWA operator. *Knowl. Based Syst.* **2008**, *21*, 847–851. [CrossRef]
5. Emrouznejad, A.; Marra, M. Ordered Weighted Averaging Operators 1988-2014: A citation-based literature survey. *Int. J. Intell. Syst.* **2014**, *29*, 994–1014. [CrossRef]
6. Filev, D.; Yager, R.R. On the issue of obtaining OWA operator weights. *Fuzzy Sets Syst.* **1988**, *94*, 157–169. [CrossRef]
7. Fullér, R.; Majlender, P. An analytic approach for obtaining maximal entropy OWA operators weights. *Fuzzy Sets Syst.* **2001**, *124*, 53–57. [CrossRef]
8. Fullér, R.; Majlender, P. On obtaining minimal variability OWA operator weights. *Fuzzy Sets Syst.* **2003**, *136*, 203–215. [CrossRef]
9. O'Hagan, M. Aggregating template or rule antecedents in real-time expert systems with fuzzy set logic. In Proceedings of the 22nd annual IEEE Asilomar Conf. on Signals, Systems, Computers, Pacific Grove, CA, USA, 31 October–2 November 1988; pp. 681–689.
10. Hong, D.H. A note on the minimal variability OWA operator weights. *Int. J. Uncertainty, Fuzziness Knowl. Based Syst.* **2006**, *14*, 747–752. [CrossRef]
11. Hong, D.H. A note on solution equivalence to general models for RIM quantifier problems. *Fuzzy Sets Syst.* **2018**, *332*, 25–28. [CrossRef]
12. Wheeden, R.L.; Zygmund, A. *Measure and Integral: An Introduction to Real Analysis*; Marcel Dekker, Inc.: New York, NY, USA, 1977.
13. Wang, Y.M.; Parkan, C. A minimax disparity approach obtaining OWA operator weights. *Inf. Sci.* **2005**, *175*, 20–29. [CrossRef]
14. Wang, Y.M.; Luo, Y.; Liu, X. Two new models for determining OWA operater weights. *Comput. Ind. Eng.* **2007**, *52*, 203–209. [CrossRef]
15. Yager, R.R. Ordered weighted averaging aggregation operators in multi-criteria decision making. *IEEE Trans. Syst. Man Cybern.* **1988**, *18*, 183–190. [CrossRef]
16. Yager, R.R. OWA aggregation over a continuous interval argument with application to decision making. *IEEE Trans. Syst. Man Cybern. Part B* **2004**, *34*, 1952–1963. [CrossRef]
17. Yager, R.R. Families of OWA operators. *Fuzzy Sets Syst.* **1993**, *59*, 125–148. [CrossRef]
18. Yager, R.R.; Filev, D. Induced ordered weighted averaging operators. *IEEE Trans. Syst. Man Cybern. Part B Cybern.* **1999**, *29*, 141–150. [CrossRef]
19. Liu, X. On the maximum entropy parameterized interval approximation of fuzzy numbers. *Fuzzy Sets Syst.* **2006**, *157*, 869–878. [CrossRef]
20. Liu, X.; Da, Q. On the properties of regular increasing monotone (RIM) quantifiers with maximum entropy. *Int. J. Gen. Syst.* **2008**, *37*, 167–179. [CrossRef]
21. Liu, X.; Lou, H. On the equivalence of some approaches to the OWA operator and RIM quantifier determination. *Fuzzy Sets Syst.* **2007**, *159*, 1673–1688. [CrossRef]
22. Hong, D.H. The relationship between the minimum variance and minimax disparity RIM quantifier problems. *Fuzzy Sets Syst.* **2011**, *181*, 50–57. [CrossRef]
23. Hong, D.H. The relationship between the maximum entropy and minimax ratio RIM quantifier problems. *Fuzzy Sets Syst.* **2012**, *202*, 110–117. [CrossRef]
24. Liu, X. A general model of parameterized OWA aggregation with given orness level. *Int. J. Approx. Reason.* **2008**, *48*, 598–627. [CrossRef]
25. Hong, D.H. The general model for least square disparity RIM quantifier problems. *Fuzzy Optim. Decis. Mak.* **2019**, submitted.
26. Hong, D.H. On proving the extended minimax disparity OWA problem. *Fuzzy Sets Syst.* **2011**, *168*, 35–46. [CrossRef]

11

Some Symmetric Identities Involving Fubini Polynomials and Euler Numbers

Zhao Jianhong [1] **and Chen Zhuoyu** [2,*]

[1] Department of Teachers Education, Lijiang Teachers College, Lijiang 674199, China; zjh3004@163.com
[2] School of Mathematics, Northwest University, Xi'an 710127, China
* Correspondence: chenzymath@163.com

Abstract: The aim of this paper is to use elementary methods and the recursive properties of a special sequence to study the computational problem of one kind symmetric sums involving Fubini polynomials and Euler numbers, and give an interesting computational formula for it. At the same time, we also give a recursive calculation method for the general case.

Keywords: Fubini polynomials; Euler numbers; symmetric identities; elementary method; computational formula

MSC: 11B83; 11B37

1. Introduction

For any integer $n \geq 0$, the Fubini polynomials $\{F_n(y)\}$ are defined by the coefficients of the generating function

$$\frac{1}{1 - y(e^t - 1)} = \sum_{n=0}^{\infty} \frac{F_n(y)}{n!} \cdot t^n, \tag{1}$$

where $F_0(y) = 1$, $F_1(y) = y$, and so on. $F_n(1) = F_n$ are called Fubini numbers. These polynomials and numbers are closely connected with the Stirling numbers. Some contents and properties of Stirling numbers can be found in reference [1]. T. Kim et al. [2] proved the identity

$$F_n(y) = \sum_{k=0}^{n} S_2(n,k)\, k!\, y^k, \ (n \geq 0),$$

where $S_2(n,k)$ are the Stirling numbers of the second kind. It not only associated Fubini polynomials with Stirling numbers, but also stressed the importance of researching Fubini polynomials.

Please note that the identity (see [3,4])

$$\frac{2e^{tx}}{1 + e^t} = \sum_{n=0}^{\infty} \frac{E_n(x)}{n!} \cdot t^n, \tag{2}$$

where $E_n(x)$ signifies the Euler polynomials.

It is distinct that if taking $y = -\frac{1}{2}$ in (1) and $x = 0$ in (2), then from (1) and (2) we can get the identity

$$E_n(0) = F_n\left(-\frac{1}{2}\right), n \in N^*0 \tag{3}$$

where $E_n(0) = E_n$ is the Euler number (see [5] for related contents).

On the other hand, two variable Fubini polynomials are defined by means of the following (see [2,6])

$$\frac{e^{xt}}{1 - y(e^t - 1)} = \sum_{n=0}^{\infty} \frac{F_n(x,y)}{n!} \cdot t^n,$$

and $F_n(y) = F_n(0, y)$ for all integers $n \geq 0$. About the properties of $F_n(x, y)$, several scholars have also researched it, especially T. Kim and others have done a large amount of vital works. For instance, they proved a series of identities linked to $F_n(x, y)$ (see [2,7]), one of which is

$$F_n(x,y) = \sum_{l=0}^{n} \binom{n}{l} x^l \cdot F_{n-l}(y), n \in N^*0.$$

These polynomials occupy indispensable positions in the theory and application of mathematics. In particular, they are widely used in combinatorial mathematics. Therefore, several scholars have researched their various properties, and acquired a series of vital results. Some involved contents can be found in references [5,7–17].

The goal of this paper is to use elementary methods and recursive properties of a special sequenc to research the computational problem of the sums

$$\sum_{a_1+a_2+\cdots+a_k=n} \frac{F_{a_1}(y)}{(a_1)!} \cdot \frac{F_{a_2}(y)}{(a_2)!} \cdots \frac{F_{a_k}(y)}{(a_k)!}, \tag{4}$$

where the summation is over all k-tuples with non-negative integer coordinates $(a_1, a_2, \cdots, a_k)$ such that $a_1 + a_2 + \cdots + a_k = n$.

About this content, it seems there is no valid method to solve the computational problem of (4). However, this problem is significant, it can reveal the structure of Fubini polynomials itself and its internal relations, at least it can reflect the combination properties of Fubini polynomials.

In this paper, we will take elementary methods and the properties of $F_n(y)$ to obtain a fascinating computational formula for (4). Simultaneously, we can also acquire a recursive calculation method for the general case. That is, we are going to prove the following major result:

Theorem 1. *For any positive integers n and k, we have the identity*

$$\sum_{a_1+a_2+\cdots+a_k=n} \frac{F_{a_1}(y)}{(a_1)!} \cdot \frac{F_{a_2}(y)}{(a_2)!} \cdots \frac{F_{a_k}(y)}{(a_k)!}$$

$$= \frac{1}{(k-1)!(y+1)^{k-1}} \cdot \frac{1}{n!} \sum_{i=0}^{k-1} C(k-1,i)F_{n+k-1-i}(y),$$

where the sequence $\{C(k,i)\}$ is defined as follows: For any positive integer k and integers $0 \leq i \leq k$, we define $C(k,0) = 1, C(k,k) = k!$ and

$$C(k+1,i+1) = C(k,i+1) + (k+1)C(k,i), \text{ for all } 0 \leq i < k,$$

providing $C(k,i) = 0$, if $i > k$.

The characteristic of this theorem is to represent a complex sum of Fubini polynomials as a linear combination of a single Fubini polynomial. Of course, our method can also be further generalized, provided a corresponding results for $F_n(x,y)$. It is just that its form is not so pretty, so we are not listing it here. If taking $k = 3, 4$ and 5, then from our theorem we may instantly deduce the following several corollaries:

Corollary 1. *For any positive integer n, we have the identity*

$$\sum_{a+b+c=n} \frac{F_a(y)}{a!} \cdot \frac{F_b(y)}{b!} \cdot \frac{F_c(y)}{c!} = \frac{1}{2 \cdot n! \cdot (y+1)^2} \left(F_{n+2}(y) + 3F_{n+1}(y) + 2F_n(y)\right).$$

Corollary 2. *For any positive integer n, we have the identity*

$$\sum_{a+b+c+d=n} \frac{F_a(y)}{a!} \cdot \frac{F_b(y)}{b!} \cdot \frac{F_c(y)}{c!} \cdot \frac{F_d(y)}{d!}$$

$$= \frac{1}{6 \cdot n! \cdot (y+1)^3} \left(F_{n+3}(y) + 6F_{n+2}(y) + 11F_{n+1}(y) + 6F_n(y)\right).$$

Corollary 3. *For any positive integer n, we have the identity*

$$\sum_{a+b+c+d+e=n} \frac{F_a(y)}{a!} \cdot \frac{F_b(y)}{b!} \cdot \frac{F_c(y)}{c!} \cdot \frac{F_d(y)}{d!} \cdot \frac{F_e(y)}{e!}$$

$$= \frac{1}{24 \cdot n! \cdot (y+1)^4} \left(F_{n+4}(y) + 10F_{n+3}(y) + 35F_{n+2}(y) + 50F_{n+1}(y) + 24F_n(y)\right).$$

If taking $y = -\frac{1}{2}$ in our theorem, then from (3) we can also infer the following:

Corollary 4. *For any positive integers n and $k \geq 2$, we have the identity*

$$\sum_{a_1+a_2+\cdots+a_k=n} \frac{E_{a_1}}{(a_1)!} \cdot \frac{E_{a_2}}{(a_2)!} \cdots \frac{E_{a_k}}{(a_k)!} = \frac{2^{k-1}}{(k-1)!} \cdot \frac{1}{n!} \sum_{i=0}^{k-1} C(k-1,i) E_{n+k-1-i}.$$

If $n = p$ is an odd prime, then taking $y = 1$ in Corollarys 1 and 2, we also have the following congruences.

Corollary 5. *For any odd prime p, we have the congruence*

$$22F_p \equiv F_{p+2} + 3F_{p+1} \pmod{p}.$$

Corollary 6. *For any odd prime p, we have the congruence*

$$186F_p \equiv F_{p+3} + 6F_{p+2} + 11F_{p+1} \pmod{p}.$$

2. A Simple Lemma

For purpose of proving our theorem, we need a uncomplicated lemma. As a matter of convenience, we first present a new sequence $\{C(k,i)\}$ as follows. For any positive integer k and integers $0 \leq i \leq k$, we define $C(k,0) = 1$, $C(k,k) = k!$ and

$C(k+1, i+1) = C(k, i+1) + (k+1)C(k,i)$, $1 \leq i \leq k$, $C(k,i) = 0$, if $i > k$.

For clarity, for $1 \leq k \leq 9$, we list values of $C(k,i)$ in the Table 1.

Table 1. Values of $C(k, i)$.

$C(k,i)$	$i=0$	$i=1$	$i=2$	$i=3$	$i=4$	$i=5$	$i=6$	$i=7$	$i=8$	$i=9$
$k=1$	1	1								
$k=2$	1	3	2							
$k=3$	1	6	11	6						
$k=4$	1	10	35	50	24					
$k=5$	1	15	85	225	274	120				
$k=6$	1	21	175	735	1624	1764	720			
$k=7$	1	28	322	1960	6769	13,132	13,068	5040		
$k=8$	1	36	546	4536	22,449	67,284	118,124	109,584	40,320	
$k=9$	1	45	870	9450	63,273	269,325	723,680	1,172,700	1,026,576	362,880

Obviously, the values of $C(k, i)$ can be easily calculated by using a computer program. Hence, for any positive integer k, the computational problem of (4) can be solved fully.

In this table of numerical values, we also find that for prime $p = 3, 5$ and 7, we have the congruence

$$C(p - 1, i) \equiv 0 \,(\mathrm{mod}\, p) \quad \text{for all } 1 \le i \le p - 2.$$

For all prime $p > 7$ is true? This is an enjoyable open problem.

If this congruence is true, then we can also deduce that for any positive integer n and odd prime p, one has the congruence

$$F_{n+p-1}(y) + F_n(y) \equiv 0 \,(\mathrm{mod}\, p).$$

Now let function $f(t) = \frac{1}{1-y(e^t-1)}$. Then we have the following

Lemma 1. *For any positive integer k, we have the identity*

$$\sum_{i=0}^{k} C(k, i) f^{(k-i)}(t) = k!(y + 1)^k f^{k+1}(t),$$

where $f^{(0)}(t) = f(t)$, $f^{(r)}(t)$ denotes the r-order derivative of $f(t)$ for variable t.

Proof. Now we prove this lemma by induction. From the definition of the derivative we acquire

$$f'(t) = \frac{ye^t}{(1 - y(e^t - 1))^2} = -f(t) + (y + 1)f^2(t) \tag{5}$$

or

$$f'(t) + f(t) = (y + 1)f^2(t). \tag{6}$$

Please note that $C(1, 0) = 1$ and $C(1, 1) = 1$, so the lemma is true for $k = 1$.
Suppose that the lemma is true for all integer $k \ge 1$. That is,

$$\sum_{i=0}^{k} C(k, i) f^{(k-i)}(t) = k!(y + 1)^k f^{k+1}(t). \tag{7}$$

Then take the derivative for t in (7) and applying (5) and (7) we obtain

$$\begin{aligned}
\sum_{i=0}^{k} C(k, i) f^{(k+1-i)}(t) &= (k + 1)!(y + 1)^k f^k(t) \cdot f'(t) \\
&= (k + 1)!(y + 1)^k f^k(t) \cdot \left(-f(t) + (y + 1)f^2(t)\right) \\
&= (k + 1)!(y + 1)^{k+1} f^{k+2}(t) - (k + 1)!(y + 1)^k f^{k+1}(t) \\
&= (k + 1)!(y + 1)^{k+1} f^{k+2}(t) - (k + 1)\left(\sum_{i=0}^{k} C(k, i) f^{(k-i)}(t)\right).
\end{aligned} \tag{8}$$

It is evident that (8) implies

$$
\begin{aligned}
&(k+1)!(y+1)^{k+1}f^{k+2}(t)\\
=\ &C(k,0)f^{(k+1)}(t)+\sum_{i=0}^{k-1}\left(C(k,i+1)+(k+1)C(k,i)\right)f^{(k-i)}(t)+(k+1)!f(t)\\
=\ &C(k,0)f^{(k+1)}(t)+\sum_{i=0}^{k-1}C(k+1,i+1)f^{(k-i)}(t)+(k+1)!f(t)\\
=\ &\sum_{i=0}^{k+1}C(k+1,i)f^{(k+1-i)}(t),
\end{aligned}
\tag{9}
$$

where we have used the identities $C(k,0)=1$ and $C(k,k)=k!$. Now the lemma follows from (9) and mathematical induction. $\square$

3. Proof of the Theorem

In this section, the proof of our theorem will be completed. Firstly, for any positive integer k, from the definition of $f(t)$ and the properties of the power series we obtain

$$
f^{(k)}(t)=\sum_{n=0}^{\infty}\frac{F_{n+k}(y)}{n!}\cdot t^n
\tag{10}
$$

and

$$
\begin{aligned}
f^k(t)&=\left(\sum_{a_1=0}^{\infty}\frac{F_{a_1}(y)}{a_1!}\cdot t^{a_2}\right)\left(\sum_{a_2=0}^{\infty}\frac{F_{a_2}(y)}{a_2!}\cdot t^{a_1}\right)\cdots\left(\sum_{a_k=0}^{\infty}\frac{F_{a_k}(y)}{a_k!}\cdot t^{a_k}\right)\\
&=\left(\sum_{a_1=0}^{\infty}\sum_{a_2=0}^{\infty}\cdots\sum_{a_k=0}^{\infty}\frac{F_{a_1}(y)}{(a_1)!}\cdot\frac{F_{a_2}(y)}{(a_2)!}\cdot\ldots\cdot\frac{F_{a_k}(y)}{(a_k)!}\cdot t^{a_1+a_2\cdots+a_k}\right)\\
&=\sum_{n=0}^{\infty}\left(\sum_{a_1+a_2+\cdots+a_k=n}\frac{F_{a_1}(y)}{(a_1)!}\cdot\frac{F_{a_2}(y)}{(a_2)!}\cdots\frac{F_{a_k}(y)}{(a_k)!}\right)\cdot t^n.
\end{aligned}
\tag{11}
$$

From (10), (11) and Lemma we acquire

$$
\begin{aligned}
&\frac{1}{(k-1)!(y+1)^{k-1}}\cdot\sum_{i=0}^{k-1}C(k-1,i)\sum_{n=0}^{\infty}\frac{F_{n+k-1-i}(y)}{n!}\cdot t^n\\
=\ &\sum_{n=0}^{\infty}\left(\sum_{a_1+a_2+\cdots+a_k=n}\frac{F_{a_1}(y)}{(a_1)!}\cdot\frac{F_{a_2}(y)}{(a_2)!}\cdots\frac{F_{a_k}(y)}{(a_k)!}\right)\cdot t^n.
\end{aligned}
\tag{12}
$$

Comparing the coefficients of t^n in (12) we have the identity

$$
\begin{aligned}
&\sum_{a_1+a_2+\cdots+a_k=n}\frac{F_{a_1}(y)}{(a_1)!}\cdot\frac{F_{a_2}(y)}{(a_2)!}\cdots\frac{F_{a_k}(y)}{(a_k)!}\\
=\ &\frac{1}{(k-1)!(y+1)^{k-1}}\cdot\frac{1}{n!}\sum_{i=0}^{k-1}C(k-1,i)F_{n+k-1-i}(y).
\end{aligned}
$$

This completes the proof of our Theorem.

Author Contributions: Writing-original draft: J.Z.; Writing-review and editing: Z.C.

Acknowledgments: The author would like to thank the referees for their very helpful and detailed comments, which have significantly improved the presentation of this paper.

References

1. Feng, R.-Q.; Song, C.-W. *Combinatorial Mathematics*; Beijing University Press: Beijing, China, 2015.
2. Kim, T.; Kim, D.S.; Jang, G.-W. A note on degenerate Fubini polynomials. *Proc. Jangjeon Math. Soc.* **2017**, *20*, 521–531.
3. Kim, T. Symmetry of power sum polynomials and multivariate fermionic p-adic invariant integral on Z_p. *Russ. J. Math. Phys.* **2009**, *16*, 93–96. [CrossRef]
4. Kim, D.S.; Park, K.H. Identities of symmetry for Bernoulli polynomials arising from quotients of Volkenborn integrals invariant under S_3. *Appl. Math. Comput.* **2013**, *219*, 5096–5104. [CrossRef]
5. Zhang, W. Some identities involving the Euler and the central factorial numbers. *Fibonacci Q.* **1998**, *36*, 154–157.
6. Kilar, N.; Simesk, Y. A new family of Fubini type numbrs and polynomials associated with Apostol-Bernoulli nujmbers and polynomials. *J. Korean Math. Soc.* **2017**, *54*, 1605–1621.
7. Kim, T.; Kim, D. S.; Jang, G.-W.; Kwon, J. Symmetric identities for Fubini polynomials. *Symmetry* **2018**, *10*, 219. [CrossRef]
8. Kim, T.; Kim, D.S. An identity of symmetry for the degernerate Frobenius-Euler polynomials. *Math. Slovaca* **2018**, *68*, 239–243. [CrossRef]
9. He, Y. Symmetric identities for Calitz's q-Bernoulli numbers and polynomials. *Adv. Differ. Equ.* **2013**, *2013*, 246. [CrossRef]
10. Rim, S.-H.; Jeong, J.-H.; Lee, S.-J.; Moon, E.-J.; Jin, J.-H. On the symmetric properties for the generalized twisted Genocchi polynomials. *ARS Comb.* **2012**, *105*, 267–272.
11. Yi, Y.; Zhang, W. Some identities involving the Fibonacci polynomials. *Fibonacci Q.* **2002**, *40*, 314–318.
12. Ma, R.; Zhang, W. Several identities involving the Fibonacci numbers and Lucas numbers. *Fibonacci Q.* **2007**, *45*, 164–170.
13. Wang, T.; Zhang, W. Some identities involving Fibonacci, Lucas polynomials and their applications. *Bull. Math. Soc. Sci. Math. Roum.* **2012**, *55*, 95–103.
14. Chen, L.; Zhang, W. Chebyshev polynomials and their some interesting applications. *Adv. Differ. Equ.* **2017**, *2017*, 303.
15. Li, X.X. Some identities involving Chebyshev polynomials. *Math. Probl. Eng.* **2015**, *2015*, 950695. [CrossRef]
16. Ma, Y.; Lv, X.-X. Several identities involving the reciprocal sums of Chebyshev polynomials. *Math. Probl. Eng.* **2017**, *2017*, 4194579. [CrossRef]
17. Clemente, C. Identities and generating functions on Chebyshev polynomials. *Georgian Math. J.* **2012**, *19*, 427–440.

12

A Note on Parametric Kinds of the Degenerate Poly-Bernoulli and Poly-Genocchi Polynomials

Taekyun Kim [1], Waseem A. Khan [2], Sunil Kumar Sharma [3],* and Mohd Ghayasuddin [4]

[1] Department of Mathematics, Kwangwoon University, Seoul 139-701, Korea; tkkim@kw.ac.kr
[2] Department of Mathematics and Natural Sciences, Prince Mohammad Bin Fahd University, P.O Box 1664, Al Khobar 31952, Kingdom of Saudi Arabia; wkhan1@pmu.edu.sa
[3] College of Computer and Information Sciences, Majmaah University, Majmaah 11952, Saudi Arabia
[4] Department of Mathematics, Integral University Campus, Shahjahanpur 242001, India; ghayas.maths@gmail.com
* Correspondence: s.sharma@mu.edu.sa

Abstract: Recently, the parametric kind of some well known polynomials have been presented by many authors. In a sequel of such type of works, in this paper, we introduce the two parametric kinds of degenerate poly-Bernoulli and poly-Genocchi polynomials. Some analytical properties of these parametric polynomials are also derived in a systematic manner. We will be able to find some identities of symmetry for those polynomials and numbers.

Keywords: degenerate poly-Bernoulli polynomials; degenerate poly-Genocchi polynomials; stirling numbers

1. Introduction

Special functions, polynomials and numbers play a prominent role in the study of many areas of mathematics, physics and engineering. In particular, the Appell polynomials and numbers are frequently used in the development of pure and applied mathematics related to functional equations in differential equations, approximation theories, interpolation problems, summation methods, quadrature rules and their multidimensional extensions (see [1]).The sequence of Appell polynomials $A_j(z)$ can be signified as follows:

$$\frac{d}{dz}A_j(z) = jA_{j-1}(z), \quad A_0(z) \neq 0, z = \eta + i\xi \in \mathbb{C}, \quad j \in \mathbb{N}, \tag{1}$$

or equivalently

$$A(z)e^{\eta z} = \sum_{j=0}^{\infty} A_j(\eta)\frac{z^j}{j!}, \tag{2}$$

where

$$A(z) = A_0 + A_1\frac{z}{1!} + A_2\frac{z^2}{2!} + \cdots + A_j\frac{z^j}{j!} + \cdots, \quad A_0 \neq 0,$$

is a formal power series with coefficients A_j known as Appell numbers.

The well known degenerate exponential function is defined by (see [2])

$$e_\mu^\eta(z) = (1 + \mu z)^{\frac{\eta}{\mu}}, \quad e_\mu(z) = e_\mu^1(z), (\mu \in \mathbb{R}). \tag{3}$$

In 1956 and 1979, Carlitz [3,4] introduced and investigated the following degenerate Bernoulli and Euler polynomials:

$$\frac{z}{e_\mu(z) - 1} e_\mu^\eta(z) = \frac{z}{(1 + \mu z)^{\frac{1}{\mu}} - 1}(1 + \mu z)^{\frac{\eta}{\mu}} = \sum_{s=0}^{\infty} \beta_s(\eta; \mu)\frac{z^s}{s!}, \tag{4}$$

and

$$\frac{2}{e_\mu(z) + 1} e_\mu^\eta(z) = \frac{2}{(1 + \mu z)^{\frac{1}{\mu}} - 1}(1 + \mu z)^{\frac{\eta}{\mu}} = \sum_{s=0}^{\infty} \mathfrak{E}_s(\eta; \mu)\frac{z^s}{s!}. \tag{5}$$

Note that

$$\lim_{\mu \longrightarrow 0} \beta_s(\eta; \mu) = B_s(\eta), \quad \lim_{\mu \longrightarrow 0} \mathfrak{E}_s(\eta; \mu) = E_s(\eta),$$

where $B_s(\eta)$ and $E_s(\eta)$ are the classical Bernoulli and Euler polynomials (see [5,6]).

Lim [7] introduced the degenerate Genocchi polynomials $G_j^{(p)}(\eta; \mu)$ of order p by means of the undermentioned generating function:

$$\left(\frac{2z}{e_\mu(z) + 1}\right)^p e_\mu^\eta(z) = \left(\frac{2z}{(1 + \mu z)^{\frac{1}{\mu}} - 1}\right)^p (1 + \mu z)^{\frac{\eta}{\mu}} = \sum_{j=0}^{\infty} G_j^{(p)}(\eta; \mu)\frac{z^j}{j!}, \tag{6}$$

so that

$$G_j^{(p)}(\eta; \mu) = \sum_{s=0}^{j} \binom{j}{s} G_s^{(p)}(\mu) \left(\frac{\eta}{\mu}\right)_{j-s}. \tag{7}$$

From Equation (6), we note that

$$\lim_{\mu \longrightarrow 0} \sum_{s=0}^{\infty} G_j^{(p)}(\eta; \mu)\frac{z^j}{j!} = \lim_{\mu \longrightarrow 0} \left(\frac{2z}{(1 + \mu z)^{\frac{1}{\mu}} - 1}\right)^p (1 + \mu z)^{\frac{\eta}{\mu}}$$

$$= \left(\frac{2z}{e^z + 1}\right)^p e^{\eta z} = \sum_{j=0}^{\infty} G_j^{(p)}(\eta)\frac{z^j}{j!},$$

where $G_j^{(p)}(\eta)$ are the generalized Genocchi polynomials of order p (see [8–11]).

The degenerate poly-Bernoulli and poly-Genocchi polynomials are defined by (see [12–14])

$$\frac{\mathrm{Li}_k(1 - e^{-z})}{e_\mu(z) - 1} e_\mu^\eta(z) = \frac{\mathrm{Li}_k(1 - e^{-z})}{(1 + \mu z)^{\frac{1}{\mu}} - 1}(1 + \mu z)^{\frac{\eta}{\mu}} = \sum_{s=0}^{\infty} B_s^{(k)}(\eta; \mu)\frac{z^s}{s!}, (k \in \mathbb{Z}), \tag{8}$$

and

$$\frac{2\mathrm{Li}_k(1 - e^{-z})}{e_\mu(z) + 1} e_\mu^\eta(z) = \frac{2\mathrm{Li}_k(1 - e^{-z})}{(1 + \mu z)^{\frac{1}{\lambda}} + 1}(1 + \mu z)^{\frac{\eta}{\mu}} = \sum_{s=0}^{\infty} G_s^{(k)}(\eta; \mu)\frac{z^s}{s!}, (k \in \mathbb{Z}). \tag{9}$$

Here, we note that (see [5,15]).

$$\lim_{\mu \longrightarrow 0} B_s^{(k)}(\eta; \mu) = B_s^{(k)}(\eta), \quad \lim_{\mu \longrightarrow 0} G_s^{(k)}(\eta; \mu) = G_s^{(k)}(\eta),$$

The Stirling numbers of the first kind are given by (see, [16–18])

$$(a)_s = a(a-1)\cdots(a-s+1) = \sum_{k=0}^{s} S^{(1)}(s,k)a^k, \, (k \geq 0), \tag{10}$$

and the Stirling numbers of the second kind are defined by (see [19,20])

$$a^s = \sum_{k=0}^{s} S^{(2)}(k,s)(a)_k. \tag{11}$$

The degenerate Stirling numbers of the of the second kind are defined by (see [10,21,22])

$$\frac{1}{k!}(e_\mu(t)-1)^k = \sum_{k=s}^{\infty} S_\mu^{(2)}(k,s)\frac{t^k}{k!}, \, (k \geq 0). \tag{12}$$

Note that $\lim_{\mu \to 0} S_\mu^{(2)}(k,s) = S^{(2)}(k,s), (s,k \geq 0)$.

In the year (2017, 2018), Jamei et al. [23,24] introduced the two parametric kinds of exponential functions as follows (see also [6,23–25]):

$$e^{\eta z}\cos \xi z = \sum_{k=0}^{\infty} C_k(\eta,\xi)\frac{z^k}{k!}, \tag{13}$$

and

$$e^{\eta z}\sin \xi z = \sum_{k=0}^{\infty} S_k(\eta,\xi)\frac{z^k}{k!}, \tag{14}$$

where

$$C_k(\eta,\xi) = \sum_{j=0}^{[\frac{k}{2}]} \binom{k}{2j} (-1)^j \eta^{k-2j}\xi^{2j}, \tag{15}$$

and

$$S_k(\eta,\xi) = \sum_{j=0}^{[\frac{k-1}{2}]} \binom{k}{2j+1} (-1)^j \eta^{k-2j-1}\xi^{2j+1}. \tag{16}$$

Recently, Kim et al. [2] introduced the following degenerate type parametric exponential functions:

$$e_\mu^\eta(z)\cos_\mu^\xi(z) = \sum_{k=0}^{\infty} C_{k,\mu}(\eta,\xi)\frac{z^k}{k!}, \tag{17}$$

and

$$e_\mu^\eta(z)\sin_\mu^\xi(z) = \sum_{k=0}^{\infty} S_{k,\mu}(\eta,\xi)\frac{z^k}{k!}, \tag{18}$$

where

$$C_{r,\mu}(\eta,\xi) = \sum_{k=0}^{[\frac{r}{2}]} \sum_{q=2k}^{r} \binom{r}{q} (-1)^k \mu^{q-2k}\xi^{2k}S^1(q,2k)(\eta)_{r-q,\mu}, \tag{19}$$

and

$$S_{r,\mu}(\eta,\xi) = \sum_{k=0}^{[\frac{r-1}{2}]} \sum_{q=2k+1}^{r} \binom{r}{q} (-1)^k \mu^{q-2k-1}\xi^{2k+1}S^1(q,2k+1)(\eta)_{r-q,\mu}. \tag{20}$$

Motivated by the importance and potential applications in certain problems in number theory, combinatorics, classical and numerical analysis and physics, several families of degenerate Bernoulli and Euler polynomials and degenerate versions of special polynomials have been recently studied by many authors, (see [3–5,11–13,16]). Recently, Kim and Kim [2] have introduced the degenerate Bernoulli and degenerate Euler polynomials of a complex variable. By separating the real and imaginary parts, they introduced the parametric kinds of these degenerate polynomials.

The main object of this article is to present the parametric kinds of degenerate poly-Bernoulli and poly-Genocchi polynomials in terms of the degenerate type parametric exponential functions. We also investigate some fundamental properties of our introduced parametric polynomials.

2. Parametric Kinds of the Degenerate Poly-Bernoulli Polynomials

In this section, we define the two parametric kinds of degenerate poly-Bernoulli polynomials by means of the two special generating functions involving the degenerate exponential as well as trigonometric functions.

It is well known that (see [2])

$$e^{(\eta + i\xi)z} = e^{\eta z}e^{i\xi z} = e^{\eta z}(\cos \xi z + i \sin \xi z), \tag{21}$$

The degenerate trigonometric functions are defined by (see [19])

$$\cos_\mu z = \frac{e_\mu^i(z) + e_\mu^{-i}(z)}{2}, \quad \sin_\mu z = \frac{e_\mu^i(z) - e_\mu^{-i}(z)}{2i}. \tag{22}$$

Note that, we have

$$\lim_{\mu \to 0} \cos_\mu z = \cos z, \quad \lim_{\mu \to 0} \sin_\mu z = \sin z.$$

In view of Equation (8), we have

$$\frac{\text{Li}_k(1 - e^{-z})}{e_\mu(z) - 1} e_\mu^{\eta + i\xi}(z) = \sum_{j=0}^{\infty} B_{j,\mu}^{(k)}(\eta + i\xi) \frac{z^j}{j!}, \tag{23}$$

and

$$\frac{\text{Li}_k(1 - e^{-z})}{e_\mu(z) - 1} e_\mu^{\eta - i\xi}(z) = \sum_{j=0}^{\infty} B_{j,\mu}^{(k)}(\eta - i\xi) \frac{z^j}{j!}. \tag{24}$$

From Equations (23) and (24), we note that

$$\frac{\text{Li}_k(1 - e^{-z})}{e_\mu(z) - 1} e_\mu^{\eta}(z) \cos_\mu^\xi(z) = \sum_{j=0}^{\infty} \left(\frac{B_{j,\mu}^{(k)}(\eta + i\xi) + B_{j,\mu}^{(k)}(\eta - i\xi)}{2} \right) \frac{z^j}{j!}, \tag{25}$$

and

$$\frac{\text{Li}_k(1 - e^{-z})}{e_\mu(z) - 1} e_\mu^{\eta}(z) \sin_\mu^\xi(z) = \sum_{j=0}^{\infty} \left(\frac{B_{j,\mu}^{(k)}(\eta + i\xi) - B_{j,\mu}^{(k)}(\eta - i\xi)}{2i} \right) \frac{z^j}{j!}. \tag{26}$$

Definition 1. *The degenerate cosine-poly-Bernoulli polynomials* $B_{p,\mu}^{(k,c)}(\eta, \xi)$ *and degenerate sine-poly-Bernoulli polynomials* $B_{p,\mu}^{(k,s)}(\eta, \xi)$ *for nonnegative integer p are defined, respectively, by*

$$\frac{\text{Li}_k(1 - e^{-z})}{e_\mu(z) - 1} e_\mu^{\eta}(z) \cos_\mu^\xi(z) = \sum_{p=0}^{\infty} B_{p,\mu}^{(k,c)}(\eta, \xi) \frac{z^p}{p!}, \tag{27}$$

and

$$\frac{\text{Li}_k(1 - e^{-z})}{e_\mu(z) - 1} e_\mu^{\eta}(z) \sin_\mu^\xi(z) = \sum_{p=0}^{\infty} B_{p,\mu}^{(k,s)}(\eta, \xi) \frac{z^p}{p!}. \tag{28}$$

For $\eta = \xi = 0$ *in Equations (27) and (28), we get*

$$B_{p,\mu}^{(k,c)}(0,0) = B_{p,\mu}^{(k)}, B_{p,\mu}^{(k,s)}(0,0) = 0, (p \geq 0).$$

Note that $\lim_{\mu \longrightarrow 0} B_{p,\mu}^{(k,c)}(\eta, \xi) = B_p^{(k,c)}(\eta, \xi)$, $\lim_{\mu \longrightarrow 0} B_{p,\mu}^{(k,s)}(\eta, \xi) = B_p^{(k,s)}(\eta, \xi)$, $(p \geq 0)$, where $B_p^{(k,c)}(\eta, \xi)$ and $B_p^{(k,s)}(\eta, \xi)$ are the new type of poly-Bernoulli polynomials.

Based on Equations (25)–(28), we determine

$$B_{p,\mu}^{(k,c)}(\eta, \xi) = \frac{B_{p,\mu}^{(k)}(\eta + i\xi) + B_{p,\mu}^{(k)}(\eta - i\xi)}{2}, \tag{29}$$

and

$$B_{p,\mu}^{(k,s)}(\eta, \xi) = \frac{B_{p,\mu}^{(k)}(\eta + i\xi) - B_{p,\mu}^{(k)}(\eta - i\xi)}{2i}. \tag{30}$$

Theorem 1. *Let $k \in \mathbb{Z}$ and $j \geq 0$. Then*

$$B_{j,\mu}^{(k)}(\eta + i\xi) = \sum_{q=0}^{j} \binom{j}{q} B_{j-q,\mu}^{(k)}(\eta)(i\xi)_{q,\mu}$$

$$= \sum_{q=0}^{j} \binom{j}{q} B_{j-q,\mu}^{(k)}(\eta + i\xi)_{q,\mu}, \tag{31}$$

and

$$B_{j,\mu}^{(k)}(\eta - i\xi) = \sum_{q=0}^{j} \binom{j}{q} B_{j-q,\mu}^{(k)}(\eta)(-1)^q(i\xi)_{q,\mu}$$

$$= \sum_{q=0}^{j} \binom{j}{q} B_{j-q,\mu}^{(k)}(\eta - i\xi)_{q,\mu}. \tag{32}$$

Proof. From Equation (23), we have

$$\sum_{j=0}^{\infty} B_{j,\mu}^{(k)}(\eta + i\xi)\frac{z^j}{j!} = \frac{\mathrm{Li}_k(1 - e^{-z})}{e_\mu(z) - 1}e_\mu^\eta(z)e_\mu^{i\xi}(z)$$

$$= \left(\sum_{j=0}^{\infty} B_{j,\mu}^{(k)}(\eta)\frac{z^j}{j!} \right) \left(\sum_{q=0}^{\infty} (i\xi)_{q,\mu}\frac{z^q}{q!} \right)$$

$$= \sum_{j=0}^{\infty} \left(\sum_{q=0}^{j} \binom{j}{q} B_{j-q,\mu}^{(k)}(\eta)(i\xi)_{q,\mu} \right) \frac{z^j}{j!}. \tag{33}$$

Similarly, we find

$$\frac{\mathrm{Li}_k(1 - e^{-z})}{e_\mu(z) - 1}e_\mu^\eta(z)e_\mu^{i\xi}(z) = \left(\sum_{j=0}^{\infty} B_{j,\mu}^{(k)}\frac{z^j}{j!} \right) \left(\sum_{q=0}^{\infty} (\eta + i\xi)_{q,\mu}\frac{z^q}{q!} \right)$$

$$= \sum_{j=0}^{\infty} \left(\sum_{q=0}^{j} \binom{j}{q} B_{j-q,\mu}^{(k)}(\eta + i\xi)_{q,\mu} \right) \frac{z^j}{j!}. \tag{34}$$

In view of Equations (33) and (34), we obtain our first claimed result shown in Equation (31). Similarly, we can establish our second result shown in Equation (32). □

Theorem 2. *The following results hold true:*

$$B_{j,\mu}^{(k,c)}(\eta,\xi) = \sum_{r=0}^{j} \binom{j}{r} B_{r,\mu}^{(k)} C_{j-r,\mu}(\eta,\xi)$$

$$= \sum_{r=0}^{[\frac{q}{2}]} \sum_{q=2r}^{j} \binom{j}{q} \mu^{q-2r}(-1)^r \xi^{2r} S^{(1)}(q,2r) B_{j-q,\mu}^{(k)}(\eta), \tag{35}$$

and

$$B_{j,\mu}^{(k,s)}(\eta,\xi) = \sum_{r=0}^{j} \binom{j}{r} B_{r,\mu}^{(k)} S_{j-r,\mu}(\eta,\xi)$$

$$= \sum_{r=0}^{[\frac{q-1}{2}]} \sum_{q=2r+1}^{j} \binom{j}{q} \mu^{q-2r-1}(-1)^r \xi^{2r+1} S^{(1)}(q,2r+1) B_{j-q,\mu}^{(k)}(\eta). \tag{36}$$

Proof. From Equations (27) and (17), we see

$$\sum_{j=0}^{\infty} B_{j,\mu}^{(k,c)}(\eta,\xi) \frac{z^j}{j!} = \frac{\mathrm{Li}_k(1-e^{-z})}{e_\mu(z)-1} e_\mu^\eta(z) \cos_\mu^\xi(z)$$

$$= \left(\sum_{r=0}^{\infty} B_{r,\mu}^{(k)} \frac{z^r}{r!} \right) \left(\sum_{j=0}^{\infty} C_{j,\mu}(\eta,\xi) \frac{z^j}{j!} \right)$$

$$= \sum_{j=0}^{\infty} \left(\sum_{r=0}^{j} \binom{j}{r} B_{r,\mu}^{(k)} C_{j-r,\mu}(\eta,\xi) \right) \frac{z^j}{j!}. \tag{37}$$

Now, by using Equations (27) and (10), we find

$$\frac{\mathrm{Li}_k(1-e^{-z})}{e_\mu(z)-1} e_\mu^\eta(z) \cos_\mu^\xi(z) = \sum_{j=0}^{\infty} B_{j,\mu}^{(k)}(\eta) \frac{z^j}{j!} \sum_{p=0}^{\infty} \sum_{r=0}^{[\frac{q}{2}]} \mu^{l-2r}(-1)^r y^{2r} S^{(1)}(q,2r) \frac{z^r}{r!}$$

$$= \sum_{j=0}^{\infty} \left(\sum_{q=0}^{j} \sum_{r=0}^{[\frac{q}{2}]} \binom{j}{q} \mu^{q-2r}(-1)^r \xi^{2r} S^{(1)}(q,2r) B_{j-r,\mu}^{(k)}(\eta) \right) \frac{z^j}{j!}$$

$$= \sum_{j=0}^{\infty} \left(\sum_{r=0}^{[\frac{q}{2}]} \sum_{q=2r}^{j} \binom{j}{q} \mu^{q-2r}(-1)^r \xi^{2r} S^{(1)}(q,2r) B_{j-q,\mu}^{(k)}(\eta) \right) \frac{z^j}{j!}. \tag{38}$$

Therefore, from Equations (37) and (38), we attain our needed result, Equation (35). Similarly, we can obtain Equation (36). □

Theorem 3. *Each of the following identities holds true:*

$$B_{r,\mu}^{(2,c)}(\eta,\xi) = \sum_{q=0}^{r} \binom{r}{q} \frac{q! B_q}{q+1} B_{r-q,\mu}^{(c)}(\eta,\xi), \tag{39}$$

and

$$B_{r,\mu}^{(2,s)}(\eta,\xi) = \sum_{q=0}^{r} \binom{r}{q} \frac{q!B_q}{q+1} B_{r-q,\mu}^{(s)}(\eta,\xi). \tag{40}$$

Proof. In view of Equation (27), we have

$$\sum_{r=0}^{\infty} B_{r,\mu}^{(k,c)}(\eta,\xi)\frac{z^r}{r!} = \frac{\mathrm{Li}_k(1-e^{-z})}{e_\mu(z)-1} e_\mu^\eta(z)\cos_\mu^\xi(z)$$

$$= \frac{e_\mu^\eta(z)\cos_\mu^\xi(z)}{e_\mu(z)-1} \int_0^z \frac{1}{e^u-1}\int_0^u \frac{1}{e^u-1}\cdots\underbrace{\frac{1}{e^u-1}\int_0^u \frac{u}{e^u-1}du\cdots du}_{(k-1)-\text{times}}. \tag{41}$$

Upon setting $k=2$, we obtain

$$\sum_{r=0}^{\infty} B_{r,\mu}^{(2,c)}(\eta,\xi)\frac{z^r}{r!} = \frac{e_\mu^\eta(z)\cos_\mu^\xi(z)}{e_\mu(z)-1}\int_0^z \frac{u}{e^u-1}du$$

$$= \left(\sum_{q=0}^{\infty}\frac{B_q z^q}{(q+1)}\right)\frac{e_\mu^\eta(z)\cos_\mu^\xi(z)}{e_\mu(z)-1}$$

$$= \left(\sum_{q=0}^{\infty}\frac{q!B_q z^q}{(q+1)q!}\right)\left(\sum_{r=0}^{\infty}B_{r,\mu}^{(c)}(\eta,\xi)\frac{z^r}{r!}\right).$$

$$= \sum_{r=0}^{\infty}\sum_{q=0}^{r}\binom{r}{q}\frac{q!B_q}{q+1}B_{r-q,\mu}^{(c)}(\eta,\xi)\frac{z^r}{r!},$$

which gives our required result, Equation (39). The proof of Equation (40) is similar; therefore, we omit the proof. □

Theorem 4. *Let $k\in\mathbb{Z}$, then*

$$B_{j,\mu}^{(k,c)}(\eta,\xi) = \sum_{r=0}^{j}\binom{j}{r}\left(\sum_{q=1}^{r+1}\frac{(-1)^{q+r+1}l!S_2(r+1,q)}{q^k(r+1)}\right)B_{j-r,\mu}^{(c)}(\eta,\xi), \tag{42}$$

and

$$B_{j,\mu}^{(k,s)}(\eta,\xi) = \sum_{r=0}^{j}\binom{j}{r}\left(\sum_{q=1}^{r+1}\frac{(-1)^{q+r+1}q!S_2(r+1,q)}{q^k(r+1)}\right)B_{j-r,\mu}^{(s)}(\eta,\xi). \tag{43}$$

Proof. From Equations (27) and (11), we see

$$\sum_{j=0}^{\infty} B_{j,\mu}^{(k,c)}(\eta,\xi)\frac{z^j}{j!} = \left(\frac{\mathrm{Li}_k(1-e^{-z})}{z}\right)\left(\frac{ze_\mu^\eta(z)\cos_\mu^\xi(z)}{e_\mu(z)-1}\right). \tag{44}$$

Now

$$\frac{1}{z}\mathrm{Li}_k(1-e^{-z}) = \frac{1}{z}\sum_{q=1}^{\infty}\frac{(1-e^{-z})^q}{q^k}$$

$$= \frac{1}{z}\sum_{q=1}^{\infty}\frac{(-1)^q}{q^k}q!\sum_{r=l}^{\infty}(-1)^r S_2(r,q)\frac{z^r}{r!}$$

$$= \frac{1}{z}\sum_{r=q}^{\infty}\sum_{q=1}^{r}\frac{(-1)^{q+r}}{q^k}q!S_2(r,q)\frac{z^r}{r!}$$

$$= \sum_{r=0}^{\infty} \left(\sum_{q=1}^{q+1} \frac{(-1)^{q+r+1}}{q^k} l! \frac{S_2(r+1,q)}{r+1} \right) \frac{z^r}{r!}. \tag{45}$$

On using Equation (45) in (44), we find

$$\sum_{j=0}^{\infty} B_{j,\mu}^{(k,c)}(\eta,\xi) \frac{z^j}{j!} = \sum_{r=0}^{\infty} \left(\sum_{q=1}^{r+1} \frac{(-1)^{q+r+1}}{q^k} l! \frac{S_2(r+1,q)}{r+1} \right) \frac{z^r}{r!} \left(\sum_{j=0}^{\infty} B_{j,\mu}^{(c)}(\eta,\xi) \frac{z^j}{j!} \right).$$

Replacing j by $j - r$ in the right side of above expression and after equating the coefficients of z^j, we obtain our needed result, Equation (42). Similarly, we can derive our second result, Equation (43). $\square$

Theorem 5. *The following recurrence relation holds true:*

$$B_{j,\mu}^{(k,c)}(\eta+1,\xi) - B_{j,\mu}^{(k,c)}(\eta,\xi)$$

$$= \sum_{r=1}^{j} \binom{j}{r} \left(\sum_{q=0}^{r-1} \frac{(-1)^{q+r+1}}{(q+1)^k} (q+1)! S_2(r,q+1) \right) C_{j-r,\mu}(\eta,\xi), \tag{46}$$

and

$$B_{j,\mu}^{(k,s)}(\eta+1,\xi) - B_{j,\mu}^{(k,s)}(\eta,\xi)$$

$$= \sum_{r=1}^{j} \binom{j}{r} \left(\sum_{q=0}^{r-1} \frac{(-1)^{q+r+1}}{(q+1)^k} (q+1)! S_2(r,q+1) \right) S_{j-r,\mu}(\eta,\xi). \tag{47}$$

Proof. In view of Equation (27), we have

$$\sum_{j=0}^{\infty} B_{j,\mu}^{(k,c)}(\eta+1,\xi) \frac{z^j}{j!} - \sum_{j=0}^{\infty} B_{j,\mu}^{(k,c)}(\eta,\xi) \frac{z^j}{j!}$$

$$= \frac{\mathrm{Li}_k(1-e^{-z})}{e_\mu(z)-1} e_\mu^{(\eta+1)}(z) \cos_\mu^\xi(z) - \frac{\mathrm{Li}_k(1-e^{-z})}{e_\mu(z)-1} e_\mu^{(\eta)}(z) \cos_\mu^\xi(z)$$

$$= \mathrm{Li}_k(1-e^{-z}) e_\mu^{(\eta)}(z) \cos_\mu^\xi(z)$$

$$= \sum_{q=0}^{\infty} \frac{(1-e^{-z})^{q+1}}{(q+1)^k} e_\mu^{(\eta)}(z) \cos_\mu^\xi(z)$$

$$= \sum_{r=1}^{\infty} \left(\sum_{q=0}^{r-1} \frac{(-1)^{q+r+1}}{(q+1)^k} (q+1)! S_2(r,q+1) \right) \frac{z^r}{r!} e_\mu^{(\eta)}(z) \cos_\mu^\xi(z)$$

$$= \left(\sum_{r=1}^{\infty} \left(\sum_{q=0}^{r-1} \frac{(-1)^{q+r+1}}{(q+1)^k} (q+1)! S_2(r,q+1) \right) \frac{z^r}{r!} \right) \left(\sum_{j=0}^{\infty} C_{j,\mu}(\eta,\xi) \frac{z^j}{j!} \right),$$

which upon replacing j by $j - r$ in the right side of above expression and after equating the coefficients of z^j, yields our first claimed result, Equation (46). Similarly, we can establish our second result, Equation (47). $\square$

Theorem 6. *Let $k \in \mathbb{Z}$ and $j \geq 0$, then we have*

$$B_{j,\mu}^{(k,c)}(\eta+\gamma,\xi) = \sum_{r=0}^{j} \binom{j}{r} B_{j-r,\mu}^{(k,c)}(\eta,\xi)(\gamma)_{r,\mu}, \tag{48}$$

and

$$B_{j,\mu}^{(k,s)}(\eta + \gamma, \xi) = \sum_{r=0}^{j} \binom{j}{r} B_{j-r,\mu}^{(k,s)}(\eta, \xi)(\gamma)_{r,\mu}. \tag{49}$$

Proof. On using Equation (27), we find

$$\sum_{j=0}^{\infty} B_{j,\mu}^{(k,c)}(\eta + \gamma, \xi) \frac{z^j}{j!} = \frac{\mathrm{Li}_k(1 - e^{-z})}{e_\mu(z) - 1} e_\mu^{(\eta + \gamma)}(z) \cos_\mu^{\xi}(z)$$

$$= \left(\sum_{j=0}^{\infty} B_{j,\mu}^{(k,c)}(\eta, \xi) \frac{z^j}{j!} \right) \left(\sum_{r=0}^{\infty} (\gamma)_{r,\mu} \frac{z^r}{r!} \right)$$

$$= \sum_{j=0}^{\infty} \left(\sum_{r=0}^{j} \binom{j}{r} B_{j-r,\mu}^{(k,c)}(\eta, \xi)(\gamma)_{r,\mu} \right) \frac{z^j}{j!}.$$

By comparing the coefficients of z^j on both sides, we obtain the result, Equation (48). The proof of Equation (49) is similar to Equation (48).

□

Theorem 7. *If $k \in \mathbb{Z}$ and $j \geq 0$, then*

$$B_{j,\mu}^{(k,c)}(\eta, \xi) = \sum_{r=0}^{j} \sum_{q=0}^{r} \binom{j}{r} (\eta)_q S_\mu^{(2)}(r, q) B_{j-r,\mu}^{(k,c)}(0, \xi), \tag{50}$$

and

$$B_{j,\mu}^{(k,s)}(\eta, \xi) = \sum_{r=0}^{j} \sum_{q=0}^{r} \binom{j}{r} (\eta)_q S_\mu^{(2)}(r, q) B_{j-r,\mu}^{(k,s)}(0, \xi). \tag{51}$$

Proof. From Equations (27) and (12), we find

$$\sum_{j=0}^{\infty} B_{j,\mu}^{(k,c)}(\eta, \xi) \frac{z^j}{j!} = \frac{\mathrm{Li}_k(1 - e^{-z})}{e_\mu(z) - 1} (e_\mu(z) - 1 + 1)^\eta \cos_\mu^{\xi}(z)$$

$$= \frac{\mathrm{Li}_k(1 - e^{-z})}{e_\mu(z) - 1} \sum_{q=0}^{\infty} \binom{\eta}{q} (e_\mu(z) - 1)^q \cos_\mu^{\xi}(z)$$

$$= \frac{\mathrm{Li}_k(1 - e^{-z})}{e_\mu(z) - 1} \cos_\mu^{\xi}(z) \sum_{q=0}^{\infty} (\eta)_q \sum_{r=q}^{\infty} S_\mu^{(2)}(r, q) \frac{z^r}{r!}$$

$$= \sum_{j=0}^{\infty} B_{j,\mu}^{(k,c)}(0, \xi) \frac{z^j}{j!} \sum_{r=0}^{\infty} \left(\sum_{q=0}^{r} (\eta)_q S_\mu^{(2)}(r, q) \right) \frac{z^r}{r!}$$

$$= \sum_{j=0}^{\infty} \left(\sum_{r=0}^{j} \sum_{q=0}^{r} \binom{j}{r} (\eta)_q S_\mu^{(2)}(r, q B_{j-r,\mu}^{(k,c)}(0, \xi) \right) \frac{z^j}{j!}.$$

On comparing the coefficients of z^j on both sides, we obtain our required result, Equation (50). The proof of Equation (51) is similar to Equation (50).

□

3. Parametric Kinds of Degenerate Poly-Genocchi Polynomials

In this section, we introduce the two parametric kinds of degenerate poly-Genocchi polynomials by defining the two special generating functions involving the degenerate exponential as well as trigonometric functions.

In view of Equation (9), we have

$$\frac{2\mathrm{Li}_k(1-e^{-z})}{e_\mu(z)+1} e_\mu^{\eta+i\xi}(z) = \sum_{j=0}^{\infty} G_{j,\mu}^{(k)}(\eta+i\xi)\frac{z^j}{j!}, \tag{52}$$

and

$$\frac{2\mathrm{Li}_k(1-e^{-z})}{e_\mu(z)+1} e_\mu^{\eta-i\xi}(z) = \sum_{j=0}^{\infty} G_{j,\mu}^{(k)}(\eta-i\xi)\frac{z^j}{j!}. \tag{53}$$

From Equations (52) and (53), we can easily get

$$\frac{2\mathrm{Li}_k(1-e^{-z})}{e_\mu(z)+1} e_\mu^{\eta}(z)\cos_\mu^{\xi}(z) = \sum_{j=0}^{\infty} \left(\frac{G_{j,\mu}^{(k)}(\eta+i\xi) + G_{j,\mu}^{(k)}(\eta-i\xi)}{2} \right)\frac{z^j}{j!}, \tag{54}$$

and

$$\frac{2\mathrm{Li}_k(1-e^{-z})}{e_\mu(z)+1} e_\mu^{\eta}(z)\sin_\mu^{\xi}(z) = \sum_{j=0}^{\infty} \left(\frac{G_{j,\mu}^{(k)}(\eta+i\xi) - G_{j,\mu}^{(k)}(\eta-i\xi)}{2i} \right)\frac{z^j}{j!}. \tag{55}$$

Definition 2. *The degenerate cosine-poly-Genocchi polynomials* $G_{j,\mu}^{(k,c)}(\eta,\xi)$ *and degenerate sine-poly-Genocchi polynomials* $G_{j,\mu}^{(k,s)}(\eta,\xi)$ *for nonnegative integer j are defined, respectively, by*

$$\frac{2\mathrm{Li}_k(1-e^{-z})}{e_\mu(z)+1} e_\mu^{\eta}(z)\cos_\mu^{\xi}(z) = \sum_{j=0}^{\infty} G_{j,\mu}^{(k,c)}(\eta,\xi)\frac{z^j}{j!}, \tag{56}$$

and

$$\frac{2\mathrm{Li}_k(1-e^{-z})}{e_\mu(z)+1} e_\mu^{\eta}(z)\sin_\mu^{\xi}(z) = \sum_{j=0}^{\infty} G_{j,\mu}^{(k,s)}(\eta,\xi)\frac{z^j}{j!}. \tag{57}$$

On setting $\eta = \xi = 0$ *in Equations (56) and (57), we get*

$$G_{j,\mu}^{(k,c)}(0,0) = G_{j,\mu}^{(k)}, \quad G_{j,\mu}^{(k,s)}(0,0) = 0, (j \geq 0).$$

Note that $\lim_{\mu\to 0} G_{j,\mu}^{(k,c)}(\eta,\xi) = G_j^{(k,c)}(\eta,\xi)$, $\lim_{\mu\to 0} G_{j,\mu}^{(k,s)}(\eta,\xi) = G_j^{(k,s)}(\eta,\xi)$, $(j \geq 0)$, *where* $G_n^{(k,c)}(\eta,\xi)$ *and* $G_j^{(k,s)}(\eta,\xi)$ *are the new type of poly-Genocchi polynomials.*

From Equations (54)–(57), we determine

$$G_{j,\mu}^{(k,c)}(\eta,\xi) = \frac{G_{j,\mu}^{(k)}(\eta+i\xi) + G_{j,\mu}^{(k)}(\eta-i\xi)}{2} \tag{58}$$

and

$$G_{j,\mu}^{(k,s)}(\eta,\xi) = \frac{G_{j,\mu}^{(k)}(\eta+i\xi) - G_{j,\mu}^{(k)}(\eta-i\xi)}{2i}. \tag{59}$$

Theorem 8. *For* $k \in \mathbb{Z}$ *and* $j \geq 0$, *we have*

$$G_{j,\mu}^{(k)}(\eta+i\xi) = \sum_{q=0}^{j} \binom{j}{q} G_{j-q,\mu}^{(k)}(\eta)(i\xi)_{q,\mu}$$

$$= \sum_{q=0}^{j} \binom{j}{q} G_{j-q,\mu}^{(k)} (\eta + i\xi)_{q,\mu}, \tag{60}$$

and

$$G_{j,\mu}^{(k)} (\eta - i\xi) = \sum_{q=0}^{j} \binom{j}{q} G_{j-q,\mu}^{(k)} (\eta)(-1)^{q} (i\xi)_{q,\mu}$$

$$= \sum_{q=0}^{j} \binom{j}{q} G_{j-q,\mu}^{(k)} (\eta - i\xi)_{q,\mu}. \tag{61}$$

Proof. On using Equation (52), we see

$$\sum_{j=0}^{\infty} G_{j,\mu}^{(k)} (\eta + i\xi) \frac{z^j}{j!} = \frac{2\mathrm{Li}_k(1 - e^{-z})}{e_\mu(z) + 1} e_\mu^\eta(z) e_\mu^{i\xi}(z)$$

$$= \left(\sum_{j=0}^{\infty} G_{j,\mu}^{(k)}(\eta) \frac{z^j}{j!} \right) \left(\sum_{q=0}^{\infty} (i\xi)_{q,\mu} \frac{z^q}{q!} \right)$$

$$= \sum_{j=0}^{\infty} \left(\sum_{q=0}^{j} \binom{j}{q} G_{j-q,\mu}^{(k)}(\eta)(i\xi)_{q,\mu} \right) \frac{z^j}{j!}. \tag{62}$$

Similarly, we find

$$\frac{2\mathrm{Li}_k(1 - e^{-z})}{e_\mu(z) + 1} e_\mu^\eta(z) e_\mu^{i\xi}(z) = \left(\sum_{j=0}^{\infty} G_{j,\mu}^{(k)} \frac{z^j}{j!} \right) \left(\sum_{q=0}^{\infty} (\eta + i\xi)_{q,\mu} \frac{z^q}{q!} \right)$$

$$= \sum_{j=0}^{\infty} \left(\sum_{q=0}^{j} \binom{j}{q} G_{j-q,\mu}^{(k)} (\eta + i\xi)_{q,\mu} \right) \frac{z^j}{j!}. \tag{63}$$

By comparing the coefficients of z^j on both sides in Equations (62) and (63), we obtain our desired result, Equation (60). The proof of Equation (61) is similar to Equation (60). $\square$

Theorem 9. *If $k \in \mathbb{Z}$ and $j \geq 0$, then*

$$G_{j,\mu}^{(k,c)} (\eta, \xi) = \sum_{r=0}^{j} \binom{j}{r} G_{r,\mu}^{(k)} C_{j-r,\mu}(\eta, \xi)$$

$$= \sum_{r=0}^{\lfloor \frac{q}{2} \rfloor} \sum_{q=2r}^{j} \binom{j}{q} \mu^{q-2r}(-1)^r \xi^{2r} S^{(1)}(q, 2r) G_{j-q,\mu}^{(k)}(\xi), \tag{64}$$

and

$$G_{j,\mu}^{(k,s)} (\eta, \xi) = \sum_{r=0}^{j} \binom{j}{r} B_{r,\mu}^{(k)} S_{j-r,\mu}(\eta, \xi)$$

$$= \sum_{r=0}^{\lfloor \frac{q-1}{2} \rfloor} \sum_{q=2r+1}^{j} \binom{j}{q} \mu^{q-2r-1}(-1)^r \xi^{2r+1} S^{(1)}(q, 2r+1) G_{j-q,\mu}^{(k)}(\eta). \tag{65}$$

Proof. From Equations (56) and (10), we see

$$\sum_{j=0}^{\infty} G_{j,\mu}^{(k,c)} (\eta, \xi) \frac{z^j}{j!} = \frac{2\mathrm{Li}_k(1 - e^{-z})}{e_\mu(z) + 1} e_\mu^\eta(t) \cos_\mu^\xi(z)$$

$$= \left(\sum_{r=0}^{\infty} G_{r,\mu}^{(k)} \frac{z^r}{r!} \right) \left(\sum_{j=0}^{\infty} C_{j,\mu}(\eta, \xi) \frac{z^j}{j!} \right)$$

$$= \sum_{j=0}^{\infty} \left(\sum_{r=0}^{j} \binom{j}{r} G_{r,\mu}^{(k)} C_{j-r,\mu}(\eta, \xi) \right) \frac{z^j}{j!}. \tag{66}$$

Similarly, we find

$$\frac{2\mathrm{Li}_k(1-e^{-z})}{e_\mu(z)+1} e_\mu^\eta(z) \cos_\mu^\xi(z) = \sum_{j=0}^{\infty} G_{j,\mu}^{(k)}(\eta) \frac{z^j}{j!} \sum_{q=0}^{\infty} \sum_{r=0}^{[\frac{q}{2}]} \mu^{q-2r}(-1)^r \xi^{2r} S^{(1)}(q, 2r) \frac{z^r}{r!}$$

$$= \sum_{j=0}^{\infty} \left(\sum_{l=0}^{j} \sum_{m=0}^{[\frac{l}{2}]} \binom{j}{l} \mu^{l-2m}(-1)^m \xi^{2m} S^{(1)}(q, 2r) G_{j-q,\mu}^{(k)}(\eta) \right) \frac{z^j}{j!}$$

$$= \sum_{j=0}^{\infty} \left(\sum_{r=0}^{[\frac{q}{2}]} \sum_{q=2r}^{j} \binom{j}{q} \mu^{q-2r}(-1)^r \xi^{2r} S^{(1)}(q, 2r) G_{j-q,\mu}^{(k)}(\eta) \right) \frac{z^j}{j!}. \tag{67}$$

By comparing the coefficients of z^j on both sides of Equations (66) and (67), we easily get our first claimed result, Equation (64). Similarly, we can establish our second needed result, Equation (65). $\square$

Theorem 10. *Let $j \geq 0$. Then, we have*

$$G_{j,\mu}^{(2,c)}(\eta, \xi) = \sum_{r=0}^{j} \binom{j}{r} \frac{r! B_r}{r+1} G_{j-r,\mu}^{(c)}(\eta, \xi), \tag{68}$$

and

$$G_{j,\mu}^{(2,s)}(\eta, \xi) = \sum_{r=0}^{j} \binom{j}{r} \frac{r! B_r}{r+1} G_{j-r,\mu}^{(s)}(\eta, \xi). \tag{69}$$

Proof. By using Equation (56), we determine

$$\sum_{j=0}^{\infty} G_{j,\mu}^{(k,c)}(\eta, \xi) \frac{z^j}{j!} = \frac{2\mathrm{Li}_k(1-e^{-z})}{e_\mu(z)+1} e_\mu^\eta(z) \cos_\mu^\xi(z)$$

$$= \frac{2e_\mu^\eta(z) \cos_\mu^\xi(z)}{e_\mu(z)+1} \underbrace{\int_0^z \frac{1}{e^u-1} \int_0^u \frac{1}{e^u-1} \cdots \frac{1}{e^u-1} \int_0^u \frac{u}{e^u-1} du \cdots du}_{(k-1)-\text{times}}. \tag{70}$$

On setting $k = 2$ in Equation (70), we find

$$\sum_{j=0}^{\infty} G_{j,\mu}^{(2,c)}(\eta, \xi) \frac{z^j}{j!} = \frac{2e_\mu^\eta(z) \cos_\mu^\xi(z)}{e_\mu(z)+1} \int_0^z \frac{u}{e^u-1} dz$$

$$= \left(\sum_{r=0}^{\infty} \frac{r! B_r z^r}{(r+1)r!} \right) \frac{2z e_\mu^\eta(z) \cos_\mu^\xi(z)}{e_\mu(z)+1}$$

$$= \left(\sum_{r=0}^{\infty} \frac{r! B_r z^r}{(r+1)r!} \right) \left(\sum_{j=0}^{\infty} G_{j,\mu}^{(c)}(\eta, \xi) \frac{z^j}{j!} \right).$$

On replacing j by $j - r$ in the above equation, we obtain

$$= \sum_{j=0}^{\infty} \sum_{r=0}^{j} \binom{j}{r} \frac{r! B_r}{r+1} G_{j-r,\mu}^{(c)}(\eta, \xi) \frac{z^j}{j!}.$$

Finally, by equating the coefficients of the like powers of z in the last expression, we get the result, Equation (68). The proof of Equation (69) is similar to Equation (68). $\square$

Theorem 11. *For $k \in \mathbb{Z}$ and $j \geq 0$, we have*

$$G_{j,\mu}^{(k,c)}(\eta, \xi) = \sum_{r=0}^{j} \binom{j}{r} \left(\sum_{q=1}^{r+1} \frac{(-1)^{q+r+1} q! S_2(r+1,q)}{q^k (r+1)} \right) G_{j-r,\mu}^{(c)}(\eta, \xi), \tag{71}$$

and

$$G_{j,\mu}^{(k,s)}(\eta, \xi) = \sum_{r=0}^{j} \binom{j}{r} \left(\sum_{q=1}^{r+1} \frac{(-1)^{q+r+1} q! S_2(r+1,q)}{q^k (r+1)} \right) G_{j-r,\mu}^{(s)}(\eta, \xi). \tag{72}$$

Proof. In view of Equations (56) and (11), we see

$$\sum_{j=0}^{\infty} G_{j,\mu}^{(k,c)}(\eta, \xi) \frac{z^j}{j!} = \left(\frac{2\mathrm{Li}_k(1 - e^{-z})}{z} \right) \left(\frac{z e_\mu^\eta(z) \cos_\mu^\xi(z)}{e_\mu(z) + 1} \right). \tag{73}$$

Now

$$\frac{1}{z} \mathrm{Li}_k(1 - e^{-z}) = \frac{1}{z} \sum_{q=1}^{\infty} \frac{(1 - e^{-z})^q}{q^k}$$

$$= \frac{1}{z} \sum_{q=1}^{\infty} \frac{(-1)^q}{q^k} q! \sum_{r=l}^{\infty} (-1)^r S_2(r,q) \frac{z^r}{r!}$$

$$= \frac{1}{z} \sum_{r=1}^{\infty} \sum_{q=1}^{r} \frac{(-1)^{q+r}}{q^k} q! S_2(r,q) \frac{t^r}{r!}$$

$$= \sum_{r=0}^{\infty} \left(\sum_{q=1}^{r+1} \frac{(-1)^{q+r+1}}{q^k} q! \frac{S_2(r+1,q)}{q+1} \right) \frac{z^r}{r!}. \tag{74}$$

Using Equation (74) in (73), we find

$$\sum_{j=0}^{\infty} G_{j,\mu}^{(k,c)}(\eta, \xi) \frac{z^j}{j!} = \sum_{r=0}^{\infty} \left(\sum_{q=1}^{r+1} \frac{(-1)^{q+r+1}}{q^k} q! \frac{S_2(r+1,q)}{r+1} \right) \frac{z^r}{r!} \left(\sum_{j=0}^{\infty} G_{j,\mu}^{(c)}(\eta, \xi) \frac{z^j}{j!} \right),$$

which on comparing the coefficients of z^j on both sides, yields our desired result, Equation (71). Similarly, we can derive our second result, Equation (72). $\square$

Theorem 12. *Let $k \in \mathbb{Z}$ and $j \geq 0$, then we have*

$$\frac{1}{2} \left[G_{j,\mu}^{(k,c)}(\eta + 1, \xi) + G_{j,\mu}^{(k,c)}(\eta, \xi) \right]$$

$$= \sum_{r=1}^{j} \binom{j}{r} \left(\sum_{q=0}^{r-1} \frac{(-1)^{q+r+1}}{(q+1)^k} (q+1)! S_2(r,q+1) \right) C_{j-r,\mu}(\eta, \xi), \tag{75}$$

and

$$\frac{1}{2} \left[G_{j,\mu}^{(k,s)}(\eta + 1, \xi) + G_{j,\mu}^{(k,s)}(\eta, \xi) \right]$$

$$= \sum_{r=1}^{j} \binom{j}{r} \left(\sum_{q=0}^{r-1} \frac{(-1)^{q+r+1}}{(q+1)^k} (q+1)! S_2(r,q+1) \right) S_{j-r,\mu}(\eta,\xi). \tag{76}$$

Proof. Taking

$$\sum_{j=0}^{\infty} G_{j,\mu}^{(k,c)}(\eta+1,\xi)\frac{z^j}{j!} + \sum_{j=0}^{\infty} G_{j,\mu}^{(k,c)}(\eta,\xi)\frac{z^j}{j!}$$

$$= \frac{2\mathrm{Li}_k(1-e^{-z})}{e_\mu(z)+1} e_\mu^{(\eta+1)}(z)\cos_\mu^\xi(z) + \frac{2\mathrm{Li}_k(1-e^{-z})}{e_\mu(z)+1} e_\mu^{(\eta)}(z)\cos_\mu^\xi(z)$$

$$= 2\mathrm{Li}_k(1-e^{-z})e_\mu^{(\eta)}(z)\cos_\mu^\xi(z)$$

$$= \sum_{q=0}^{\infty} \frac{(1-e^{-z})^{q+1}}{(q+1)^k} 2e_\mu^\eta(z)\cos_\mu^{(\xi)}(z)$$

$$= \sum_{r=1}^{\infty} \left(\sum_{q=0}^{r-1} \frac{(-1)^{q+r+1}}{(q+1)^k} (q+1)! S_2(r,q+1) \right) \frac{z^r}{r!} 2e_\mu^x(z)\cos_\mu^{(\xi)}(z)$$

$$= 2\left(\sum_{r=1}^{\infty} \left(\sum_{q=0}^{r-1} \frac{(-1)^{q+r+1}}{(q+1)^k} (q+1)! S_2(r,q+1) \right) \frac{z^r}{r!} \right) \left(\sum_{j=0}^{\infty} C_{j,\mu}(\eta,\xi)\frac{z^j}{j!} \right).$$

On replacing j by $j-r$ in the right side of the above equation, and after comparing the coefficients of z^j on both sides, we acquire the desired result, Equation (75). Similarly, we can obtain the result, Equation (76). □

Theorem 13. *For $k \in \mathbb{Z}$ and $j \geq 0$, we have*

$$G_{j,\mu}^{(k,c)}(\eta+\alpha,\xi) = \sum_{m=0}^{j} \binom{j}{m} G_{j-m,\mu}^{(k,c)}(\eta,\xi)(\alpha)_{m,\mu}, \tag{77}$$

and

$$G_{j,\mu}^{(k,s)}(\eta+\alpha,\xi) = \sum_{m=0}^{j} \binom{j}{m} G_{j-m,\mu}^{(k,s)}(\eta,\xi)(\alpha)_{m,\mu}. \tag{78}$$

Proof. By using Equation (56), we have

$$\sum_{j=0}^{\infty} G_{j,\mu}^{(k,c)}(\eta+\alpha,\xi)\frac{z^j}{j!} = \frac{2\mathrm{Li}_k(1-e^{-z})}{e_\mu(z)+1} e_\mu^{(\eta+\alpha)}(z)\cos_\mu^{(\xi)}(z)$$

$$= \left(\sum_{j=0}^{\infty} G_{j,\mu}^{(k,c)}(\eta,\xi)\frac{z^j}{j!} \right) \left(\sum_{m=0}^{\infty} (\alpha)_{m,\mu}\frac{z^m}{m!} \right)$$

$$= \sum_{j=0}^{\infty} \left(\sum_{m=0}^{j} \binom{j}{m} G_{j-m,\mu}^{(k,c)}(\eta,\xi)(\alpha)_{m,\mu} \right) \frac{z^j}{j!}.$$

By comparing the coefficients of z^j on both sides in the last expression, we acquire our desired result, Equation (77). Similarly, we can derive our second result, Equation (78).
  □

Theorem 14. *If $k \in \mathbb{Z}$ and $j \geq 0$, then*

$$G_{j,\mu}^{(k,c)}(\eta,\xi) = \sum_{r=0}^{j} \sum_{q=0}^{r} \binom{j}{r} (\eta)_l S_\mu^{(2)}(r,q) G_{j-r,\mu}^{(k,c)}(0,\xi), \tag{79}$$

and

$$G_{j,\mu}^{(k,s)}(\eta,\xi) = \sum_{r=0}^{j}\sum_{q=0}^{r} \begin{pmatrix} j \\ r \end{pmatrix} (\eta)_l S_\mu^{(2)}(r,q) G_{j-r,\mu}^{(k,s)}(0,\xi). \tag{80}$$

Proof. From Equations (56) and (12), we have

$$\sum_{j=0}^{\infty} G_{j,\mu}^{(k,c)}(\eta,\xi)\frac{z^j}{j!} = \frac{2\mathrm{Li}_k(1-e^{-z})}{e_\mu(z)+1}(e_\mu(z)-1+1)^\eta \cos_\mu^\xi(z)$$

$$= \frac{2\mathrm{Li}_k(1-e^{-z})}{e_\mu(z)+1}\sum_{q=0}^{\infty}\begin{pmatrix}\eta\\q\end{pmatrix}(e_\mu(z)-1)^q \cos_\mu^\xi(z)$$

$$= \frac{2\mathrm{Li}_k(1-e^{-z})}{e_\mu(z)+1}\cos_\mu^\xi(z)\sum_{q=0}^{\infty}(\eta)_q\sum_{r=q}^{\infty}S_\mu^{(2)}(r,q)\frac{z^r}{r!}$$

$$= \sum_{j=0}^{\infty}G_{j,\mu}^{(k,c)}(0,\xi)\frac{z^j}{j!}\sum_{r=0}^{\infty}\left(\sum_{q=0}^{r}(\eta)_q\,S_\mu^{(2)}(r,q)\right)\frac{z^r}{r!}$$

$$= \sum_{j=0}^{\infty}\left(\sum_{r=0}^{j}\sum_{q=0}^{r}\begin{pmatrix}j\\r\end{pmatrix}(\eta)_q S_\mu^{(2)}(r,q)G_{j-r,\mu}^{(k,c)}(0,\xi)\right)\frac{z^j}{j!}.$$

Finally, by comparing the coefficients of z^j on both sides in the last expression, we arrive at our claimed result, Equation (79). Similarly, we can establish our second result, Equation (80). □

4. Conclusions

In the present article, we have considered the parametric kinds of degenerate poly-Bernoulli and poly-Genocchi polynomials by making use of the degenerate type exponential as well as trigonometric functions. We have also derived some analytical properties of our newly introduced parametric polynomials by using the series manipulation technique. Furthermore, it is noticed that, if we consider any Appell polynomials of a complex variable (as discussed in the present article), then we can easily define its parametric kinds by separating the complex variable into real and imaginary parts.

Author Contributions: All authors contributed equally to the manuscript and typed, read, and approved the final manuscript. All authors have read and agreed to the published version of the manuscript.

Abbreviations

The following abbreviations are used in this manuscript:

MKdV   modified Korteweg–de Vries equation

References

1. Avram, F.; Taqqu, M.S. Noncentral limit theorems and Appell polynomials. *Ann. Probab.* **1987**, *15*, 767–775. [CrossRef]
2. Kim, D.S.; Kim, T.; Lee, H. A note on degenerate Euler and Bernoulli polynomials of complex variable. *Symmetry* **2019**, *11*, 1168; doi:10.3390/sym11091168. [CrossRef]
3. Carlitz, L. Degenerate Stirling Bernoulli and Eulerian numbers. *Util. Math.* **1979**, *15*, 51–88.
4. Carlitz, L. A degenerate Staud-Clausen theorem. *Arch. Math.* **1956**, *7*, 28–33. [CrossRef]
5. Haroon, H.; Khan, W.A. Degenerate Bernoulli numbers and polynomials associated with degenerate Hermite polynomials. Commun. *Korean Math. Soc.* **2018**, *33*, 651–669.

6. Masjed-Jamei, M.; Beyki, M.R.; Koepf, W. A new type of Euler polynomials and numbers. *Mediterr. J. Math.* **2018**, *15*, 138. [CrossRef]
7. Lim, D. Some identities of degenerate Genocchi polynomials. *Bull. Korean Math. Soc.* **2016**, *53*, 569–579. [CrossRef]
8. Khan, W.A. A note on Hermite-based poly-Euler and multi poly-Euler polynomials. *Palest. J. Math.* **2017**, *6*, 204–214.
9. Khan, W.A. A note on degenerate Hermite poly-Bernoulli numbers and polynomials. *J. Class. Anal.* **2016**, *8*, 65–76. [CrossRef]
10. Kim, D. A note on the degenerate type of complex Appell polynomials. *Symmetry* **2019**, *11*, 1339. [CrossRef]
11. Ryoo, C.S.; Khan, W.A. On two bivariate kinds of poly-Bernoulli and poly-Genocchi polynomials. *Mathematics* **2020**, *8*, 417. [CrossRef]
12. Sharma, S.K. A note on degenerate poly-Genocchi polynomials. *Int. J. Adv. Appl. Sci.* **2020**, *7*, 1–5. [CrossRef]
13. Kim, D.S.; Kim, T. A note on degenerate poly-Bernoulli numbers polynomials. *Adv. Diff. Equat.* **2015**, *2015*, 258. [CrossRef]
14. Sharma, S.K.; Khan, W.A.; Ryoo, C.S. A parametric kind of the degenerate Fubini numbers and polynomials. *Mathematics* **2020**, *8*, 405. [CrossRef]
15. Kim, T.; Jang, Y.S.; Seo, J.J. A note on poly-Genocchi numbers and polynomials. *Appl. Math. Sci.* **2014**, *8*, 4475–4781. [CrossRef]
16. Kim, T.; Jang, G.-W. A note on degenerate gamma function and degenerate Stirling numbers of the second kind. *Adv. Stud. Contemp. Math.* **2018**, *28*, 207–214.
17. Kim, T. A note on degenerate Stirling polynomials of the second kind. *Proc. Jangjeon Math. Soc.* **2017**, *20*, 319–331.
18. Kim, T.; Yao, Y.; Kim, D.S.; Jang, G.-W. Degenerate *r*-Stirling numbers and *r*-Bell polynomials. *Russ. J. Math. Phys.* **2018**, *25*, 44–58. [CrossRef]
19. Kim, T.; Ryoo, C.S. Some identities for Euler and Bernoulli polynomials and their zeros. *Axioms* **2018**, *7*, 56. [CrossRef]
20. Kim, T.; Kim, D. S.; Kim, H. Y.; Jang, L.-C. Degenerate poly-Bernoulli number and polynomials. *Informatica* **2020**, *31*, 2–8.
21. Kim, T.; Kim, D.S.; Kwon, H.-I. A note on degenerate Stirling numbers and their applications. *Proc. Jangjeon Math. Soc.* **2018**, *21*, 195–203.
22. Kim, D. A class of Sheffer sequences of some complex polynomials and their degenerate types. *Mathematics* **2019**, *7*, 1064. [CrossRef]
23. Masjed-Jamei, M.; Beyki, M.R.; Koepf, W. An extension of the Euler-Maclaurin quadrature formula using a parametric type of Bernoulli polynomials. *Bull. Sci. Math.* **2019**, *156*, 102798. [CrossRef]
24. Masjed-Jamei, M.; Koepf, W. Symbolic computation of some power trigonometric series. *J. Symb. Comput.* **2017**, *80*, 273–284. [CrossRef]
25. Masjed-Jamei, M.; Beyki, M.R.; Omey, E. On a parametric kind of Genocchi polynomials. *J. Inq. Spec. Funct.* **2018**, *9*, 68–81.

13

New Families of Three-Variable Polynomials Coupled with Well-Known Polynomials and Numbers

Can Kızılateş [1,*]**, Bayram Çekim** [2]**, Naim Tuğlu** [2] **and Taekyun Kim** [3]

[1] Faculty of Art and Science, Department of Mathematics, Zonguldak Bülent Ecevit University, Zonguldak 67100, Turkey

[2] Faculty of Science, Department of Mathematics, Gazi University, Teknikokullar, Ankara 06500, Turkey; bayramcekim@gazi.edu.tr (B.Ç.); naimtuglu@gazi.edu.tr (N.T.)

[3] Department of Mathematics, Kwangwoon University, Seoul 139-701, Korea; tkkim@kw.ac.kr

* Correspondence: cankizilates@gmail.com

Abstract: In this paper, firstly the definitions of the families of three-variable polynomials with the new generalized polynomials related to the generating functions of the famous polynomials and numbers in literature are given. Then, the explicit representation and partial differential equations for new polynomials are derived. The special cases of our polynomials are given in tables. In the last section, the interesting applications of these polynomials are found.

Keywords: Fibonacci polynomials; Lucas polynomials; trivariate Fibonacci polynomials; trivariate Lucas polynomials; generating functions

MSC: 11B39; 11B37; 05A19

---

1. Introduction

In literature, the Fibonacci and Lucas numbers have been studied extensively and some authors tried to enhance and derive some directions to mathematical calculations using these special numbers [1–3]. By favour of the Fibonacci and Lucas numbers, one of these directions verges on the tribonacci and the tribonacci-Lucas numbers. In fact, M. Feinberg in 1963 has introduced the tribonacci numbers and then derived some properties for these numbers in [4–7]. Elia in [4] has given and investigated the tribonacci-Lucas numbers. The tribonacci numbers T_n for any integer $n > 2$ are defined via the following recurrence relation

$$T_n = T_{n-1} + T_{n-2} + T_{n-3}, \tag{1}$$

with the initial values $T_0 = 0$, $T_1 = 1$, and $T_2 = 1$. Similarly, by way of the initial values $K_0 = 3$, $K_1 = 1$, and $K_2 = 3$, the tribonacci-Lucas numbers K_n are given by the recurrence relation

$$K_n = K_{n-1} + K_{n-2} + K_{n-3}. \tag{2}$$

By dint of the above extensions, the tribonacci and tribonacci-Lucas numbers are introduced with the help of the following generating functions, respectively:

$$\sum_{n=0}^{\infty} T_n t^n = \frac{t}{1 - t - t^2 - t^3}, \tag{3}$$

and

$$\sum_{n=0}^{\infty} K_n t^n = \frac{3 - 2t - t^2}{1 - t - t^2 - t^3}. \tag{4}$$

Moreover, some authors define a large class of polynomials by using the Fibonacci and the tribonacci numbers [6–9]. Firstly, the well-known Fibonacci polynomials are defined via the recurrence relation

$$F_{n+1}(x) = xF_n(x) + F_{n-1}(x),$$

with $F_0(x) = 0$, $F_1(x) = 1$. The well-known Lucas polynomials are defined with the help of the recurrence relation

$$L_{n+1}(x) = xL_n(x) + L_{n-1}(x),$$

with $L_0(x) = 2$, $L_1(x) = x$.

Fibonacci or Fibonacci-like polynomials have been studied by many mathematicians for many years. Recently, in [10], Kim et al. kept in mind the sums of finite products of Fibonacci polynomials and of Chebyshev polynomials of the second kind and obtained Fourier series expansions of functions related to them. In [11], Kim et al. studied the convolved Fibonacci numbers by using the generating functions of them and gave some new identities for the convolved Fibonacci numbers. In [12], Wang and Zhang studied some sums of powers Fibonacci polynomials and Lucas polynomials. In [13], Wu and Zhang obtained the several new identities involving the Fibonacci polynomials and Lucas polynomials.

Afterwards, by giving the Pell and Jacobsthal polynomials, in 1973, Hoggatt and Bicknell [6] introduced the tribonacci polynomials. The tribonacci polynomials are defined by the recurrence relation for $n \geq 0$,

$$t_{n+3}(x) = x^2 t_{n+2}(x) + x t_{n+1}(x) + t_n(x), \tag{5}$$

where $t_0(x) = 0$, $t_1(x) = 1$, and $t_2(x) = x^2$. The tribonacci-Lucas polynomials are defined by the recurrence relation for $n \geq 0$,

$$k_{n+3}(x) = x^2 k_{n+2}(x) + x k_{n+1}(x) + k_n(x), \tag{6}$$

where $k_0(x) = 3$, $k_1(x) = x^2$, and $k_2(x) = x^4 + 2x$, respectively. Here we note that $t_n(1) = T_n$ which is the tribonacci numbers and $k_n(1) = K_n$ which is the tribonacci-Lucas numbers. Also for these polynomials, we have the generating function as follows:

$$\sum_{n=0}^{\infty} t_n(x) t^n = \frac{t}{1 - x^2 t - x t^2 - t^3}, \tag{7}$$

and

$$\sum_{n=0}^{\infty} k_n(x) t^n = \frac{3 - 2x^2 t - x t^2}{1 - x^2 t - x t^2 - t^3}. \tag{8}$$

On the other hand, some authors try to define the second and third variables of these polynomials with the help of these numbers. For example [8], for integer $n > 2$, the recurrence relations of the trivariate Fibonacci and Lucas polynomials are as follows:

$$H_n(x, y, z) = xH_{n-1}(x, y, z) + yH_{n-2}(x, y, z) + zH_{n-3}(x, y, z), \tag{9}$$

with $H_0(x, y, z) = 0$, $H_1(x, y, z) = 1$, $H_2(x, y, z) = x$ and

$$K_n(x, y, z) = xK_{n-1}(x, y, z) + yK_{n-2}(x, y, z) + zK_{n-3}(x, y, z), \tag{10}$$

with $K_0(x, y, z) = 3$, $K_1(x, y, z) = x$, $K_2(x, y, z) = x^2 + 2y$, respectively. Also for these, we have the generating functions as follows:

$$\sum_{n=0}^{\infty} H_n(x, y, z) t^n = \frac{t}{1 - xt - yt^2 - zt^3}, \tag{11}$$

and

$$\sum_{n=0}^{\infty} K_n(x,y,z)t^n = \frac{3 - 2xt - yt^2}{1 - xt - yt^2 - zt^3}. \tag{12}$$

After that, Ozdemir and Simsek [14] give the family of two-variable polynomials, reducing some well-known polynomials and obtaining some properties of these polynomials. In light of these polynomials, we introduce the families of three-variable polynomials with the new generalized polynomials reduced to the generating functions of the famous polynomials and numbers in literature. Then, we obtain the explicit representations and partial differential equations for new polynomials. The special cases of our polynomials are given in tables. Also the last section, we give the interesting applications of these polynomials.

2. The New Generalized Polynomials: Definitions and Properties

Now, we introduce the original and wide generating functions reduce the well-known polynomials and the well-known numbers such as the trivariate Fibonacci and Lucas polynomials, the tribonacci and the tribonacci-Lucas polynomials, the tribonacci and the tribonacci-Lucas numbers, and so on.

Firstly, some properties of these functions are investigated. Then, in the case of the new generating function, we give some properties the particular well-known polynomials as tables.

Via the following generating functions, a new original and wide family of three-variable polynomials denoted by $S_j := S_j(x,y,z;k,m,n,c)$ is defined as follows:

$$T := M(t;x,y,z;k,m,n,c) = \sum_{j=0}^{\infty} S_j t^j = \frac{1}{1 - x^k t - y^m t^{m+n} - z^c t^{m+n+c}}, \tag{13}$$

where $k,m,n,c \in \mathbb{N} - \{0\}$, and $\left| x^k t + y^m t^{m+n} + z^c t^{m+n+c} \right| < 1$. Now we derive the explicit representation for polynomials S_j. By means of Taylor series of the generating function of the right hand side of (13), we can write

$$T = \sum_{j=0}^{\infty} S_j t^j = \sum_{j=0}^{\infty} \left(x^k t + y^m t^{m+n} + z^c t^{m+n+c} \right)^j.$$

After that, using the binomial expansion and taking $j + s$ instead of j, we get

$$T = \sum_{j=0}^{\infty} \sum_{s=0}^{\infty} \binom{j+s}{s} \left(x^k t \right)^j \left(t^{m+n} \right)^s \left(y^m + z^c t^c \right)^s.$$

Lastly, using the expansion of $(y^m + z^c t^c)^s$, taking $u + s$ instead of s, taking $j - (m+n+c)u$ instead of j and taking $j - (m+n)s$ instead of j, respectively, we have

$$T = \sum_{j=0}^{\infty} \sum_{s=0}^{\left\lfloor \frac{j}{n+m} \right\rfloor} \sum_{u=0}^{\left\lfloor \frac{j-(m+n)s}{n+m+c} \right\rfloor} \binom{j - (n+m-1)s - (n+m+c-1)u}{s+u} \binom{s+u}{u} \left(x^k \right)^{j-(n+m)(s+u)-cu} z^{cu} y^{ms} t^j.$$

Thus after the equalization of coefficients of t^j, we obtain

$$S_j = \sum_{s=0}^{\left\lfloor \frac{j}{n+m} \right\rfloor} \sum_{u=0}^{\left\lfloor \frac{j-(m+n)s}{n+m+c} \right\rfloor} \binom{j-(n+m-1)s-(n+m+c-1)u}{s+u} \binom{s+u}{u} \left(x^k \right)^{j-(n+m)(s+u)-cu} z^{cu} y^{ms}. \tag{14}$$

Note that for $z = 0$, our polynomials reduces to the polynomials Equation (4) [14].

Remark 1. *As a similar to Theorem 2.3 in [14], we can write the following relation*

$$S_j(2x, -1, 0; 1, 1, 1, c) = \sum_{r=0}^{j} P_{j-r}(x) P_r(x),$$

where $P_r(x)$ are the Legendre polynomials.

To obtain other wide family of well-known polynomials, we define the second new generating function for the family of the polynomials $W_j := W_j(x, y, z; k, m, n, c)$ as follows

$$R := R(t; x, y, z; k, m, n, c) = M(t; x, y, z; k, m, n, c)t^n$$

$$= \frac{t^n}{1 - x^k t - y^m t^{m+n} - z^c t^{m+n+c}}$$

$$= \sum_{j=0}^{\infty} W_j t^j, \tag{15}$$

where $k, m, n, c \in \mathbb{N} - \{0\}$, and $\left| x^k t + y^m t^{m+n} + z^c t^{m+n+c} \right| < 1$. Similarly for $z = 0$, our polynomials reduces to the polynomials in (5) in [14]. Now we give some special case. Firstly taking $k = m = n = c = 1$ in (15), we give the generating function

$$\frac{t}{1 - xt - yt^2 - zt^3} = \sum_{j=0}^{\infty} W_j(x, y, z; 1, 1, 1, 1) t^j,$$

where $W_j(x, y, z; 1, 1, 1, 1) = H_j(x, y, z)$, which are trivariate Fibonacci polynomials in (11). Secondly, writing $k = m = n = c = 1$ and $x \to x^2$, $y \to x$, $z \to 1$, we have the generating function

$$\frac{t}{1 - x^2 t - xt^2 - t^3} = \sum_{j=0}^{\infty} W_j(x^2, x, 1; 1, 1, 1, 1) t^j,$$

where $W_j(x^2, x, 1; 1, 1, 1, 1) = t_j(x)$ which are the tribonacci polynomials in (7). In the above generating function, for $x = 1$, we find the generating function of the tribonacci numbers in (3). Now, we give other special cases as the following table related to (15).

Now, we define a new family of the polynomials denoted by $K_j := K_j(x, y, z; k, m, n, c)$ via the generating function

$$\sum_{j=0}^{\infty} K_j t^j = \frac{\alpha(t; x, y) - \beta(t; x, y)t^n}{1 - x^k t - y^m t^{m+n} - z^c t^{m+n+c}}, \tag{16}$$

where $k, m, n, c \in \mathbb{N} - \{0\}$, $\alpha(t; x, y)$ and $\beta(t; x, y)$ are arbitrary polynomials depending on t, x, y and $\left| x^k t + y^m t^{m+n} + z^c t^{m+n+c} \right| < 1$. Thirdly, via (16), we give

$$3M(t; x, y, z; 1, 1, 1, 1) - 2xR(t; x, y, z; 1, 1, 1, 1) - ytR(t; x, y, z; 1, 1, 1, 1) = \frac{3 - 2xt - yt^2}{1 - xt - yt^2 - zt^3}$$

$$= \sum_{j=0}^{\infty} K_j(x, y, z) t^j,$$

where $K_j(x, y, z)$ are the trivariate Lucas polynomials in (12). Due to the last equation, we have the polynomial representation

$$3S_j(x, y, z; 1, 1, 1, 1) - 2xW_j(x, y, z; 1, 1, 1, 1) - ytW_j(x, y, z; 1, 1, 1, 1) = K_j(x, y, z). \tag{17}$$

In (17) substituting $x \to x^2, y \to x, z \to 1$, we get

$$3S_j(x^2, x, 1; 1, 1, 1, 1) - 2xW_j(x^2, x, 1; 1, 1, 1, 1) - ytW_j(x^2, x, 1; 1, 1, 1, 1) = k_j(x),$$

where $k_j(x)$ are the tribonacci-Lucas polynomials in (6). In the above representation, for $x = 1$, we find the generating function of the tribonacci-Lucas numbers in (4).

Now, we give other special cases as Table 1 and Table 2 related to (15) and (16) respectively.

Table 1. Special cases of W_j.

x	y	z	k	m	n	c	**Special Case**
x	y	z	1	1	1	1	Trivariate Fibonacci Polynomials [8]
x^2	x	1	1	1	1	1	tribonacci Polynomials [8]
x	y	0	1	1	1	c	Bivariate Fibonacci Polynomials [9]
x	1	0	1	p	1	c	Fibonacci $p-$Polynomials [9]
$2x$	1	0	1	p	1	c	Pell $p-$Polynomials [9]
x	1	0	1	1	1	c	Fibonacci Polynomials [9]
$2x$	1	0	1	1	1	c	Pell Polynomials [9]
1	$2y$	0	k	1	1	c	Jacobsthal Polynomials [9]
$3x$	-2	0	1	1	1	c	Fermat Polynomials [15]
x	-2	0	1	1	1	c	First kind of Fermat–Horadam Polynomials [16]
x	$-\alpha$	0	1	1	1	c	Second kind of Dickson Polynomials [17]
$x+2$	-1	0	1	1	1	c	Morgan–Voyce Polynomials [18]
$x+1$	$-x$	0	1	1	1	c	Delannoy Polynomials [19]
$h(x)$	1	0	1	1	1	c	$h(x)-$Fibonacci Polynomials [2]
$p(x)$	$q(x)$	0	1	1	1	c	$(p, q)-$Fibonacci Polynomials [15]
1	1	0	k	1	1	c	Fibonacci Numbers [9]
2	1	0	1	1	1	c	Pell Numbers [9]
1	2	0	k	1	1	c	Jacobsthal Numbers [9]

Table 2. Special cases of K_j

α	β	x	y	z	k	m	n	c	**Special Case**
3	$2x + yt$	x	y	z	1	1	1	1	Trivariate Lucas Polynomials [8]
3	$2x^2 + xt$	x^2	x	1	1	1	1	1	tribonacci-Lucas Polynomials [8]
2	xz	x	y	0	1	1	1	c	Bivariate Lucas Polynomials [9]
$p+1$	px	x	1	0	1	p	1	c	Lucas $p-$Polynomials [9]
0	-1	$2x$	1	0	1	p	1	c	Pell Lucas $p-$Polynomials [9]
2	x	x	1	0	1	1	1	c	Lucas Polynomials [9]
2	$2x$	$2x$	1	0	1	1	1	c	Pell Lucas Polynomials [9]
2	1	1	$2y$	0	k	1	1	c	Jacobsthal Lucas Polynomials [9]
2	$3x$	$3x$	-2	0	1	1	1	c	Fermat Lucas Polynomials [15]
2	x	x	-2	0	1	1	1	c	Second kind of Fermat–Horadam P. [16]
2	x	x	$-\alpha$	0	1	1	1	c	First kind of Dickson Polynomials [17]
2	$x+2$	$x+2$	-1	0	1	1	1	c	Morgan–Voyce Polynomials [18]
2	$x+1$	$x+1$	$-x$	0	1	1	1	c	Corona Polynomials [19]
2	$h(x)$	$h(x)$	1	0	1	1	1	c	$h(x)-$Lucas Polynomials [2]
2	$p(x)$	$p(x)$	$q(x)$	0	1	1	1	c	$(p, q)-$Lucas Polynomials [15]
2	1	1	1	0	k	1	1	c	Lucas Numbers [9]
2	2	2	1	0	1	1	1	c	Pell–Lucas Numbers [9]
2	1	1	2	0	k	1	1	c	Jacobsthal–Lucas Numbers [9]
t	t	2	2	-1	1	1	1	1	Squares of Fibonacci Numbers [1]

3. Partial Differential Equations for Polynomials in (13)

With the help of the derivatives of these generating functions with regard to some variable and algebraic arrangements, we derive some partial differential equations for new polynomials. Taking the derivative with regard to x, y, z, t of the generating function in (13), respectively, they hold

$$\frac{\partial}{\partial x} M = k x^{k-1} t M^2, \tag{18}$$

$$\frac{\partial}{\partial y} M = m y^{m-1} t^{n+m} M^2, \tag{19}$$

$$\frac{\partial}{\partial z} M = c z^{c-1} t^{n+m+c} M^2, \tag{20}$$

$$\frac{\partial}{\partial t} M = \left(x^k + y^m (n+m) t^{n+m-1} + z^c (n+m+c) t^{n+m+c-1} \right) M^2. \tag{21}$$

From (13) and (18), we get the following theorem.

Theorem 1. *For $j \geq 0$, we have the first relation as follows:*

$$\frac{\partial}{\partial x} S_j = k x^{k-1} \sum_{l=0}^{j-1} S_{j-l-1} S_l.$$

Combining (13) and (19), we have the next theorem.

Theorem 2. *For $j \geq m + n$, we have the second relation as follows:*

$$\frac{\partial}{\partial y} S_j = \sum_{l=0}^{j-m-n} m y^{m-1} S_{j-m-n-l} S_l.$$

With the help of considering (13) and (20), we get the next result.

Theorem 3. *For $j \geq m + n + c$, we have the third relation as follows:*

$$\frac{\partial}{\partial z} S_j = c z^{c-1} \sum_{l=0}^{j-m-n-c} S_{j-m-n-c-l} S_l.$$

Lastly, by means of (13) and (21), we get the following result.

Theorem 4.

(i) *For $m + n - 1 \leq j \leq m + n + c - 1$, then we obtain*

$$(j+1) S_{j+1} = x^k \sum_{l=0}^{j} S_{j-l} S_l + y^m (n+m) \sum_{l=0}^{j-m-n+1} S_l S_{j-m-n-l+1}.$$

(ii) *For $j \leq m + n - 1$, then we derive*

$$(j+1) S_{j+1} = x^k \sum_{l=0}^{j} S_{j-l} S_l.$$

(iii)   *For $j \geq m + n + c - 1$, then we get*

$$(j+1)S_{j+1} = x^k \sum_{l=0}^{j} S_{j-l}S_l + y^m(n+m) \sum_{l=0}^{j-m-n+1} S_l S_{j-m-n-l+1}$$

$$+ z^c(n+m+c) \sum_{l=0}^{j-m-n-c+1} S_l S_{j-m-n-c-l+1}.$$

After that, using the partial differential equations in (18)–(21), we get the new partial differential equation for S_j.

Theorem 5. *For $j \geq 0$, we have*

$$jS_j = \frac{x}{k}\frac{\partial}{\partial x}S_j + \left(\frac{n+m}{m}\right)y\frac{\partial}{\partial y}S_j + \left(\frac{n+m+c}{c}\right)z\frac{\partial}{\partial z}S_j.$$

Proof. Combining (18)–(21), we get

$$\frac{\partial}{\partial t}M - \frac{x}{kt}\frac{\partial}{\partial x}M = \left(\frac{n+m}{m}\right)\frac{y}{t}\frac{\partial}{\partial y}M + \left(\frac{n+m+c}{c}\right)\frac{z}{t}\frac{\partial}{\partial z}M.$$

In the above, using (13), we get the desired result.   □

4. Some Applications of Generating Functions

In this section, by using these functions, some identities connected with these polynomials are derived. Furthermore, in the special case, we show that these identities reduce to the well-known sum identities connected with the well-known numbers in literature.

Case 1.   Taking $t = \frac{1}{a}$ in (15) for $|a| > 1$, we get the following equation

$$\sum_{j=0}^{\infty} \frac{W_j}{a^j} = \frac{a^{m+c}}{a^{m+n+c} - x^k a^{m+n+c-1} - y^m a^c - z^c}. \tag{22}$$

(i)   Substituting $a = 2$, $x \to x^2$, $y \to x$, $z \to 1$ and $k = m = n = c = 1$ in (22), we obtain the relation for the tribonacci polynomials as

$$\sum_{j=0}^{\infty} \frac{t_j(x)}{2^j} = \frac{4}{7 - 4x^2 - 2x}. \tag{23}$$

Writing $x = 1$ in (23), we have

$$\sum_{j=0}^{\infty} \frac{T_j}{2^j} = 4,$$

where T_j are the tribonacci numbers.

(ii)   Taking $a = 10$, $x \to x^2$, $y \to x$, $z \to 1$ and $k = m = n = c = 1$ in (22), we get

$$\sum_{j=0}^{\infty} \frac{T_j(x)}{10^{j+2}} = \frac{1}{999 - 100x^2 - 10x}, \tag{24}$$

and writing $x = 1$ (24), we get for the tribonacci numbers

$$\sum_{j=0}^{\infty} \frac{T_j}{10^{j+2}} = \frac{1}{889}.$$

(iii) Substituting $x \to x$, $y \to 1$, $z \to 0$, $a = 2$, and $k = m = n = c = 1$ into (22), we get for the Fibonacci polynomials

$$\sum_{j=0}^{\infty} \frac{F_j(x)}{2^j} = \frac{2}{3 - 2x}, \tag{25}$$

which was given in [14]. Then taking $x = 1$ in (25), we have for Fibonacci numbers

$$\sum_{j=0}^{\infty} \frac{F_j}{2^j} = 2,$$

which was given in [14].

(iv) Substituting $x \to x$, $y \to 1$, $z \to 0$, $a = 3$, and $k = m = n = c = 1$ in (22), we get for the Fibonacci polynomials

$$\sum_{j=0}^{\infty} \frac{F_j(x)}{3^{j+1}} = \frac{1}{8 - 3x}. \tag{26}$$

Taking $x = 1$ in (26), we get for the Fibonacci numbers

$$\sum_{j=0}^{\infty} \frac{F_j}{3^{j+1}} = \frac{1}{5} = \frac{1}{F_5},$$

was given in page 424 in [1].

(v) Substituting $x \to x$, $y \to 1$, $z \to 0$, $a = 8$, and $k = m = n = c = 1$ in (22), we get for the Fibonacci polynomials

$$\sum_{j=0}^{\infty} \frac{F_j(x)}{8^{j+1}} = \frac{1}{63 - 3x}. \tag{27}$$

Taking $x = 1$ in (27), we get for the Fibonacci numbers

$$\sum_{j=0}^{\infty} \frac{F_j}{8^{j+1}} = \frac{1}{55} = \frac{1}{F_{10}},$$

was given in page 424 in [1].

(vi) Substituting $x \to x$, $y \to 1$, $z \to 0$, $a = -10$, and $k = m = n = c = 1$ in (22), we get for the Fibonacci polynomials

$$\sum_{j=0}^{\infty} \frac{F_j(x)}{(-10)^{j+1}} = \frac{1}{99 + 10x}. \tag{28}$$

Taking $x = 1$ in (28), we get for the Fibonacci numbers

$$\sum_{j=0}^{\infty} \frac{F_j}{(-10)^{j+1}} = \frac{1}{109},$$

was given in page 427 in [1].

(vii) Substituting $x \to 2x$, $y \to 1$, $z \to 0$, $a = 3$, and $k = m = n = c = 1$ in (22), we get

$$\sum_{j=0}^{\infty} \frac{P_j(x)}{3^{j+1}} = \frac{1}{8 - 6x}, \tag{29}$$

where $P_j(x)$ are the Pell polynomials. Then taking $x = 1$ in (29), we have

$$\sum_{j=0}^{\infty} \frac{P_j}{3^{j+1}} = \frac{1}{2},$$

where P_j are the Pell numbers.

(viii) Substituting $x \to 1$, $y \to 2y$, $z \to 0$, $a = 3$, and $k = m = n = c = 1$ in (22), we get

$$\sum_{s=0}^{\infty} \frac{J_s(x)}{3^{s+1}} = \frac{1}{6 - 2y}, \tag{30}$$

where $J_s(x)$ are the Jacobsthal polynomials. Then taking $y = 1$ in (30), we have

$$\sum_{s=0}^{\infty} \frac{J_s}{3^{s+1}} = \frac{1}{4},$$

where J_s are the Jacobsthal numbers.

Case 2. Taking $t = \frac{1}{a}$ in (16) for $|a| > 1$, we get the following equation

$$\sum_{j=0}^{\infty} \frac{K_j}{a^j} = \frac{a^{m+n+c}\alpha(t; x, y) - a^{m+c}\beta(t; x, y)}{a^{m+n+c} - x^k a^{m+n+c-1} - y^m a^c - z^c}. \tag{31}$$

(i) Substituting $x \to x^2$, $y \to x$, $z \to 1$, $a = 2$, and $k = m = n = c = 1$, $\alpha(t; x, y) = 3$, $\beta(t; x, y) = 2x^2 + xt$ in (31), we get

$$\sum_{j=0}^{\infty} \frac{k_j(x)}{2^j} = \frac{24 - 8x^4 - 2x^2}{7 - 4x^2 - 2x}, \tag{32}$$

where $k_j(x)$ are the tribonacci-Lucas polynomials. Then taking $x = 1$ in (32), we have

$$\sum_{j=0}^{\infty} \frac{k_j}{2^{j+1}} = 7,$$

where $k_j(x)$ are the tribonacci-Lucas numbers.

(ii) Substituting $x \to x$, $y \to 1$, $z \to 0$, $a = 2$, and $k = m = n = c = 1$, $\alpha(t; x, y) = 2$, $\beta(t; x, y) = x$ in (31), we get

$$\sum_{j=0}^{\infty} \frac{L_j(x)}{2^{j+1}} = \frac{4 - x}{3 - 2x}, \tag{33}$$

where $L_j(x)$ are the Lucas polynomials. Then taking $x = 1$ in (33), we have

$$\sum_{j=0}^{\infty} \frac{L_j}{2^{j+1}} = 3,$$

where L_j are the Lucas numbers.

(iii) Substituting $x \to x$, $y \to 1$, $z \to 0$, $a = 10$, and $k = m = n = c = 1$, $\alpha(t; x, y) = 2$, $\beta(t; x, y) = x$ in (31), we get

$$\sum_{j=0}^{\infty} \frac{L_j(x)}{10^j} = \frac{200 - 10x}{99 - 10x}, \tag{34}$$

where $L_j(x)$ are the Lucas polynomials. Then taking $x = 1$ in (34), for L_j are the Lucas numbers, we have

$$\sum_{j=0}^{\infty} \frac{L_j}{10^{j+1}} = \frac{19}{89} = \frac{L_6 - L_1}{F_{11}}$$

was given in page 427 in [1].

(iv) Substituting $x \to x$, $y \to 1$, $z \to 0$, $a = 3$, and $k = m = n = c = 1$, $\alpha(t; x, y) = 2$, $\beta(t; x, y) = x$ in (31), we get

$$\sum_{j=0}^{\infty} \frac{L_j(x)}{3^{j+1}} = \frac{6 - x}{8 - 3x}, \tag{35}$$

and taking $x = 1$ in (35), we have

$$\sum_{j=0}^{\infty} \frac{L_j}{3^{j+1}} = 1.$$

(v) Substituting $x \to x$, $y \to 1$, $z \to 0$, $a = 8$, and $k = m = n = c = 1$, $\alpha(t; x, y) = 2$, $\beta(t; x, y) = x$ in (31), we get

$$\sum_{j=0}^{\infty} \frac{L_j(x)}{8^{j+1}} = \frac{16 - x}{63 - 8x}, \tag{36}$$

and taking $x = 1$ in (36), we have

$$\sum_{j=0}^{\infty} \frac{L_j}{8^{j+1}} = \frac{3}{11} = \frac{L_2}{L_5}.$$

(vi) Substituting $x \to x$, $y \to 1$, $z \to 0$, $a = -10$, and $k = m = n = c = 1$, $\alpha(t; x, y) = 2$, $\beta(t; x, y) = x$ in (31), we get

$$\sum_{j=0}^{\infty} \frac{L_j(x)}{(-10)^{j+1}} = \frac{-20 - x}{99 + 10x}. \tag{37}$$

Taking $x = 1$ in (37), we have

$$\sum_{j=0}^{\infty} \frac{L_j}{(-10)^{j+1}} = \frac{-21}{109},$$

was given in page 427 in [1].

(vii) Substituting $x \to 2x$, $y \to 1$, $z \to 0$, $a = 5$, and $k = m = n = c = 1$, $\alpha(t; x, y) = 2$, $\beta(t; x, y) = 2x$ in (31), we get

$$\sum_{j=0}^{\infty} \frac{Q_j(x)}{5^{j+1}} = \frac{5 - x}{12 - 5x}, \tag{38}$$

where $Q_j(x)$ are the Pell Lucas polynomials. Then taking $x = 1$ in (38), we have

$$\sum_{j=0}^{\infty} \frac{Q_j}{5^{j+1}} = \frac{4}{7},$$

where Q_j are the Pell Lucas numbers.

(viii) Substituting $x \to 1$, $y \to 2y$, $z \to 0$, $a = 3$, and $k = m = n = c = 1$, $\alpha(t; x, y) = 2$, $\beta(t; x, y) = 1$ in (31), we get

$$\sum_{s=0}^{\infty} \frac{j_s(y)}{3^{s+1}} = \frac{5}{6 - 2y}, \tag{39}$$

where $j_s(y)$ are the Jocabsthal Lucas polynomials. Then taking $y = 1$ in (39), we have

$$\sum_{s=0}^{\infty} \frac{j_s}{3^{s+1}} = \frac{5}{4},$$

where j_s is Jocabsthal Lucas number.

(ix)   Substituting $x \to 2$, $y \to 2$, $z \to -1$, $a = 4$, and $k = m = n = c = 1$, $\alpha(t; x, y) = t$, $\beta(t; x, y) = t$ in (31), for the square of Fibonacci numbers F_j, we get

$$\sum_{j=0}^{\infty} \frac{F_j^2}{4^j} = \frac{12}{25},$$

was given in page 439 in [1].

Let us give Tables 3 and 4 containing the obtained formulas for simplify reading.

Table 3. Special cases of Equation (22) for $k = m = n = c = 1$.

a	x	y	z	Formulas
2	x^2	x	1	$\sum_{j=0}^{\infty} \frac{t_j(x)}{2^j} = \frac{4}{7-4x^2-2x}$
2	1	1	1	$\sum_{j=0}^{\infty} \frac{T_j}{2^j} = 4$
10	x^2	x	1	$\sum_{j=0}^{\infty} \frac{T_j(x)}{10^{j+2}} = \frac{1}{999-100x^2-10x}$
10	1	1	1	$\sum_{j=0}^{\infty} \frac{T_j}{10^{j+2}} = \frac{1}{889}$
2	x	1	0	$\sum_{j=0}^{\infty} \frac{F_j(x)}{2^j} = \frac{2}{3-2x}$
2	1	1	0	$\sum_{j=0}^{\infty} \frac{F_j}{2^j} = 2$
3	x	1	0	$\sum_{j=0}^{\infty} \frac{F_j(x)}{3^{j+1}} = \frac{1}{8-3x}$
3	1	1	0	$\sum_{j=0}^{\infty} \frac{F_j}{3^{j+1}} = \frac{1}{5} = \frac{1}{F_5}$
8	x	1	0	$\sum_{j=0}^{\infty} \frac{F_j(x)}{8^{j+1}} = \frac{1}{63-3x}$
8	1	1	0	$\sum_{j=0}^{\infty} \frac{F_j}{8^{j+1}} = \frac{1}{55} = \frac{1}{F_{10}}$
-10	x	1	0	$\sum_{j=0}^{\infty} \frac{F_j(x)}{(-10)^{j+1}} = \frac{1}{99+10x}$
-10	1	1	0	$\sum_{j=0}^{\infty} \frac{F_j}{(-10)^{j+1}} = \frac{1}{109}$
3	$2x$	1	0	$\sum_{j=0}^{\infty} \frac{P_j(x)}{3^{j+1}} = \frac{1}{8-6x}$
3	2	1	0	$\sum_{j=0}^{\infty} \frac{P_j}{3^{j+1}} = \frac{1}{2}$
3	1	$2y$	0	$\sum_{s=0}^{\infty} \frac{J_s(x)}{3^{s+1}} = \frac{1}{6-2y}$
3	1	2	0	$\sum_{s=0}^{\infty} \frac{J_s}{3^{s+1}} = \frac{1}{4}$

Table 4. Special cases of Equation (31) for $k = m = n = c = 1$.

a	x	y	z	α	β	Formulas
2	x^2	x	1	3	$2x^2 + xt$	$\sum\limits_{j=0}^{\infty} \frac{k_j(x)}{2^j} = \frac{24 - 8x^4 - 2x^2}{7 - 4x^2 - 2x}$
2	1	1	1	3	$2 + t$	$\sum\limits_{j=0}^{\infty} \frac{k_j}{2^{j+1}} = 7$
2	x	1	0	2	x	$\sum\limits_{j=0}^{\infty} \frac{L_j(x)}{2^{j+1}} = \frac{4-x}{3-2x}$
2	1	1	0	2	1	$\sum\limits_{j=0}^{\infty} \frac{L_j}{2^{j+1}} = 3$
10	x	1	0	2	x	$\sum\limits_{j=0}^{\infty} \frac{L_j(x)}{10^j} = \frac{200 - 10x}{99 - 10x}$
10	1	1	0	2	1	$\sum\limits_{j=0}^{\infty} \frac{L_j}{10^{j+1}} = \frac{19}{89} = \frac{L_6 - L_1}{F_{11}}$
3	x	1	0	2	x	$\sum\limits_{j=0}^{\infty} \frac{L_j(x)}{3^{j+1}} = \frac{6-x}{8-3x}$
3	1	1	0	2	1	$\sum\limits_{j=0}^{\infty} \frac{L_j}{3^{j+1}} = 1$
8	x	1	0	2	x	$\sum\limits_{j=0}^{\infty} \frac{L_j(x)}{8^{j+1}} = \frac{16-x}{63-8x}$
8	1	1	0	2	1	$\sum\limits_{j=0}^{\infty} \frac{L_j}{8^{j+1}} = \frac{3}{11} = \frac{L_2}{L_5}$
-10	x	1	0	2	x	$\sum\limits_{j=0}^{\infty} \frac{L_j(x)}{(-10)^{j+1}} = \frac{-20-x}{99+10x}$
-10	1	1	0	2	1	$\sum\limits_{j=0}^{\infty} \frac{L_j}{(-10)^{j+1}} = \frac{-21}{109}$
5	$2x$	1	0	2	$2x$	$\sum\limits_{j=0}^{\infty} \frac{Q_j(x)}{5^{j+1}} = \frac{5-x}{12-5x}$
5	2	1	0	2	2	$\sum\limits_{j=0}^{\infty} \frac{Q_j}{5^{j+1}} = \frac{4}{7}$
3	1	$2y$	0	2	1	$\sum\limits_{s=0}^{\infty} \frac{j_s(y)}{3^{s+1}} = \frac{5}{6-2y}$
3	1	2	0	2	1	$\sum\limits_{s=0}^{\infty} \frac{j_s}{3^{s+1}} = \frac{5}{4}$
4	2	2	-1	$1/4$	$1/4$	$\sum\limits_{j=0}^{\infty} \frac{F_j^2}{4^j} = \frac{12}{25}$

5. Conclusions

In the present paper, we considered the families of three-variable polynomials with the generalized polynomials reduce to generating function of the polynomials and numbers in the literature. In Section 2, we gave special polynomials and numbers as the tables related to (15) and (16). Then we obtained the explicit representations and partial differential equations for new polynomials. In the last section, we gave the interesting sum identities related to the well-known numbers and polynomials in the literature.

For all of the resuts, if the appropriate values given in the tables are taken, many infinite sums including various polynomials are obtained.

In recent years, some authors use the well-known polynomials and numbers in the applications of ordinary and fractional differential equations and difference equations (for example [20–23]). Therefore, our new families of three variables polynomials could been used for future works of some application areas such as mathematical modelling, physics, engineering, and applied sciences.

Author Contributions: All authors contributed equally to this work. All authors read and approved the final manuscript.

Acknowledgments: The authors would like to express their sincere gratitude to the referees for their valuable comments which have significantly improved the presentation of this paper.

References

1.   Koshy, T. *Fibonacci and Lucas Numbers with Applications*; John Wiley and Sons Inc.: New York, NY, USA, 2001.
2.   Nalli, A.; Haukkanen, P. On generalized Fibonacci and Lucas polynomials. *Chaos Solitons Fractals* **2009**, *42*, 3179–3186. [CrossRef]
3.   Vajda, S. *Fibonacci and Lucas Numbers, and the Golden Section, Theory and Applications*; Ellis Horwood Limited: Chichester, UK, 1989.
4.   Elia, M. Derived sequences, the tribonacci recurrence and cubic forms. *Fibonacci Q.* **2001**, *39*, 107–109.
5.   Feinberg, M. Fibonacci-tribonacci. *Fibonacci Q.* **1963**, *3*, 70–74.
6.   Hoggatt, V.E., Jr. Bicknell, M. Generalized Fibonacci polynomials. *Fibonacci Q.* **1973**, *11*, 457–465.
7.   Tan, M.; Zhang, Y. A note on bivariate and trivariate Fibonacci polynomials. *Southeast Asian Bull. Math.* **2005**, *29*, 975–990.
8.   Kocer, E.G.; Gedikli, H. Trivariate Fibonacci and Lucas polynomials. *Konuralp J. Math.* **2016**, *4*, 247–254.
9.   Tuglu, N.; Kocer, E.G.; Stakhov, A. Bivariate Fibonacci like $p-$polynomials. *Appl. Math. Comput.* **2011**, *217*, 10239–10246. [CrossRef]
10.   Kim, T.; Kim, D.S.; Dolgy, D.V.; Park, J.W. Sums of finite products of Chebyshev polynomials of the second kind and of Fibonacci polynomials. *J. Inequal. Appl.* **2018**, *148*. [CrossRef] [PubMed]
11.   Kim, T.; Dolgy, D.V.; Kim, D.S.; Seo, J.J. Convolved Fibonacci numbers and their applications. *arXiv* **2017**, arXiv:1607.06380 .
12.   Wang, T.; Zhang, W. Some identities involving Fibonacci, Lucas polynomials and their applications. *Bull. Math. Soc. Sci. Math. Roum.* **2012**, *55*, 95–103.
13.   Wu, Z.; Zhang, W. Several identities involving the Fibonacci polynomials and Lucas polynomials. *J. Inequal. Appl.* **2013**, *2013*, 14. [CrossRef]
14.   Ozdemir, G.; Simsek, Y. Generating functions for two-variable polynomials related to a family of Fibonacci type polynomials and numbers. *Filomat* **2016**, *30*, 969–975. [CrossRef]
15.   Lee, G.Y.; Asci, M. Some properties of the $(p,q)-$Fibonacci and $(p,q)-$Lucas polynomials. *J. Appl. Math.* **2012**, *2012*, 264842. [CrossRef]
16.   Horadam, A.F. Chebyshev and Fermat polynomials for diagonal functions. *Fibonacci Q.* **1979**, *17*, 328–333.
17.   Lidl, R.; Mullen, G.; Tumwald. G. Dickson Polynomials. *Pitman Monographs and Surveys in Pure and Applied Mathematics*; Longman Scientific and Technical: Essex, UK, 1993; Volume 65,
18.   Shannon, A.G.; Horadam, A.F. Some relationships among Vieta, Morgan-Voyce and Jacobsthal polynomials. In *Applications of Fibonacci Numbers*; Howard, F.T., Ed.; Kluwer Academic Publishers: Dordrecht, The Netherlands, 1999; pp. 307–323.
19.   Cheon, G.-S.; Kim, H.; Shapiro, L.W. A generalization of Lucas polynomial sequence. *Discret. Appl. Math.* **2009**, *157*, 920–927. [CrossRef]
20.   Bulut, H; Zhang, Pandir, Y.; Baskonus, H.M. Symmetrical hyperbolic Fibonacci function solutions of generalized Fisher equation with fractional order. *AIP Conf. Proc.* **2013**, *1558*, 1914–1918.
21.   Mirzaee, F.; Hoseini, S.F. Solving singularly perturbed differential-difference equations arising in science and engineering whit Fibonacci polynomials. *Results Phys.* **2013**, *3*, 134–141. [CrossRef]
22.   Mirzaee, F.; Hoseini, S.F. Solving systems of linear Fredholm integro-differential equations with Fibonacci polynomials. *Ain. Shams Eng.* **2014**, *5*, 271–283. [CrossRef]
23.   Kurt, A.; Yalinbash, S.; Sezer, M. Fibonacci-collocation method for solving high-order linear Fredholm integro-differential-difference equations. *Int. J. Math. Math. Sci.* **2013**, *2013*, 486013. [CrossRef]

14

Some Identities of Ordinary and Degenerate Bernoulli Numbers and Polynomials

Dmitry V. Dolgy [1], **Dae San Kim** [2], **Jongkyum Kwon** [3,*] **and Taekyun Kim** [4]

[1]   Hanrimwon, Kwangwoon University, Seoul 139-701, Korea
[2]   Department of Mathematics, Sogang University, Seoul 121-742, Korea
[3]   Department of Mathematics Education and ERI, Gyeongsang National University, Jinju, Gyeongsangnamdo 52828, Korea
[4]   Department of Mathematics, Kwangwoon University, Seoul 139-701, Korea
*   Correspondence: mathkjk26@gnu.ac.kr

Abstract: In this paper, we investigate some identities on Bernoulli numbers and polynomials and those on degenerate Bernoulli numbers and polynomials arising from certain p-adic invariant integrals on $\mathbb{Z}_p$. In particular, we derive various expressions for the polynomials associated with integer power sums, called integer power sum polynomials and also for their degenerate versions. Further, we compute the expectations of an infinite family of random variables which involve the degenerate Stirling polynomials of the second and some value of higher-order Bernoulli polynomials.

Keywords: Bernoulli polynomials; degenerate Bernoulli polynomials; random variables; p-adic invariant integral on $\mathbb{Z}_p$; integer power sums polynomials; Stirling polynomials of the second kind; degenerate Stirling polynomials of the second kind

1. Introduction

We begin this section by reviewing some known facts. In more detail, we recall the integral equation for the p-adic invariant integral of a uniformly differentiable function on $\mathbb{Z}_p$ and its generalizations, the expression in terms of some values of Bernoulli polynomials for the integer power sums, and the p-adic integral representaions of Bernoulli polynomials and of their generating functions.

Throughout this paper, $\mathbb{Z}_p$, $\mathbb{Q}_p$ and $\mathbb{C}_p$ will denote the ring of p-adic integers, the field of p-adic rational numbers and the completion of the algebraic closure of $\mathbb{Q}_p$, respectively. The p-adic norm is normalized as $|p|_p = \frac{1}{p}$. Let f be a uniformly differentiable function on $\mathbb{Z}_p$. Then the p-adic invariant integral of f (also called the Volkenborn integral of f) on $\mathbb{Z}_p$ is defined by

$$
\begin{aligned}
I_0(f) = \int_{\mathbb{Z}_p} f(x) d\mu_0(x) &= \lim_{N \to \infty} \frac{1}{p^N} \sum_{x=0}^{p^N-1} f(x) \\
&= \lim_{N \to \infty} \sum_{x=0}^{p^N-1} f(x) \mu_0(x + p^N \mathbb{Z}_p).
\end{aligned}
\tag{1}
$$

Here we note that $\mu_0(x + p^N \mathbb{Z}_p) = \frac{1}{p^N}$ is a distribution but not a measure. The existence of such integrals for uniformly differentiable functions on $\mathbb{Z}_p$ is detailed in [1,2]. It can be seen from (1) that

$$
I_0(f_1) = I_0(f) + f'(0),
\tag{2}
$$

where $f_1(x) = f(x+1)$, and $f'(0) = \frac{df(x)}{dx}|_{x=0}$, (see [1,2]).

In general, by induction and with $f_n(x) = f(x+n)$, we can show that

$$I_0(f_n) = I_0(f) + \sum_{k=0}^{n-1} f'(k), \quad (n \in \mathbb{N}), \tag{3}$$

As is well known, the Bernoulli polynomials are given by the generating function (see [3–5])

$$\frac{t}{e^t - 1} e^{xt} = \sum_{n=0}^{\infty} B_n(x) \frac{t^n}{n!}, \tag{4}$$

When $x = 0$, $B_n = B_n(0)$ are called the Bernoulli numbers.
From (4), we note that (see [3–5])

$$B_n(x) = \sum_{l=0}^{n} \binom{n}{l} B_l x^{n-l}, \quad (n \geq 0), \tag{5}$$

and

$$B_0 = 1, \quad \sum_{k=0}^{n} \binom{n}{k} B_k - B_n = \begin{cases} 1, & \text{if } n = 1, \\ 0, & \text{if } n > 1, \end{cases}$$

Let (see [6–13])

$$S_p(n) = \sum_{k=1}^{n} k^p, \quad (n, p \in \mathbb{N}). \tag{6}$$

The generating function of $S_p(n)$ is given by

$$\sum_{p=0}^{\infty} S_p(n) \frac{t^p}{p!} = \sum_{k=1}^{n} e^{kt} = \frac{1}{t} \left(\frac{t}{e^t - 1} \left(e^{(n+1)t} - e^t \right) \right)$$

$$= \sum_{p=0}^{\infty} \left(\frac{B_{p+1}(n+1) - B_{p+1}(1)}{p+1} \right) \frac{t^p}{p!}. \tag{7}$$

Thus, by (7), we get

$$S_p(n) = \frac{B_{p+1}(n+1) - B_{p+1}(1)}{p+1}, \quad (n, p \in \mathbb{N}). \tag{8}$$

From (2), we have

$$\int_{\mathbb{Z}_p} e^{(x+y)t} d\mu_0(y) = \frac{t}{e^t - 1} e^{xt} = \sum_{n=0}^{\infty} B_n(x) \frac{t^n}{n!}. \tag{9}$$

By (9), we get (see [11,12])

$$\int_{\mathbb{Z}_p} (x+y)^n d\mu_0(y) = B_n(x), \quad (n \geq 0), \tag{10}$$

From (8) and (10), we can derive the following equation.

$$\int_{\mathbb{Z}_p} (x+k+1)^{p+1} d\mu_0(x) - \int_{\mathbb{Z}_p} x^{p+1} d\mu_0(x) = (p+1) \sum_{n=1}^{k} n^p, \quad (p \in \mathbb{N}). \tag{11}$$

Thus, by (6) and (11), and for $p \in \mathbb{N}$, we get

$$S_p(k) = \frac{1}{p+1} \left\{ \int_{\mathbb{Z}_p} (x+k+1)^{p+1} d\mu_0(x) - \int_{\mathbb{Z}_p} x^{p+1} d\mu_0(x) \right\}. \tag{12}$$

The purpose of this paper is to investigate some identities on Bernoulli numbers and polynomials and those on degenerate Bernoulli numbers and polynomials arising from certain p-adic invariant integrals on $\mathbb{Z}_p$.

The outline of this paper is as in the following. After reviewing well-known necessary results in Section 1, we will derive some identities on Bernoulli polynomials and numbers in Section 2. In particular, we will introduce the integer power sum polynomials and derive several expressions for them. In Section 3, we will obtain some identities on degenerate Bernoulli numbers and polynomials. Especially, we will introduce the degenerate integer power sum polynomials, a degenerate version of the integer power sum polynomials and deduce various representations of them. In the final Section 4, we will consider an infinite family of random variables and compute their expectations to see that they involve the degenerate Stirling polynomials of the second and some value of higher-order Bernoulli polynomials.

2. Some Identities of Bernoulli Numbers and Polynomials

For $p \in \mathbb{N}$, we observe that

$$(j+1)^{p+1} - j^{p+1} = \sum_{i=0}^{p+1} \binom{p+1}{i} j^i - j^{p+1}$$

$$= (p+1)j^p + \sum_{i=1}^{p-1} \binom{p+1}{i} j^i + 1. \tag{13}$$

Thus, we get

$$(n+1)^{p+1} = \sum_{j=0}^{n} \left\{ (j+1)^{p+1} - j^{p+1} \right\} = (p+1) \sum_{j=0}^{n} j^p + \sum_{i=1}^{p-1} \binom{p+1}{i} \sum_{j=0}^{n} j^i + (n+1). \tag{14}$$

From (14), we have

$$S_p(n) = \frac{1}{p+1} \left\{ (n+1)^{p+1} - (n+1) - \sum_{i=1}^{p-1} \binom{p+1}{i} S_i(n) \right\}. \tag{15}$$

Therefore, by (15), we obtain the following lemma.

Lemma 1. *For $n, p \in \mathbb{N}$, we have*

$$\int_{\mathbb{Z}_p} (x+n+1)^{p+1} d\mu_0(x) - \int_{\mathbb{Z}_p} x^{p+1} d\mu_0(x)$$

$$= (n+1)^{p+1} - (n+1) - \sum_{i=1}^{n-1} \binom{p+1}{i} \frac{1}{i+1} \tag{16}$$

$$\times \left\{ \int_{\mathbb{Z}_p} (x+n+1)^{i+1} d\mu_0(x) - \int_{\mathbb{Z}_p} x^{i+1} d\mu_0(x) \right\}.$$

From Lemma 1, we note the following.

Corollary 1. *For $n, p \in \mathbb{N}$, we have*

$$B_{p+1}(n+1) - B_{p+1} = (n+1)^{p+1} - (n+1) - \sum_{i=1}^{p-1} \binom{p+1}{i} \frac{1}{i+1} \left(B_{i+1}(n+1) - B_{i+1} \right). \tag{17}$$

For $n \in \mathbb{N}_0 = \mathbb{N} \cup \{0\}$, by (1), we get

$$\int_{\mathbb{Z}_p} \left(y + 1 - x \right)^n d\mu_0(y) = (-1)^n \int_{\mathbb{Z}_p} (y + x)^n d\mu_0(y). \tag{18}$$

From (18), we note that

$$B_n(1-x) = (-1)^n B_n(x), \quad (n \geq 0). \tag{19}$$

Now, we observe that, for $n \geq 1$,

$$
\begin{aligned}
B_n(2) &= \sum_{l=0}^{n} \binom{n}{l} B_l(1) = B_0 + \binom{n}{1} B_1(1) + \sum_{l=2}^{n} \binom{n}{l} B_l(1) \\
&= B_0 + \binom{n}{1} B_1 + n + \sum_{l=2}^{n} \binom{n}{l} B_l = n + \sum_{l=0}^{n} \binom{n}{l} B_l \\
&= n + B_n(1).
\end{aligned}
\tag{20}
$$

Thus we have completed the proof for the next lemma.

Lemma 2. *For any $n \in \mathbb{N}_0$, the following identity is valid:*

$$B_n(2) = n + B_n + \delta_{n,1}, \tag{21}$$

where $\delta_{n,1}$ is the Kronecker's delta.

For any $n, m \in \mathbb{N}$ with $n, m \geq 2$, we have

$$
\begin{aligned}
\int_{\mathbb{Z}_p} x^m(-1+x)^n d\mu_0(x) &= \sum_{i=0}^{n} \binom{n}{i} (-1)^{n-i} \int_{\mathbb{Z}_p} x^{m+i} d\mu_0(x) \\
&= \sum_{i=0}^{n} \binom{n}{i} (-1)^{n-i} B_{m+i} \\
&= (-1)^{n-m} \sum_{i=0}^{n} \binom{n}{i} B_{m+i}.
\end{aligned}
\tag{22}
$$

On the other hand,

$$
\begin{aligned}
\int_{\mathbb{Z}_p} x^m(x-1)^n d\mu_0(x) &= \sum_{i=0}^{m} \binom{m}{i} \int_{\mathbb{Z}_p} (x-1)^{n+i} d\mu_0(x) \\
&= \sum_{i=0}^{m} \binom{m}{i} (-1)^{n+i} \int_{\mathbb{Z}_p} (x+2)^{n+i} d\mu_0(x) \\
&= \sum_{i=0}^{m} \binom{m}{i} (-1)^{n+i} \left(B_{n+i} + n + i \right) \\
&= \sum_{i=0}^{m} \binom{m}{i} (-1)^{n+i} B_{n+i} \\
&= \sum_{i=0}^{m} \binom{m}{i} B_{n+i}.
\end{aligned}
\tag{23}
$$

Therefore, by (22) and (23), we obtain the following theorem.

Theorem 1. *For any $m, n \in \mathbb{N}$ with $m, n \geq 2$, the following symmetric identity holds:*

$$(-1)^n \sum_{i=0}^{n} \binom{n}{i} B_{m+i} = (-1)^m \sum_{i=0}^{m} \binom{m}{i} B_{n+i}. \tag{24}$$

From (5), we note that

$$B_n(1) = \sum_{l=0}^{n} \binom{n}{l} B_l, \quad (n \geq 0).$$

For $n \geq 2$, we have

$$B_n = B_n(1) = \sum_{l=0}^{n} \binom{n}{l} B_l = \sum_{l=0}^{n} \binom{n}{l} B_{n-l}. \tag{25}$$

Now, we define the *integer power sum polynomials* by

$$S_p(n|x) = \sum_{k=0}^{n} (k+x)^p, \quad (n, p \in \mathbb{N}_0). \tag{26}$$

Note that $S_p(n|0) = S_p(n), \quad (n \in \mathbb{N}_0, p \in \mathbb{N})$.
For $N \in \mathbb{N}_0$, we have

$$t \sum_{k=0}^{N} e^{(k+x)t} = \int_{\mathbb{Z}_p} e^{(N+1+x+y)t} d\mu_0(y) - \int_{\mathbb{Z}_p} e^{(x+y)t} d\mu_0(y). \tag{27}$$

Then it is immediate to see from (27) that we have

$$\sum_{k=0}^{N} e^{(k+x)t} = \sum_{n=0}^{\infty} \frac{1}{n+1} \left\{ \int_{\mathbb{Z}_p} (N+1+x+y)^{n+1} d\mu_0(y) - \int_{\mathbb{Z}_p} (x+y)^{n+1} d\mu_0(y) \right\} \frac{t^n}{n!}. \tag{28}$$

Now, we see that (28) is equivalent to the next theorem.

Theorem 2. *For $n, N \in \mathbb{N}_0$, we have*

$$S_n(N|x) = \frac{1}{n+1} \left\{ B_{n+1}(x+N+1) - B_{n+1}(x) \right\}. \tag{29}$$

Let $\triangle$ denote the difference operator given by

$$\triangle f(x) = f(x+1) - f(x). \tag{30}$$

Then, by (30) and induction, we get

$$\triangle^n f(x) = \sum_{k=0}^{n} \binom{n}{k} (-1)^{n-k} f(x+k), \quad (n \geq 0). \tag{31}$$

Now, we can deduce the Equation (32) from (27) as in the following:

$$
\begin{aligned}
\sum_{k=0}^{N} e^{(k+x)t} &= \frac{1}{t} e^{xt} (e^{(N+1)t} - 1) \int_{\mathbb{Z}_p} e^{yt} d\mu_0(y) \\
&= \frac{1}{e^t - 1} \left(\sum_{m=0}^{N+1} \binom{N+1}{m} (e^t - 1)^m - 1 \right) e^{xt} \\
&= \frac{1}{e^t - 1} \sum_{m=1}^{N+1} \binom{N+1}{m} (e^t - 1)^m e^{xt} \\
&= \sum_{m=0}^{N} \binom{N+1}{m+1} (e^t - 1)^m e^{xt} \\
&= \sum_{n=0}^{\infty} \left\{ \sum_{m=0}^{N} \binom{N+1}{m+1} \sum_{k=0}^{m} \binom{m}{k} (-1)^{m-k} (k+x)^n \right\} \frac{t^n}{n!} \\
&= \sum_{n=0}^{\infty} \left\{ \sum_{k=0}^{N} \sum_{m=k}^{N} \binom{N+1}{m+1} \binom{m}{k} (-1)^{m-k} (k+x)^n \right\} \frac{t^n}{n!}.
\end{aligned}
\tag{32}
$$

Therefore, (31) and (32) together yield the next theorem.

Theorem 3. *For $n, N \geq 0$, we have*

$$
S_n(N|x) = \sum_{m=0}^{N} \binom{N+1}{m+1} \triangle^m x^n = \sum_{k=0}^{N} (k+x)^n T(N,k),
\tag{33}
$$

where $T(N,k) = \sum_{m=k}^{N} \binom{N+1}{m+1} \binom{m}{k} (-1)^{m-k}$.
In particular, we have

$$
S_0(N|x) = \sum_{k=0}^{N} T(N,k) = N + 1.
$$

We recall here that the Stirling polynomials of the second kind $S_2(n,k|x)$ are given by (see [14])

$$
\frac{1}{k!} (e^t - 1)^k e^{xt} = \sum_{n=k}^{\infty} S_2(n,k|x) \frac{t^n}{n!}.
\tag{34}
$$

Note here that $S_2(n,k|0) = S_2(n,k)$ are Stirling numbers of the second kind. Then, we can show that, for integers $n, m \geq 0$, we have

$$
\frac{1}{m!} \triangle^m x^n = \begin{cases} S_2(n,m|x), & \text{if } n \geq m, \\ 0, & \text{if } n < m. \end{cases}
\tag{35}
$$

We can see this, for example, by taking $\lambda \to 0$ in (51).

Remark 1. *Combing (33) and (35), we obtain*

$$
S_n(N|x) = \sum_{m=0}^{\min\{N,n\}} \binom{N+1}{m+1} m! S_2(n,m|x).
$$

For any $m, k \in \mathbb{N}$ with $m - k \geq 2$, we observe that

$$
\begin{aligned}
\int_{\mathbb{Z}_p} x^{m-k} d\mu_0(x) &= \int_{\mathbb{Z}_p} (x+1)^{m-k} d\mu_0(x) \\
&= \sum_{j=0}^{m-k} \binom{m-k}{m-k-j} \int_{\mathbb{Z}_p} x^{m-k-j} d\mu_0(x) \\
&= \sum_{j=k}^{m} \binom{m-k}{m-j} \int_{\mathbb{Z}_p} x^{m-j} d\mu_0(x) \\
&= \frac{1}{\binom{m}{k}} \sum_{j=k}^{m} \binom{m}{j} \binom{j}{k} \int_{\mathbb{Z}_p} x^{m-j} d\mu_0(x).
\end{aligned}
\tag{36}
$$

Thus we have shown the following result.

Theorem 4. *For any $m, k \in \mathbb{N}$ with $m - k \geq 2$, the following holds true:*

$$
\binom{m}{k} \int_{\mathbb{Z}_p} x^{m-k} d\mu_0(x) = \sum_{j=k}^{m} \binom{m}{j} \binom{j}{k} \int_{\mathbb{Z}_p} x^{m-j} d\mu_0(x).
\tag{37}
$$

From (10) and (37), we derive the following corollary.

Corollary 2. *For $m, k \in \mathbb{N}$ with $m - k \geq 2$, we have*

$$
\binom{m}{k} B_{m-k} = \sum_{j=k}^{m} \binom{m}{j} \binom{j}{k} B_{m-j}.
\tag{38}
$$

3. Some Identities of Degenerate Bernoulli Numbers and Polynomials

In this section, we assume that $0 \neq \lambda \in \mathbb{C}_p$ with $|\lambda|_p < p^{-\frac{1}{p-1}}$. The degenerate exponential function is defined as (see [3,13])
$$
e_\lambda^x(t) = (1 + \lambda t)^{\frac{x}{\lambda}}.
$$

Note that $\lim_{\lambda \to 0} e_\lambda^x(t) = e^{xt}$. In addition, we denote $(1 + \lambda t)^{\frac{1}{\lambda}} = e_\lambda^1(t)$ simply by $e_\lambda(t)$.
As is well known, the degenerate Bernoulli polynomials are defined by Carlitz as

$$
\frac{t}{e_\lambda(t) - 1} e_\lambda^x(t) = \frac{t}{(1+\lambda t)^{\frac{1}{\lambda}} - 1} (1 + \lambda t)^{\frac{x}{\lambda}} = \sum_{n=0}^{\infty} \beta_{n,\lambda}(x) \frac{t^n}{n!}.
\tag{39}
$$

When $x = 0$, $\beta_{n,\lambda} = \beta_{n,\lambda}(0)$ are called the degenerate Bernoulli numbers, (see [3,15]).
From (39), we note that (see [3])

$$
\beta_{n,\lambda}(x) = \sum_{l=0}^{n} \binom{n}{l} (x)_{n-l,\lambda} \beta_{l,\lambda},
\tag{40}
$$

where $(x)_{0,\lambda} = 1$, $(x)_{n,\lambda} = x(x - \lambda) \cdots (x - (n-1)\lambda)$, $(n \geq 1)$.
By (39) and (40), we get

$$
\beta_{n,\lambda}(1) - \beta_{n,\lambda} = \delta_{n,1}.
\tag{41}
$$

Now, we observe that

$$\sum_{k=0}^{N} e_\lambda^{k+x}(t) = \frac{e_\lambda^{N+1}(t) - 1}{e_\lambda(t) - 1} e_\lambda^x(t) = \frac{1}{t}\left\{ \frac{t}{e_\lambda(t) - 1}\left(e_\lambda^{N+1+x}(t) - e_\lambda^x(t) \right) \right\}$$

$$= \frac{1}{t}\sum_{n=0}^{\infty} \left(\beta_{n,\lambda}(N+1+x) - \beta_{n,\lambda}(x) \right)\frac{t^n}{n!} \tag{42}$$

$$= \sum_{n=0}^{\infty} \left(\frac{\beta_{n+1,\lambda}(N+1+x) - \beta_{n+1,\lambda}(x)}{n+1} \right)\frac{t^n}{n!}, \quad (n \in \mathbb{N}_0).$$

On the other hand,

$$\sum_{k=0}^{N} e_\lambda^{k+x}(t) = \sum_{n=0}^{\infty}\left(\sum_{k=0}^{N}(k+x)_{n,\lambda} \right)\frac{t^n}{n!}. \tag{43}$$

Let us define a degenerate version of the integer power sum polynomials, called the *degenerate integer power sum polynomials*, by

$$S_{p,\lambda}(n|x) = \sum_{k=0}^{n}(k+x)_{p,\lambda}, \quad (n \geq 0). \tag{44}$$

Note that $\lim_{\lambda \to 0} S_{p,\lambda}(n|x) = S_p(n|x)$, $(n \geq 0)$.
Therefore, by (42) and (43), we obtain the following theorem.

Theorem 5. *For $n, N \in \mathbb{N}_0$, we have*

$$S_{n,\lambda}(N|x) = \frac{1}{n+1}\left(\beta_{n+1,\lambda}(N+1+x) - \beta_{n+1,\lambda}(x) \right). \tag{45}$$

Now, we observe that

$$\sum_{k=0}^{N} e_\lambda^{x+k}(t) = \frac{1}{e_\lambda(t) - 1}\left(e_\lambda^{N+1}(t) - 1 \right)e_\lambda^x(t)$$

$$= \frac{1}{e_\lambda(t) - 1}\left((e_\lambda(t) - 1 + 1)^{N+1} - 1 \right)e_\lambda^x(t)$$

$$= \frac{1}{e_\lambda(t) - 1}\sum_{m=1}^{N+1}\binom{N+1}{m}(e_\lambda(t) - 1)^m e_\lambda^x(t)$$

$$= \sum_{m=0}^{N}\binom{N+1}{m+1}(e_\lambda(t) - 1)^m e_\lambda^x(t) \tag{46}$$

$$= \sum_{n=0}^{\infty}\left(\sum_{m=0}^{N}\binom{N+1}{m+1}\sum_{k=0}^{m}\binom{m}{k}(-1)^{m-k}(k+x)_{n,\lambda} \right)\frac{t^n}{n!}.$$

$$= \sum_{n=0}^{\infty}\left(\sum_{k=0}^{N}\sum_{m=k}^{N}\binom{N+1}{m+1}\binom{m}{k}(-1)^{m-k}(k+x)_{n,\lambda} \right)\frac{t^n}{n!}.$$

Therefore, (31) and (46) together give the next result.

Theorem 6. *For any $n, N \in \mathbb{N}_0$, the following identity holds:*

$$S_{n,\lambda}(N|x) = \sum_{m=0}^{N}\binom{N+1}{m+1}\triangle^m (x)_{n,\lambda} = \sum_{k=0}^{N}(k+x)_{n,\lambda}T(N,k), \tag{47}$$

where $T(N,k) = \sum_{m=k}^{N} \binom{N+1}{m+1}\binom{m}{k}(-1)^{m-k}.$

As is known, the degenerate Stirling polynomials of the second kind are defined by Kim as (see [14])

$$(x+y)_{n,\lambda} = \sum_{k=0}^{n} S_{2,\lambda}(n,k|x)(y)_k, \tag{48}$$

where $(x)_0 = 1, (x)_n = x(x-1)\cdots(x-n+1), \ (n \geq 1).$

From (48), we can derive the generating function for $S_{2,\lambda}(n,k|x), \ (n,k \geq 0),$ as follows:

$$\frac{1}{k!}(e_\lambda(t)-1)^k e_\lambda^x(t) = \sum_{n=k}^{\infty} S_{2,\lambda}(n,k|x)\frac{t^n}{n!}. \tag{49}$$

When $x = 0, \ S_{2,\lambda}(n,k|0) = S_{2,\lambda}(n,k)$ are called the degenerate Stirling numbers of the second kind.

By (49), we get

$$
\begin{aligned}
\sum_{n=m}^{\infty} S_{2,\lambda}(n,m|x)\frac{t^n}{n!} &= \frac{1}{m!}(e_\lambda(t)-1)^m e_\lambda^x(t) \\
&= \frac{1}{m!}\sum_{k=0}^{m}\binom{m}{k}(-1)^{m-k}e_\lambda^{k+x}(t) \\
&= \sum_{n=0}^{\infty}\left(\frac{1}{m!}\sum_{k=0}^{m}\binom{m}{k}(-1)^{m-k}(x+k)_{n,\lambda}\right)\frac{t^n}{n!} \\
&= \sum_{n=0}^{\infty}\left(\frac{1}{m!}\triangle^m (x)_{n,\lambda}\right)\frac{t^n}{n!}.
\end{aligned}
\tag{50}
$$

Now, comparison of the coefficients on both sides of (50) yield following theorem.

Theorem 7. *For any $n, m \geq 0$, the following identity holds:*

$$\frac{1}{m!}\triangle^m (x)_{n,\lambda} = \begin{cases} S_{2,\lambda}(n,m|x), & \textit{if } n \geq m, \\ 0, & \textit{if } n < m. \end{cases} \tag{51}$$

Remark 2. *Combing (47) and (51), we obtain*

$$S_{n,\lambda}(N|x) = \sum_{m=0}^{\min\{N,n\}} \binom{N+1}{m+1}m!S_{2,\lambda}(n,m|x).$$

From (30) and proceeding by induction, we have

$$(1+\triangle)^k f(x) = \sum_{m=0}^{k}\binom{k}{m}\triangle^m f(x) = f(x+k), \ (k \geq 0). \tag{52}$$

By (52), we get

$$\sum_{k=0}^{N}(x+k)_{n,\lambda} = \sum_{k=0}^{N}(1+\triangle)^k (x)_{n,\lambda}. \tag{53}$$

It is known that Daehee numbers are given by the generating function

$$\frac{\log(1+t)}{t} = \sum_{n=0}^{\infty} D_n \frac{t^n}{n!}, \ (\text{see } [1,4,6]). \tag{54}$$

From (2), we have

$$
\begin{aligned}
\int_{\mathbb{Z}_p} e_\lambda^{x+y}(t)d\mu_0(y) &= \frac{\frac{1}{\lambda}\log(1+\lambda t)}{e_\lambda(t)-1}e_\lambda^x(t) \\
&= \frac{\log(1+\lambda t)}{\lambda t}\frac{t}{e_\lambda(t)-1}e_\lambda^x(t) \\
&= \sum_{l=0}^{\infty} D_l \frac{\lambda^l t^l}{l!} \sum_{m=0}^{\infty} \beta_{m,\lambda}(x)\frac{t^m}{m!} \\
&= \sum_{n=0}^{\infty}\left(\sum_{l=0}^{n}\binom{n}{l}\lambda^l D_l \beta_{n-l,\lambda}(x)\right)\frac{t^n}{n!}.
\end{aligned}
$$

(55)

From (55), we have

$$
\int_{\mathbb{Z}_p}(x+y)_{n,\lambda}d\mu_0(y) = \sum_{l=0}^{n}\binom{n}{l}\lambda^l D_l \beta_{n-l,\lambda}(x), \quad (n\geq 0).
$$

4. Further Remark

A random variable X is a real-valued function defined on a sample space. We say that X is a continuous random variable if there exists a nonnegative function f, defined on $(-\infty,\infty)$, having the property that for any set B of real numbers (see [16,17])

$$
P\{X\in B\} = \int_B f(x)dx.
$$

(56)

The function f is called the probability density function of random variable X.

Let X be a uniform random variable on the interval (α,β). Then the probability density function f of X is given by

$$
f(x) = \begin{cases} \frac{1}{\beta-\alpha}, & \text{if } \alpha < x < \beta, \\ 0, & \text{otherwise.} \end{cases}
$$

(57)

Let X be a continuous random variable with the probability density function f. Then the expectation of X is defined by

$$
E[X] = \int_{-\infty}^{\infty} xf(x)dx.
$$

For any real-valued function $g(x)$, we have (see [16])

$$
E[g(X)] = \int_{-\infty}^{\infty} g(x)f(x)dx.
$$

(58)

Assume that $X_1, X_2, \cdots, X_k$ are independent uniform random variables on $(0,1)$. Then we have

$$
E[e_\lambda^{x+X_1+X_2+\cdots+X_k}(t)] = e_\lambda^x(t) E[e_\lambda^{X_1}(t)] E[e_\lambda^{X_2}(t)] \cdots E[e_\lambda^{X_k}(t)]
$$

$$
= e_\lambda^x(t) \underbrace{\frac{\lambda}{\log(1+\lambda t)}(e_\lambda(t)-1) \times \cdots \times \frac{\lambda}{\log(1+\lambda t)}(e_\lambda(t)-1)}_{k-times}
$$

$$
= \left(\frac{\lambda t}{\log(1+\lambda t)} \right)^k \frac{k!}{t^k} \frac{1}{k!} (e_\lambda(t)-1)^k e_\lambda^x(t) \tag{59}
$$

$$
= \frac{k!}{t^k} \sum_{l=0}^\infty B_l^{(l-k+1)}(1)\lambda^l \frac{t^l}{l!} \sum_{m=k}^\infty S_{2,\lambda}(m,k \mid x) \frac{t^m}{m!}
$$

$$
= \frac{k!}{t^k} \sum_{n=k}^\infty \left(\sum_{m=k}^n \binom{n}{m} S_{2,\lambda}(m,k \mid x) B_{n-m}^{(n-m-k+1)}(1)\lambda^{n-m} \right) \frac{t^n}{n!},
$$

where $B_n^{(\alpha)}(x)$ are the Bernoulli polynomials of order α, given by (see [4,7,8])

$$
\left(\frac{t}{e^t-1} \right)^\alpha e^{xt} = \sum_{n=0}^\infty B_n^{(\alpha)}(x) \frac{t^n}{n!}, \tag{60}
$$

and we used the well-known formula

$$
\left(\frac{t}{\log(1+t)} \right)^n (1+t)^{x-1} = \sum_{k=0}^\infty B_k^{(k-n+1)}(x) \frac{t^k}{k!}. \tag{61}
$$

From (59), we note that

$$
\binom{n}{k} E[(x+X_1+X_2+\cdots+X_k)_{n-k,\lambda}]
$$

$$
= \sum_{m=k}^n \binom{n}{m} S_{2,\lambda}(m,k \mid x) B_{n-m}^{(n-m-k+1)}(1)\lambda^{n-m}. \tag{62}
$$

5. Conclusions

It is well-known and classical that the first n positive integer power sums can be given by an expression involving some values of Bernoulli polynomials. Here we investigated some identities on Bernoulli numbers and polynomials and those on degenerate Bernoulli numbers and polynomials, which can be deduced from certain p-adic invariant integrals on $\mathbb{Z}_p$.

In particular, we introduced the integer power sum polynomials associated with integer power sums and obtained various expressions of them. Namely, they can be given in terms of Bernoulli polynomials, difference operators, and of the Stirling polynomials of the second kind. In addition, we introduced a degenerate version of the integer power sum polynomials, called the degenerate integer power sum polynomials and were able to find several representations of them. In detail, they can be represented in terms of Carlitz degenerate Bernoulli polynomials, difference operators, and of the degenerate Stirling numbers of the second kind.

In the final section, we considered an infinite family of random variables and proved that the expectations of them are expressed in terms of the degenerate Stirling polynomials of the second and some value of higher-order Bernoulli polynomials.

Most of the results in Sections 1 and 2 are reviews of known results, other than that, we demonstrated the usefulness of the p-adic invariant integrals in the study of integer power sum polynomials. However, we emphasize that the results in Sections 3 and 4 are new. In particular, we showed that the degenerate Stirling polynomials of the second kind, introduced as a degenerate version of the Stirling polynomials of the second kind, appear naturally and meaningfully in the context of calculations of an infinite family of random variables (see (62)). We also showed that they appear in an expression of the degenerate integer power sum polynomials (Remark 2) which is a degenerate version of the integer power sum polynomials (see (26)).

We have witnessed in recent years that studying various degenerate versions of some old and new polynomials, initiated by Carlitz in the classical papers [3,15], is very productive and promising (see [3,5,14,15,18,19] and references therein). Lastly, we note that this idea of considering degenerate versions of some polynomials extended even to transcendental functions like the gamma functions (see [19]).

Author Contributions: All authors contributed equally to the manuscript, and typed, read and approved the final manuscript.

References

1. Kim, T. q-Volkenborn integration. *Russ. J. Math. Phys.* **2002**, *9*, 288–299.
2. Schikhof, W.H. *Ultrametric Calculus: An Introduction to p-Adic Analysis*; Cambridge Studies in Advanced Mathematics, 4; Cambridge University Press: Cambridge, UK, 1984.
3. Carlitz, L. Degenerate Stirling, Bernoulli and Eulerian numbers. *Utilitas Math. J.* **1979**, *15*, 51–88.
4. El-Desoulky, B.S.; Mustafa, A. New results on higher-order Daehee and Bernoulli numbers and polynomials. *Adv. Differ. Equ.* **2016**, *2016*, 32. [CrossRef]
5. Kim, T.; Kim, D.S. Identities for degenerate Bernoulli polynomials and Korobov polynomials of the first kind. *Sci. China Math.* **2019**, *62*, 999–1028. [CrossRef]
6. Araci, S.; Özer, O. Extened q-Dedekind-type Daehee-Changhee sums associated with q-Euler polynomials. *Adv. Differ. Equ.* **2015**, *2015*, 272. [CrossRef]
7. Kim, T. Symmetry of power sum polynomials and multivariate fermionic p-adic invariant integral on $\mathbb{Z}_p$. *Russ. J. Math. Phys.* **2009**, *16*, 93–96. [CrossRef]
8. Kim, T. Sums of powers of consecutive q-integers. *Adv. Stud. Contemp. Math. (Kyungshang)* **2004**, *9*, 15–18.
9. Kim, T. A note on exploring the sums of powers of consecutive q-integers. *Adv. Stud. Contemp. Math. (Kyungshang)* **2005**, *11*, 137–140.
10. Kim, T. On the alternating sums of powers of consecutive integers. *J. Anal. Comput.* **2005**, *1*, 117–120.
11. Rim, S.-H.; Kim, T.; Ryoo, C.S. On the alternating sums of powers of consecutive q-integers. *Bull. Korean Math. Soc.* **2006**, *43*, 611–617. [CrossRef]
12. Ryoo, C.S.; Kim, T. Exploring the q-analogues of the sums of powers of consecutive integers with Mathematica. *Adv. Stud. Contemp. Math. (Kyungshang)* **2009**, *18*, 69–77.
13. Simsek, Y.; Kim, D.S.; Kim, T.; Rim, S.-H. A note on the sums of powers of consecutive q-integers. *J. Appl. Funct. Differ. Equ.* **2016**, *1*, 81–88.
14. Kim, T. A note on degenerate Stirling polynomials of the second kind. *Proc. Jangjeon Math. Soc.* **2017**, *20*, 319–331.
15. Carlitz, L. A degenerate Staudt-Clausen theorem. *Arch. Math. (Basel)* **1956**, *7*, 28–33. [CrossRef]
16. Kim, T.; Yao, Y.; Kim, D.S.; Kwon, H.I. Some identities involving special numbers and moments of random variables. *Rocky Mt. J. Math.* **2019**, *49*, 521–538. [CrossRef]
17. Liu, C.; Bao, W. Application of probabilistic method on Daehee sequences. *Eur. J. Pure Appl. Math.* **2018**, *11*, 69–78. [CrossRef]
18. Kim, T.; Yao, Y.; Kim, D.S.; Jang, G.-W. Degenerate r-Stirling numbers and r-Bell polynomials. *Russ. J. Math. Phys.* **2018**, *25*, 44–58. [CrossRef]
19. Kim, T.; Kim, D.S. Degenerate Laplace transform and degenerate gamma function. *Russ. J. Math. Phys.* **2017**, *24*, 241–248. [CrossRef]

15

A New Sequence and its Some Congruence Properties

Wenpeng Zhang and Xin Lin *

School of Mathematics, Northwest University, Xi'an 710127, China; wpzhang@nwu.edu.cn
* Correspondence: estelle-xin@hotmail.com

Abstract: The aim of this paper is to study the congruence properties of a new sequence, which is closely related to Fubini polynomials and Euler numbers, using the elementary method and the properties of the second kind Stirling numbers. As results, we obtain some interesting congruences for it. This solves a problem proposed in a published paper.

Keywords: Fubini polynomials; Euler numbers; congruence; elementary method

MSC: 11B83; 11B37

1. Introduction

Let $n \geq 0$ be an integer, the famous Fubini polynomials $F_n(y)$ are defined according to the coefficients of following generating function:

$$\frac{1}{1 - y(e^t - 1)} = \sum_{n=0}^{\infty} \frac{F_n(y)}{n!} \cdot t^n, \tag{1}$$

where $F_0(y) = 1$, $F_1(y) = y$, and so on.

These polynomials are closely related to the Stirling numbers and Euler numbers. For example, if $y = -\frac{1}{2}$, then (1) becomes

$$\frac{2}{1 + e^t} = \sum_{n=0}^{\infty} \frac{E_n}{n!} \cdot t^n, \tag{2}$$

where E_n denotes the Euler numbers.

At the same time, the Fubini polynomials with two variables can also be defined by the following identity (see [1,2]):

$$\frac{e^{xt}}{1 - y(e^t - 1)} = \sum_{n=0}^{\infty} \frac{F_n(x, y)}{n!} \cdot t^n,$$

and $F_n(y) = F_n(0, y)$ for all integers $n \geq 0$. Many scholars have studied the properties of $F_n(x, y)$, and have obtained many important works. For example, T. Kim et al. proved a series of identities related to $F_n(x, y)$ (see [2,3]), one of which is

$$F_n(x, y) = \sum_{l=0}^{n} \binom{n}{l} x^l \cdot F_{n-l}(y), \ n \geq 0.$$

Zhao Jianhong and Chen Zhuoyu [4] studied the computational problem of the sums

$$\sum_{a_1 + a_2 + \cdots + a_k = n} \frac{F_{a_1}(y)}{(a_1)!} \cdot \frac{F_{a_2}(y)}{(a_2)!} \cdots \frac{F_{a_k}(y)}{(a_k)!},$$

where the summation in the formula above denotes all k-dimension non-negative integer coordinates $(a_1, a_2, \cdots, a_k)$ such that $a_1 + a_2 + \cdots + a_k = n$. They proved the identity

$$\sum_{a_1+a_2+\cdots+a_k=n} \frac{F_{a_1}(y)}{(a_1)!} \cdot \frac{F_{a_2}(y)}{(a_2)!} \cdots \frac{F_{a_k}(y)}{(a_k)!}$$
$$= \frac{1}{(k-1)!(y+1)^{k-1}} \cdot \frac{1}{n!} \sum_{i=0}^{k-1} C(k-1,i) F_{n+k-1-i}(y), \tag{3}$$

where the sequence $C(k,i)$ is defined for positive integer k and i with $0 \leq i \leq k$, $C(k,0) = 1$, $C(k,k) = k!$ and

$$C(k+1, i+1) = C(k, i+1) + (k+1)C(k,i), \text{ for all } 0 \leq i < k,$$

providing $C(k,i) = 0$, if $i > k$.

For clarity, for $1 \leq k \leq 9$, we list values of $C(k,i)$ in the following Table 1.

Table 1. Values of $C(k,i)$.

$C(k,i)$	$i=0$	$i=1$	$i=2$	$i=3$	$i=4$	$i=5$	$i=6$	$i=7$	$i=8$	$i=9$
$k=1$	1	1								
$k=2$	1	3	2							
$k=3$	1	6	11	6						
$k=4$	1	10	35	50	24					
$k=5$	1	15	85	225	274	120				
$k=6$	1	21	175	735	1624	1764	720			
$k=7$	1	28	322	1960	6769	13,132	13,068	5040		
$k=8$	1	36	546	4536	22,449	67,284	118,124	109,584	40,320	
$k=9$	1	45	870	9450	63,273	269,325	723,680	1,172,700	1,026,576	362,880

Meanwhile, Zhao Jianhong and Chen Zhuoyu [4] proposed some conjectures related to the sequence. We believe that this sequence is meaningful because it satisfies some very interesting congruence properties, such as

$$C(p-2, i) \equiv 1 \,(\text{mod } p) \tag{4}$$

for all odd primes p and integers $0 \leq i \leq p - 2$. The equivalent conclusion is

$$C(p-1, i) \equiv 0 \,(\text{mod } p) \tag{5}$$

for all odd primes p and positive integers $1 \leq i \leq p - 2$. Since some related content can be found in references [5–15], we will not go through all of them here.

The aim of this paper is to prove congruence (5) by applying the elementary method and the properties of the second kind Stirling numbers. That is, we will solve the conjectures in [4], which are listed in the following.

Theorem 1. *Let p be an odd prime. For any integer $1 \leq i \leq p - 2$, we have congruence*

$$C(p-1, i) \equiv 0 \,(\text{mod } p).$$

From this theorem and (3), we can deduce following three corollaries:

Corollary 1. *For any positive integer n and odd prime p, we have*

$$F_{n+p-1}(y) - F_n(y) \equiv 0 \,(\text{mod } p).$$

Corollary 2. *For any positive integer n and odd prime p, we have*

$$E_{n+p-1} - E_n \equiv 0 \,(\text{mod } p).$$

Corollary 3. *For any odd prime p, we have the congruences*

$$2E_p \equiv -1 \,(\text{mod } p), \quad 4E_{p+2} \equiv 1 \,(\text{mod } p), \quad and \quad 2E_{p+4} \equiv -1 \,(\text{mod } p).$$

Note. Since E_n is a rational number, we can denote $E_n = \dfrac{U_n}{V_n}$, where U_n and V_n are integers with $(U_n, V_n) = 1$. Based on this, in our paper, the expression $E_n \equiv 0 \,(\text{mod } p)$ means $p \mid U_n$, while $p \nmid V_n$.

2. Several Lemmas

Lemma 1. *For any positive integer k, we have the identity*

$$k!y(y+1)^{k-1} = \sum_{i=0}^{k-1} C(k-1, i) F_{k-i}(y).$$

Proof. Taking $n = 1$ in (3), and noting that $F_0(y) = 1$, $F_1(y) = y$, and the equation $a_1 + a_2 + \cdots + a_k = 1$ holds if and only if one of a_i is 1, others are 0. The number of the solutions of this equation is $\binom{k}{1} = k$. So, from (3), we have

$$\sum_{a_1+a_2+\cdots+a_k=1} \frac{F_{a_1}(y)}{(a_1)!} \cdot \frac{F_{a_2}(y)}{(a_2)!} \cdots \frac{F_{a_k}(y)}{(a_k)!} = \binom{k}{1} y = ky$$

$$= \frac{1}{(k-1)!(y+1)^{k-1}} \cdot \sum_{i=0}^{k-1} C(k-1, i) F_{k-i}(y)$$

or identity

$$k!y(y+1)^{k-1} = \sum_{i=0}^{k-1} C(k-1, i) F_{k-i}(y),$$

which proves Lemma 1. $\quad\square$

Lemma 2. *For any positive integer n, we have the identity*

$$F_n(y) = \sum_{k=0}^{n} S(n, k) \, k! \, y^k, \quad (n \geq 0),$$

where $S(n, k)$ are the second kind Stirling numbers, which are defined for any integer k, n with $0 \leq k \leq n$ as:

$$S(n, k) = kS(n-1, k) + S(n-1, k-1)$$

where $S(0, 0) = 1$, $S(n, 0) = 0$ and $S(0, k) = 0$ for $n, k > 0$.

Proof. See Reference [2]. $\quad\square$

Lemma 3. *For any positive integers n and k, we have*

$$S(n, k) = \frac{1}{k!} \sum_{j=0}^{k} \binom{k}{j} j^n (-1)^{k-j}.$$

Proof. See Theorem 4.3.12 of [16]. $\quad\square$

Lemma 4. *For any odd prime p and positive integer $2 \leq k \leq p-1$, we have the congruence*

$$k!S(p,k) \equiv 0 \,(\text{mod } p).$$

Proof. From the definition and properties of $S(n,k)$, we have $S(n,k) = 0$, if $k > n$. For any integers $0 \leq j \leq p-1$, from the famous Fermat's little theorem, we have the congruence $j^p \equiv j \,(\text{mod } p)$. From this congruence and Lemma 3, we have

$$k!S(p,k) = \sum_{j=0}^{k} \binom{k}{j} j^p (-1)^{k-j} \equiv \sum_{j=0}^{k} \binom{k}{j} j(-1)^{k-j} \equiv k!S(1,k) \equiv 0 \,(\text{mod } p),$$

if $k \geq 2$. This completes the proof of Lemma 4. $\square$

3. Proof of the Theorem

In this section, we will prove Theorem by mathematical induction. Taking $k = p$ in Lemma 1 and noting that $C(p-1,0) = 1$ and $C(p-1,p-1) = (p-1)!$, we have:

$$p!y(y+1)^{p-1} = \sum_{i=0}^{p-1} C(p-1,i)F_{p-i}(y)$$

$$= F_p(y) + y(p-1)! + \sum_{i=1}^{p-2} C(p-1,i)F_{p-i}(y).$$

Note that $(p-1)! + 1 \equiv 0 \,(\text{mod } p)$, which implies

$$F_p(y) - y + \sum_{i=1}^{p-2} C(p-1,i)F_{p-i}(y) \equiv 0 \,(\text{mod } p). \tag{6}$$

From (6), we have the congruence

$$y - F_p(y) \equiv \sum_{i=1}^{p-2} C(p-1,i)F_{p-i}(y) \,(\text{mod } p). \tag{7}$$

From Lemma 2, we have

$$F_p(y) = \sum_{k=0}^{p} S(p,k)\, k!\, y^k \tag{8}$$

and

$$F_p^{(p-1)}(0) = S(p,p-1)\,(p-1)! \cdot (p-1)!, \tag{9}$$

where $F_n^{(k)}(y)$ denotes the k-order derivative of $F_n(y)$ for variable y.

$$F_{p-1}^{(p-1)}(0) = S(p-1,p-1)\,(p-1)! \cdot (p-1)! = (p-1)! \cdot (p-1)!. \tag{10}$$

Then, applying Lemma 3 and Lemma 4 and noting that $S(1,p-1) = 0$, we have

$$\begin{aligned}
(p-1)!S(p,p-1) &\equiv \sum_{j=0}^{p-1} \binom{p-1}{j} j^p (-1)^{p-1-j} \equiv \sum_{j=0}^{p-1} \binom{p-1}{j} j(-1)^{p-1-j} \\
&\equiv (p-1)!S(1,p-1) \equiv 0 \,(\text{mod } p).
\end{aligned} \tag{11}$$

Combining (7), (9), (10), and (11), we have:

$$0 \equiv -S(p, p-1)(p-1)!(p-1)! \equiv C(p-1,1)(p-1)! \cdot (p-1)! \,(\mathrm{mod}\ p) \tag{12}$$

or

$$C(p-1,1) \equiv 0 \,(\mathrm{mod}\ p). \tag{13}$$

That is, the theorem is true for $i = 1$.

Assume that the theorem is true for all $1 \leq i \leq s$. That is,

$$C(p-1,i) \equiv 0 \,(\mathrm{mod}\ p)$$

for $1 \leq i \leq s < p - 1$. It is clear that if $s = p - 2$, then the theorem is true.

If $1 < s < p - 2$, then from (7) we have the congruence

$$y - F_p(y) \equiv \sum_{i=s+1}^{p-2} C(p-1,i)F_{p-i}(y) \,(\mathrm{mod}\ p). \tag{14}$$

In congruence (14), taking the $(p-s-1)$-order derivative with respect to t, then let $y = 0$, applying Lemma 2, we have:

$$
\begin{aligned}
&-S(p, p-s-1)(p-s-1)! \cdot (p-s-1)! \\
\equiv\ & C(p-1, s+1)(p-s-1)!(p-s-1)! \,(\mathrm{mod}\ p).
\end{aligned}
\tag{15}
$$

Note that $((p-s-1)!, p) = 1$, from Lemma 4 and (15) we have the congruence

$$C(p-1, s+1)(p-s-1)! \equiv -(p-s-1)!S(p, p-s-1) \equiv 0 \,(\mathrm{mod}\ p),$$

which implies

$$C(p-1, s+1) \equiv 0 \,(\mathrm{mod}\ p).$$

That is, the theorem is true for $i = s + 1$. Now the proof of the theorem completes by mathematical induction.

Now, we prove Corollary 1. For any integer $n \geq 0$, taking $k = p$ in (3) and noting that

$$n! \sum_{a_1+a_2+\cdots+a_p=n} \frac{F_{a_1}(y)}{(a_1)!} \cdot \frac{F_{a_2}(y)}{(a_2)!} \cdots \frac{F_{a_p}(y)}{(a_p)!} \equiv 0 \,(\mathrm{mod}\ p),$$

we have

$$\sum_{i=0}^{p-1} C(p-1,i)F_{n+p-1-i}(y) \equiv 0 \,(\mathrm{mod}\ p). \tag{16}$$

From our theorem, we have

$$\sum_{i=1}^{p-2} C(p-1,i)F_{n+p-1-i}(y) \equiv 0 \,(\mathrm{mod}\ p). \tag{17}$$

Note that $C(p-1,0) = 1$, $C(p-1,p-1) = (p-1)!$. Combining (16) and (17), we can deduce the congruence

$$F_{n+p-1}(y) - F_n(y) \equiv 0 \,(\mathrm{mod}\ p).$$

Now the proof of Corollary 1 completes. Since Corollarys 2 and 3 are the special situation of Corollary 1, we will not prove Corollarys 2 and 3 here.

Author Contributions: Conceptualization, W.Z.; Methodology, W.Z. and X.L.; Software, X.L.; Validation, W.Z. and X.L.; Formal Analysis, W.Z.; Investigation, X.L.; Resources, W.Z.; Data Curation, X.L.; Writing Original Draft Preparation, W.Z.; Writing Review & Editing, X.L.; Visualization, W.Z.; Supervision, W.Z.; Project Administration, X.L.; Funding Acquisition, W.Z. All authors have read and approved the final manuscript.

Acknowledgments: The authors would like to thank the reviewers for their very detailed and helpful comments, which have significantly improved the presentation of this paper.

References

1. Kilar, N.; Simsek, Y. A new family of Fubini type numbrs and polynomials associated with Apostol-Bernoulli nujmbers and polynomials. *J. Korean Math. Soc.* **2017**, *54*, 1605–1621.
2. Kim, T.; Kim, D.S.; Jang, G.-W. A note on degenerate Fubini polynomials. *Proc. Jiangjeon Math. Soc.* **2017**, *20*, 521–531.
3. Kim, T.; Kim, D.S.; Jang, G.-W.; Kwon, J. Symmetric identities for Fubini polynomials. *Symmetry* **2018**, *10*, 219.
4. Zhao, J.-H.; Chen, Z.-Y. Some symmetric identities involving Fubini polynomials and Euler numbers. *Symmetry* **2018**, *10*, 303.
5. Chen, L.; Zhang, W.-P. Chebyshev polynomials and their some interesting applications. *Adv. Differ. Equ.* **2017**, *2017*, 303.
6. Clemente, C. Identities and generating functions on Chebyshev polynomials. *Georgian Math. J.* **2012**, *19*, 427–440.
7. He, Y. Symmetric identities for Calitz's *q*-Bernoulli numbers and polynomials. *Adv. Differ. Equ.* **2013**, *2013*, 246.
8. Kim, T. Symmetry of power sum polynomials and multivariate fermionic *p*-adic invariant integral on Z_p. *Russ. J. Math. Phys.* **2009**, *16*, 93–96.
9. Kim, T.; Kim, D.S. An identity of symmetry for the degernerate Frobenius-Euler polynomials. *Math. Slovaca* **2018**, *68*, 239–243.
10. Li, X.-X. Some identities involving Chebyshev polynomials. *Math. Probl. Eng.* **2015**, *2015*, 950695.
11. Rim, S.-H.; Jeong, J.-H.; Lee, S.-J.; Moon, E.-J.; Jin, J.-H. On the symmetric properties for the generalized twisted Genocchi polynomials. *ARS Comb.* **2012**, *105*, 267–272.
12. Wang, T.-T.; Zhang, W.-P. Some identities involving Fibonacci, Lucas polynomials and their applications. *Bull. Math. Soc. Sci. Math. Roum.* **2012**, *55*, 95–103.
13. Yi, Y.; Zhang, W.P. Some identities involving the Fibonacci polynomials. *Fibonacci Q.* **2002**, *40*, 314–318.
14. Zhang, W.-P. Some identities involving the Euler and the central factorial numbers. *Fibonacci Q.* **1998**, *36*, 154–157.
15. Kim, D.S.; Park, K.H. Identities of symmetry for Bernoulli polynomials arising from quotients of Volkenborn integrals invariant under S_3. *Appl. Math. Comput.* **2013**, *219*, 5096–5104.
16. Feng R.-Q.; Song C.-W. *Combinatorial Mathematics*; Beijing University Press: Beijing, China, 2015.

Connection Problem for Sums of Finite Products of Legendre and Laguerre Polynomials

Taekyun Kim [1], Kyung-Won Hwang [2,*], Dae San Kim [3] and Dmitry V. Dolgy [4]

[1] Department of Mathematics, Kwangwoon University, Seoul 01897, Korea; tkkim@kw.ac.kr
[2] Department of Mathematics, Dong-A University, Busan 49315, Korea
[3] Department of Mathematics, Sogang University, Seoul 04107, Korea; dskim@sogang.ac.kr
[4] Institute of National Science, Far Eastern Federal University, Vladivostok 690950, Russia; dvdolgy@gmail.com
* Correspondence: khwang@dau.ac.kr

Abstract: The purpose of this paper is to represent sums of finite products of Legendre and Laguerre polynomials in terms of several orthogonal polynomials. Indeed, by explicit computations we express each of them as linear combinations of Hermite, generalized Laguerre, Legendre, Gegenbauer and Jacobi polynomials, some of which involve terminating hypergeometric functions $_1F_1$ and $_2F_1$.

Keywords: Legendre polynomials; Laguerre polynomials; generalized Laguerre polynomials; Gegenbauer polynomials; hypergeometric functions $_1F_1$ and $_2F_1$

1. Preliminaries

Here, after fixing some notations that will be needed throughout this paper, we will review briefly some basic facts about orthogonal polynomials relevant to our discussion. As general references on orthogonal polynomials, we recommend the reader to refer to [1,2].

As is well known, the falling factorial sequence $(x)_n$ and the rising factorial sequence $\langle x \rangle_n$ are respectively defined by

$$(x)_n = x(x-1)\ldots(x-n+1), \quad (n \geq 1), (x)_0 = 1, \tag{1}$$

$$\langle x \rangle_n = x(x+1)\ldots(x+n-1), \quad (n \geq 1), \langle x \rangle_0 = 1. \tag{2}$$

The two factorial sequences are related by

$$(-1)^n (x)_n = \langle -x \rangle_n, \quad (-1)^n \langle x \rangle_n = (-x)_n. \tag{3}$$

$$\frac{(2n-2s)!}{(n-s)!} = \frac{2^{2n-2s}(-1)^s \left\langle \frac{1}{2} \right\rangle_n}{\left\langle \frac{1}{2} - n \right\rangle_s}, \quad (n \geq s \geq 0). \tag{4}$$

$$\frac{(2n+2s)!}{(n+s)!} = 2^{2n+2s} \left\langle \frac{1}{2} \right\rangle_n \left\langle n + \frac{1}{2} \right\rangle_s, \quad (n,s \geq 0). \tag{5}$$

$$\Gamma(n + \frac{1}{2}) = \frac{(2n)!\sqrt{\pi}}{2^{2n}n!}, \quad (n \geq 0), \tag{6}$$

$$\frac{\Gamma(x+1)}{\Gamma(x+1-n)} = (x)_n, \quad \frac{\Gamma(x+n)}{\Gamma(x)} = \langle x \rangle_n, \quad (n \geq 0), \tag{7}$$

$$B(x,y) = \int_0^1 t^{x-1}(1-t)^{y-1}dt = \frac{\Gamma(x)\Gamma(y)}{\Gamma(x+y)}, \quad (\operatorname{Re} x, \operatorname{Re} y > 0), \tag{8}$$

where $\Gamma(x)$ and $B(x, y)$ denote respectively the gamma and beta functions.

The hypergeometric function is defined by

$$_pF_q = (a_1, \ldots, a_p; b_1, \ldots, b_q; x) = \sum_{n=0}^{\infty} \frac{\langle a_1 \rangle_n \cdots \langle a_p \rangle_n}{\langle b_1 \rangle_n \cdots \langle b_q \rangle_n} \frac{x^n}{n!}. \tag{9}$$

Now, we are ready to recall some relevant facts about Legendre polynomials $P_n(x)$, Laguerre polynomials $L_n(x)$, Hermite polynomials $H_n(x)$, generalized (extended) Laguerre polynomials $L_n^\alpha(x)$, Gegenbauer polynomials $C_n^{(\lambda)}(x)$, and Jacobi polynomials $P_n^{(\alpha,\beta)}(x)$. All the facts stated here can also be found in [3–8].Interested readers may refer to [1,2,9–13] for full accounts of orthogonal polynomials and also to [14,15] for papers discussing relevant orthogonal polynomials.

The above-mentioned orthogonal polynomials are given, in terms of generating functions, by

$$F(t, x) = (1 - 2xt + t^2)^{-\frac{1}{2}} = \sum_{n=0}^{\infty} P_n(x) t^n, \tag{10}$$

$$G(t, x) = (1 - t)^{-1} exp\left(-\frac{xt}{1-t}\right) = \sum_{n=0}^{\infty} L_n(x) t^n, \tag{11}$$

$$e^{2xt - t^2} = \sum_{n=0}^{\infty} H_n(x) \frac{t^n}{n!}, \tag{12}$$

$$(1 - t)^{-\alpha - 1} exp\left(-\frac{xt}{1-t}\right) = \sum_{n=0}^{\infty} L_n^\alpha(x) t^n, \tag{13}$$

$$\frac{1}{(1 - 2xt + t^2)^\lambda} = \sum_{n=0}^{\infty} C_n^{(\lambda)}(x) t^n, \quad \left(\lambda > -\frac{1}{2}, \lambda \neq 0, |t| < 1, |x| \leq 1\right), \tag{14}$$

$$\frac{\alpha + \beta}{R(1 - t + +R)^\alpha (1 + t + R)^\beta} = \sum_{n=0}^{\infty} P_n^{(\alpha,\beta)}(x) t^n,$$
$$(R = \sqrt{1 - 2xt + t^2}, \alpha, \beta > -1). \tag{15}$$

In terms of explicit expressions, those orthogonal polynomials are given explicitly as follows:

$$P_n(x) = {_2F_1}\left(-n, n + 1; 1; \frac{1 - x}{2}\right)$$
$$= \frac{1}{2^n} \sum_{l=0}^{[\frac{n}{2}]} (-1)^l \binom{n}{l} \binom{2n - 2l}{n} x^{n-2l}, \tag{16}$$

$$L_n(x) = {_1F_1}(-n; 1; x)$$
$$= \sum_{l=0}^{n} (-1)^{n-l} \binom{n}{l} \frac{1}{(n-l)!} x^{n-l}, \tag{17}$$

$$H_n(x) = n! \sum_{l=0}^{[\frac{n}{2}]} \frac{(-1)^l}{l!(n - 2l)!} (2x)^{n-2l}, \tag{18}$$

$$L_n^\alpha(x) = \frac{\langle \alpha + 1 \rangle_n}{n!} {_1F_1}(-n; \alpha + 1; x)$$
$$= \sum_{l=0}^{n} \frac{(-1)^l \binom{n+\alpha}{n-l}}{l!} x^l, \tag{19}$$

$$C_n^\lambda(x) = \binom{n + 2\lambda - 1}{n} {}_2F_1\left(-n, n + 2\lambda; \lambda + \frac{1}{2}; \frac{1 - x}{2}\right)$$

$$= \sum_{k=0}^{\lfloor\frac{n}{2}\rfloor} (-1)^k \frac{\Gamma(n - k + \lambda)}{\Gamma(\lambda)k!(n - 2k)!} (2x)^{n-2k}, \tag{20}$$

$$P_n^{(\alpha,\beta)}(x) = \frac{\langle\alpha + 1\rangle_n}{n!} {}_2F_1\left(-n, 1 + \alpha + \beta + n; \alpha + 1; \frac{1 - x}{2}\right)$$

$$= \sum_{k=0}^{n} \binom{n + \alpha}{n - k}\binom{n + \beta}{k}\left(\frac{x - 1}{2}\right)^k\left(\frac{x + 1}{2}\right)^{n-k}. \tag{21}$$

For Legendre, Gegenbauer and Jacobi polynomials, we have Rodrigues' formulas, and for Hermite and generalized Laguerre polynomials, we have Rodrigues-type formulas.

$$H_n(x) = (-1)^n e^{x^2} \frac{d^n}{dx^n} e^{-x^2}, \tag{22}$$

$$L_n^\alpha(x) = \frac{1}{n!} x^{-\alpha} e^x \frac{d^n}{dx^n}(e^{-x} x^{n+\alpha}), \tag{23}$$

$$P_n(x) = \frac{1}{2^n n!} \frac{d^n}{dx^n}(x^2 - 1)^n, \tag{24}$$

$$(1 - x^2)^{\lambda - \frac{1}{2}} C_n^{(\lambda)}(x) = \frac{(-2)^n}{n!} \frac{\langle\lambda\rangle_n}{\langle n + 2\lambda\rangle_n} \frac{d^n}{dx^n}(1 - x^2)^{n+\lambda-\frac{1}{2}}, \tag{25}$$

$$(1 - x)^\alpha(1 + x)^\beta P_n^{(\alpha,\beta)}(x) = \frac{(-1)^n}{2^n n!} \frac{d^n}{dx^n}(1 - x)^{n+\alpha}(1 + x)^{n+\beta}. \tag{26}$$

The orthogonal polynomials in Equations (22)–(26) satisfy the following orthogonality relations with respect to various weight functions.

$$\int_{-\infty}^{\infty} e^{-x^2} H_n(x)H_m(x)dx = 2^n n!\sqrt{\pi}\delta_{m,n}, \tag{27}$$

$$\int_0^{\infty} x^\alpha e^{-x} L_n^\alpha(x)L_m^\alpha(x)dx = \frac{1}{n!}\Gamma(\alpha + n + 1)\delta_{m,n}, \tag{28}$$

$$\int_{-1}^{1} P_n(x)P_m(x)dx = \frac{2}{2n + 1}\delta_{m,n}, \tag{29}$$

$$\int_{-1}^{1}(1 - x^2)^{\lambda - \frac{1}{2}} C_n^{(\lambda)}(x)C_m^{(\lambda)}(x)dx = \frac{\pi 2^{1-2\lambda}\Gamma(n + 2\lambda)}{n!(n + \lambda)\Gamma(\lambda)^2}\delta_{m,n}, \tag{30}$$

$$\int_{-1}^{1}(1 - x)^\alpha(1 + x)^\beta P_n^{(\alpha,\beta)}(x)P_m^{(\alpha,\beta)}(x)dx = \frac{2^{\alpha+\beta+1}\Gamma(n + \alpha + 1)\Gamma(n + \beta + 1)}{(2n + \alpha + \beta + 1)\Gamma(n + \alpha + \beta + 1)\Gamma(n + 1)}\delta_{m,n}. \tag{31}$$

2. Introduction

In this paper, we will consider two sums of finite products

$$\gamma_{n,r}(x) = \sum_{i_1 + \cdots + i_{2r+1} = n} P_{i_1}(x)P_{i_2}(x)\ldots P_{i_{2r+1}}(x), \quad (n, r \geq 0), \tag{32}$$

in terms of Legendre polynomials and

$$\varepsilon_{n,r}(x) = \sum_{i_1 + \cdots + i_{r+1} = n} L_{i_1}\left(\frac{x}{r + 1}\right) L_{i_2}\left(\frac{x}{r + 1}\right) \ldots L_{i_{r+1}}\left(\frac{x}{r + 1}\right), \quad (n, r \geq 0), \tag{33}$$

in terms of Laguerre polynomials. We represent each of them as linear combinations of Hermite, extended Laguerre, Legendre, Gegenbauer, and Jacobi polynomials (see Theorems 1 and 2). It is amusing to note here that, for some of these expressions, the coefficients involve certain terminating hypergeometric functions $_2F_1$ and $_1F_1$. These representations are obtained by carrying out explicit computations with the help of Propositions 1 and 2. We observe here that the formulas in Proposition 1 can be derived from the orthogonalities in Equation (27)–(31), Rodrigues' and Rodrigues-type formulas in Equation (22)–(26), and integration by parts.

Our study of such representation problems can be justified by the following. Firstly, the present research can be viewed as a generalization of the classical connection problems. Indeed, the classical connection problems are concerned with determining the coefficients in the expansion of a product of two polynomials in terms of any given sequence of polynomials (see [1,2]).

Secondly, studying such kinds of sums of finite products of special polynomials can be well justified also by the following example. Let us put

$$\alpha_m(x) = \sum_{k=1}^{m-1} \frac{1}{k(m-k)} B_k(x) B_{m-k}(x), \ (m \geq 2),$$

where $B_n(x)$ are the Bernoulli polynomials. Then we can express $\alpha_m(x)$ as linear combinations of Bernoulli polynomials, for example from the Fourier series expansion of the function closely related to that. Indeed, we can show that

$$\sum_{k=1}^{m-1} \frac{1}{2k(2m-2k)} B_{2k}(x) B_{2m-2k}(x) + \frac{2}{2m-1} B_1(x) B_{2m-1}(x) \tag{34}$$

$$= \frac{1}{m} \sum_{k=1}^{m} \frac{1}{2k} \binom{2m}{2k} B_{2k} B_{2m-2k}(x) + \frac{1}{m} H_{2m-1} B_{2m}(x) + \frac{2}{2m-1} B_{2m-1} B_1(x),$$

where $H_m = \sum_{j=1}^{m} \frac{1}{j}$ are the harmonic numbers.

Further, some simple modification of this gives us the famous Faber-Pandharipande-Zagier identity and a slightly different variant of the Miki's identity by letting respectively $x = \frac{1}{2}$ and $x = 0$ in (34). We note here that all the other known derivations of F-P-Z and Miki's identity are quite involved, while our proof of Miki's and Faber-Pandharipande-Zagier identities follow from the polynomial identity (34), which in turn follows immediately the Fourier series expansion of $\alpha_m(x)$. Indeed, Miki makes use of a formula for the Fermat quotient $\frac{a^p - a}{p}$ modulo p^2, Shiratani-Yokoyama employs p-adic analysis, Gessel's proof is based on two different expressions for Stirling numbers of the second kind $S_2(n, k)$, and Dunne-Schubert exploits the asymptotic expansion of some special polynomials coming from the quantum field theory computations. For some details on these, we let the reader refer to the introduction in [16] and the papers therein.

The next two theorems are the main results of this paper.

Theorem 1. *For any nonnegative integers n and r, we have the following representation.*

$$\sum_{i_1+i_2+\cdots+i_{2r+1}=n} P_{i_1}(x) P_{i_2}(x) \ldots P_{i_{2r+1}}(x)$$

$$= \frac{2^r (n+r-\frac{1}{2})_{n+r}}{(2r-1)!!} \sum_{j=0}^{[\frac{n}{2}]} \frac{_1F_1(-j; \frac{1}{2}-n-r; -1)}{j!(n-2j)!} H_{n-2j}(x) \tag{35}$$

$$= \frac{1}{(2r-1)!! 2^{n+r}} \sum_{k=0}^{n} \frac{(-1)^k}{\Gamma(\alpha+k+1)}$$

$$\times \sum_{l=0}^{[\frac{n-k}{2}]} \frac{(-1)^l (2n+2r-2l)! \Gamma(n-2l+\alpha+1)}{l!(n+r-l)!(n-k-2l)!} L_k^{\alpha}(x) \tag{36}$$

$$= \frac{2^{r-1}(n+r-\frac{1}{2})_{n+r}}{(2r-1)!!}$$

$$\times \sum_{j=0}^{[\frac{n}{2}]} \frac{(2n+1-4j)_2F_1(-j, j-n-\frac{1}{2}; \frac{1}{2}-n-r; 1)}{j!(n-j+\frac{1}{2})_{n-j+1}} P_{n-2j}(x) \tag{37}$$

$$= \frac{2^r \Gamma(\lambda)(n+r-\frac{1}{2})_{n+r}}{(2r-1)!!}$$

$$\times \sum_{j=0}^{[\frac{n}{2}]} \frac{(n+\lambda-2j)_2F_1(-j, j-n-r; \frac{1}{2}-n-r; 1)}{\Gamma(n+\lambda-j+1)j!} C_{n-2j}^{(\lambda)}(x) \tag{38}$$

$$= \frac{(-1)^n}{(2r-1)!!2^{n+r}} \sum_{k=0}^{n} \frac{\Gamma(k+\alpha+\beta+1)(-2)^k}{\Gamma(2k+\alpha+\beta+1)}$$

$$\times \sum_{l=0}^{[\frac{n-k}{2}]} \frac{(-1)^l(2n+2r-2l)!}{l!(n+r-l)!(n-k-2l)!} \tag{39}$$

$$\times {}_2F_1(2l+k-n, k+\beta+1; 2k+\alpha+\beta+2; 2)P_k^{(\alpha,\beta)}(x).$$

Here $(2r-1)!!$ is the double factorial given by

$$(2r-1)!! = (2r-1)(2r-3)\ldots 1, \quad (r \geq 1), (-1)!! = 1. \tag{40}$$

Remark 1. *An alternative expression for (36) is given by*

$$\gamma_{n,r}(x) = \frac{1}{\Gamma(\alpha+1)(2r-1)!!2^{n+r}}$$

$$\times \sum_{l=0}^{[\frac{n}{2}]} \frac{(-1)^l(2n+2r-2l)!\Gamma(n-2l+\alpha+1)}{l!(n+r-l)!(n-2l)!} \sum_{k=0}^{n-2l} \frac{\langle 2l-n \rangle_k}{\langle \alpha+1 \rangle_k} L_k^\alpha(x). \tag{41}$$

Theorem 2. *For any nonnegative integers n and r, we have the following representation.*

$$\sum_{i_1+i_2+\cdots+i_{r+1}=n} L_{i_1}\left(\frac{x}{r+1}\right) L_{i_2}\left(\frac{x}{r+1}\right) \ldots L_{i_{r+1}}\left(\frac{x}{r+1}\right)$$

$$= (n+r)! \sum_{k=0}^{n} \frac{(-\frac{1}{2})^k}{k!} \sum_{j=0}^{[\frac{n-k}{2}]} \frac{(\frac{1}{4})^j}{j!(n-k-2j)!(r+k+2j)!} H_k(x) \tag{42}$$

$$= (n+r)! \sum_{k=0}^{n} \frac{{}_2F_1(k-n, k+\alpha+1; r+k+1; 1)}{(n-k)!(r+k)!} L_k^\alpha(x) \tag{43}$$

$$= (n+r)! \sum_{k=0}^{n} 2^{k+1}(2k+1)$$

$$\times \sum_{j=0}^{[\frac{n-k}{2}]} \frac{(k+j+1)!}{j!(n-k-2j)!(r+k+2j)!(2k+2j+2)!} P_k(x) \tag{44}$$

$$= (n+r)!\Gamma(\lambda) \sum_{k=0}^{n} \left(-\frac{1}{2}\right)^k (k+\lambda)$$

$$\times \sum_{j=0}^{[\frac{n-k}{2}]} \frac{(\frac{1}{4})^j}{j!(n-k-2j)!(r+k+2j)!\Gamma(k+j+\lambda+1)} C_k^{(\lambda)}(x) \tag{45}$$

$$= (n+r)! \sum_{k=0}^{n} \frac{\Gamma(k+\alpha+\beta+1)(-2)^k}{\Gamma(2k+\alpha+\beta+1)}$$

$$\times \sum_{l=0}^{n-k} \frac{{}_2F_1(k-n+l, k+\beta+1; 2k+\alpha+\beta+2; 2)}{l!(n+r-l)!(n-k-l)!} P_k^{(\alpha,\beta)}(x). \tag{46}$$

Remark 2. *An alternative expression for* (42) *is as follows:*

$$\varepsilon_{n,r}(x) = (n+r)! \sum_{j=0}^{[\frac{n}{2}]} \frac{(\frac{1}{4})^j}{j!(n-2j)!(r+2j)!} \sum_{k=0}^{n-2j} \frac{(\frac{1}{2})^k \langle 2j-n \rangle_k}{k! \langle r+2j+1 \rangle_k} H_k(x). \tag{47}$$

Before we move on to the next section, we would like to mention some of the related previous works. In [16–18], sums of finite products of Bernoulli, Euler and Genocchi polynomials were represented as linear combinations of Bernoulli polynomials. These were derived from the Fourier series expansions for the functions closely related to those sums of finite products. In addition, in [9] the same had been done for sums of finite products of Chebyshev polynomials of the second kind and of Fibonacci polynomials.

On the other hand, in terms of all kinds of Chebyshev polynomials, sums of finite products of Chebyshev polynomials of the second, third and fourth kinds and of Fibonacci, Legendre and Laguerre polynomials were expressed in [11,12,19]. Further, by the orthogonal polynomials in Equations (16), and (18)–(21), sums of finite products of Chebyshev polynomials of the second kind and Fibonacci polynomials were represented in [13].

Finally, the reader may want to see [20,21] for some other aspects of Legendre and Laguerre polynomials.

3. Proof of Theorem 1

We will first state Propositions 1 and 2 that will be needed in showing Theorems 1 and 2.

The results in the next proposition can be derived from the orthogonalities in (27)–(31), Rodrigues' and Rodrigues-type formulas in (22)–(26), and integration by parts, as we mentioned earlier. The facts (a), (b), (c), (d) and (e) in Proposition 1 are respectively from (3.7) of [5], (2.3) of [7] (see also (2.4) of [3]), (2.3) of [6], (2.3) of [4] and (2.7) of [8].

Proposition 1. *For any polynomial* $q(x) \in \mathbb{R}[x]$ *of degree* n, *the following hold.*

(a)

$$q(x) = \sum_{k=0}^{n} C_{k,1} H_k(x), \text{ where } C_{k,1} = \frac{(-1)^k}{2^k k! \sqrt{\pi}} \int_{-\infty}^{\infty} q(x) \frac{d^k}{dx^k} e^{-x^2} dx,$$

(b)

$$q(x) = \sum_{k=0}^{n} C_{k,2} L_k^\alpha(x), \text{ where } C_{k,2} = \frac{1}{\Gamma(\alpha+k+1)} \int_0^{\infty} q(x) \frac{d^k}{dx^k} (e^{-x} x^{k+\alpha}) dx,$$

(c)

$$q(x) = \sum_{k=0}^{n} C_{k,3} P_k(x), \text{ where } C_{k,3} = \frac{2k+1}{2^{k+1} k!} \int_{-1}^{1} q(x) \frac{d^k}{dx^k} (x^2-1)^k dx,$$

(d)

$$q(x) = \sum_{k=0}^{n} C_{k,4} C_k^{(\lambda)}(x), \text{ where}$$

$$C_{k,4} = \frac{(k+\lambda)\Gamma(\lambda)}{(-2)^k \sqrt{\pi} \Gamma(k+\lambda+\frac{1}{2})} \int_{-1}^{1} q(x) \frac{d^k}{dx^k} (1-x^2)^{k+\lambda-\frac{1}{2}} dx,$$

(e)

$$q(x) = \sum_{k=0}^{n} C_{k,5} P_k^{(\alpha,\beta)}(x), \text{ where}$$

$$C_{k,5} = \frac{(-1)^k (2k + \alpha + \beta + 1)\Gamma(k + \alpha + \beta + 1)}{2^{\alpha+\beta+k+1}\Gamma(\alpha + k + 1)\Gamma(\beta + k + 1)}$$

$$\times \int_{-1}^{1} q(x) \frac{d^k}{dx^k}(1 - x)^{k+\alpha}(1 + x)^{k+\beta} dx.$$

Proposition 2. *The following proposition was stated in [16].*
For any nonnegative integers m and k, the following identities hold.

(a)

$$\int_{-\infty}^{\infty} x^m e^{-x^2} dx = \begin{cases} 0, & \text{if } m \equiv 1 \ (mod \ 2), \\ \frac{m!\sqrt{\pi}}{(\frac{m}{2})!2^m}, & \text{if } m \equiv 0 \ (mod \ 2), \end{cases}$$

(b)

$$\int_{-1}^{1} x^m (1 - x^2)^k dx = \begin{cases} 0, & \text{if } m \equiv 1 \ (mod \ 2), \\ \frac{2^{2k+2} k! m! (k + \frac{m}{2} + 1)!}{(\frac{m}{2})!(2k+m+2)!}, & \text{if } m \equiv 0 \ (mod \ 2), \end{cases}$$

$$= 2^{2k+1} k! \sum_{s=0}^{m} \binom{m}{s} 2^s (-1)^{m-s} \frac{(k+s)!}{(2k+s+1)!},$$

(c)

$$\int_{-1}^{1} x^m (1 - x^2)^{k+\lambda-\frac{1}{2}} dx = \begin{cases} 0, & \text{if } m \equiv 1 \ (mod \ 2), \\ \frac{\Gamma(k+\lambda+\frac{1}{2})\Gamma(\frac{m}{2}+\frac{1}{2})}{\Gamma(k+\lambda+\frac{m}{2}+1)}, & \text{if } m \equiv 0 \ (mod \ 2), \end{cases}$$

(d)

$$\int_{-1}^{1} x^m (1 - x)^{k+\alpha}(1 + x)^{k+\beta} dx = 2^{2k+\alpha+\beta+1} \sum_{s=0}^{m} \binom{m}{s}(-1)^{m-s} 2^s$$

$$\times \frac{\Gamma(k + \alpha + 1)\Gamma(k + \beta + s + 1)}{\Gamma(2k + \alpha + \beta + s + 2)}.$$

Differentiation of (10) gives us the following lemma.

Lemma 1. *For any nonnegative integers n and r, we have the following identity.*

$$\sum_{i_1+i_2+\ldots i_{2r+1}=n} P_{i_1}(x), P_{i_2}(x), \ldots, P_{i_{2r+1}}(x) = \frac{1}{(2r-1)!!} P_{n+r}^{(r)}(x), \tag{48}$$

where the sum is over all nonnegative integers $i_1, i_2, \ldots, i_{2r+1}$, with $i_1 + i_2 + \ldots i_{2r+1} = n$.
By taking rth derivative of (16), we have

$$P_n^{(r)}(x) = \frac{1}{2^n} \sum_{l=0}^{[\frac{n-r}{2}]} (-1)^l \binom{n}{l}\binom{2n - 2l}{n}(n - 2l)_r x^{n-2l-r}. \tag{49}$$

Actually, we need the following particular case of (49).

$$P_{n+r}^{(r+k)}(x) = \frac{1}{2^{n+r}} \sum_{l=0}^{[\frac{n-k}{2}]} (-1)^l \binom{n+r}{l}\binom{2n + 2r - 2l}{n+r}$$

$$\times (n + r - 2l)_{r+k} x^{n-2l-k}. \tag{50}$$

Here we are going to show (35), (36) and (38), leaving the other two (37) and (39) as exercises.

With $\gamma_{n,r}(x)$ as in (32), let us put

$$\gamma_{n,r}(x) = \sum_{k=0}^{n} C_{k,1} H_k(x). \tag{51}$$

Then, from (a) of Proposition 1, (48), (50), and by integrating by parts k times, we have

$$\begin{aligned}
C_{k,1} &= \frac{(-1)^k}{2^k k! \sqrt{\pi}} \int_{-\infty}^{\infty} \gamma_{n,r}(x) \frac{d^k}{dx^k} e^{-x^2} dx \\
&= \frac{(-1)^k}{2^k k! \sqrt{\pi}(2r-1)!!} \int_{-\infty}^{\infty} P_{n+r}^{(r)}(x) \frac{d^k}{dx^k} e^{-x^2} dx \\
&= \frac{1}{2^k k! \sqrt{\pi}(2r-1)!!} \int_{-\infty}^{\infty} P_{n+r}^{(r+k)}(x) e^{-x^2} dx \\
&= \frac{1}{2^{k+n+r} k! \sqrt{\pi}(2r-1)!!} \\
&\quad \times \sum_{l=0}^{[\frac{n-k}{2}]} (-1)^l \binom{n+r}{l} \binom{2n+2r-2l}{n+r} (n+r-2l)_{r+k} \\
&\quad \times \int_{-\infty}^{\infty} x^{n-2l-k} e^{-x^2} dx.
\end{aligned} \tag{52}$$

From (52) and making use of (a) of Proposition 2, we obtain

$$\begin{aligned}
C_{k,1} &= \frac{1}{2^{k+n+r} k! \sqrt{\pi}(2r-1)!!} \\
&\quad \times \sum_{l=0}^{[\frac{n-k}{2}]} (-1)^l \binom{n+r}{l} \binom{2n+2r-2l}{n+r} (n+r-2l)_{r+k} \\
&\quad \times \begin{cases} 0, & \text{if } k \not\equiv n \ (mod\ 2), \\ \frac{(n-k-2l)! \sqrt{\pi}}{2^{n-k-2l}(\frac{n-k}{2}-l)!}, & \text{if } k \equiv n \ (mod\ 2). \end{cases}
\end{aligned} \tag{53}$$

Now, from (51) and (53) and after some simplifications,

$$\begin{aligned}
\gamma_{n,r}(x) &= \frac{1}{2^{2n+r}(2r-1)!!} \sum_{\substack{0 \le k \le n \\ k \equiv n \ (mod\ 2)}} \frac{1}{k!} \\
&\quad \times \sum_{l=0}^{[\frac{n-k}{2}]} \frac{(-4)^l (2n+2r-2l)!}{l!(n+r-l)!(\frac{n-k}{2}-l)!} H_k(x) \\
&= \frac{1}{2^{2n+r}(2r-1)!!} \sum_{j=0}^{[\frac{n}{2}]} \frac{1}{j!(n-2j)!} H_{n-2j}(x) \\
&\quad \times \sum_{l=0}^{j} \frac{(-4)^l (j)_l (2n+2r-2l)!}{l!(n+r-l)!} \\
&= \frac{2^r (n+r-\frac{1}{2})_{n+r}}{(2r-1)!!} \sum_{j=0}^{[\frac{n}{2}]} \frac{1}{j!(n-2j)!} H_{n-2j}(x) \\
&\quad \times \sum_{l=0}^{j} \frac{(-1)^l \langle -j \rangle_l}{l! \langle \frac{1}{2}-n-r \rangle_l} \\
&= \frac{2^r (n+r-\frac{1}{2})_{n+r}}{(2r-1)!!} \sum_{j=0}^{[\frac{n}{2}]} \frac{{}_1F_1(-j;\frac{1}{2}-n-r;-1)}{j!(n-2j)!} H_{n-2j}(x).
\end{aligned} \tag{54}$$

This shows (35) in Theorem 1.

Next, we put

$$\gamma_{n,r}(x) = \sum_{k=0}^{n} C_{k,2}\, L_k^{\alpha}(x).$$
(55)

Then, from (b) of Proposition 1, (48), (50) and integration by parts k times, we get

$$
\begin{aligned}
C_{k,2} &= \frac{(-1)^k}{\Gamma(\alpha+k+1)(2r-1)!!} \int_0^{\infty} P_{n+r}^{(r+k)}(x)e^{-x}x^{k+\alpha}dx \\
&= \frac{(-1)^k}{\Gamma(\alpha+k+1)(2r-1)!!\,2^{n+r}} \sum_{l=0}^{[\frac{n-k}{2}]} (-1)^l \binom{n+r}{l}\binom{2n+2r-2l}{n+r} \\
&\quad \times (n+r-2l)_{r+k}\,\Gamma(n-2l+\alpha+1) \\
&= \frac{(-1)^k}{\Gamma(\alpha+k+1)(2r-1)!!\,2^{n+r}} \\
&\quad \times \sum_{l=0}^{[\frac{n-k}{2}]} \frac{(-1)^l(2n+2r-2l)!\,\Gamma(n-2l+\alpha+1)}{l!(n+r-l)!(n-k-2l)!}.
\end{aligned}
$$
(56)

Combining (55) and (56), and changing order of summation, we immediately have

$$
\begin{aligned}
\gamma_{n,r}(x) &= \frac{1}{(2r-1)!!\,2^{n+r}} \sum_{k=0}^{n} \frac{(-1)^k}{\Gamma(\alpha+k+1)} \\
&\quad \times \sum_{l=0}^{[\frac{n-k}{2}]} \frac{(-1)^l(2n+2r-2l)!\,\Gamma(n-2l+\alpha+1)}{l!(n+r-l)!(n-k-2l)!} L_k^{\alpha}(x) \\
&= \frac{1}{\Gamma(\alpha+1)(2r-1)!!\,2^{n+r}} \sum_{l=0}^{[\frac{n}{2}]} \frac{(-1)^l(2n+2r-2l)!\,\Gamma(n-2l+\alpha+1)}{l!(n+r-l)!(n-2l)!} \\
&\quad \times \sum_{k=0}^{n-2l} \frac{\langle 2l-n\rangle_k}{\langle \alpha+1\rangle_k} L_k^{\alpha}(x).
\end{aligned}
$$
(57)

This yields (36) in Theorem 1.

Finally, we let

$$\gamma_{n,r}(x) = \sum_{k=0}^{n} C_{k,4}\, C_k^{(\lambda)}(x)$$
(58)

Then, from (d) of Proposition 1, (48), (50), integration by parts k times and making use of (c) of Proposition 2, we have

$$
\begin{aligned}
C_{k,4} &= \frac{(k+\lambda)\,\Gamma(\lambda)(-1)^k}{(-2)^k\sqrt{\pi}\,\Gamma(k+\lambda+\frac{1}{2})(2r-1)!!} \\
&\quad \times \int_{-1}^{1} P_{n+r}^{(r+k)}(x)(1-x^2)^{k+\lambda-\frac{1}{2}}dx \\
&= \frac{(k+\lambda)\Gamma(\lambda)}{2^{k+n+r}\sqrt{\pi}\,\Gamma(k+\lambda+\frac{1}{2})(2r-1)!!} \\
&\quad \times \sum_{l=0}^{[\frac{n-k}{2}]} (-1)^l \binom{n+r}{l}\binom{2n+2r-2l}{n+r}(n+r-2l)_{r+k} \\
&\quad \times \begin{cases} 0, & \text{if } k \not\equiv n \ (mod\ 2), \\ \frac{\Gamma(k+\lambda+\frac{1}{2})\Gamma(\frac{n-k+1}{2}-l)}{\Gamma(\frac{n+k}{2}+\lambda-l+1)}, & \text{if } k \equiv n \ (mod\ 2). \end{cases}
\end{aligned}
$$
(59)

From (58) and (59), exploiting (3), (4), (6) and (7), and after some simplifications, we finally derive

$$
\gamma_{n,r}(x) = \frac{\Gamma(\lambda)}{\sqrt{\pi}(2r-1)!!2^{n+r}} \sum_{\substack{0 \le k \le n \\ k \equiv n \ (mod \ 2)}} \frac{(k+\lambda)}{2^k}
$$

$$
\times \sum_{l=0}^{[\frac{n-k}{2}]} \frac{(-1)^l(2n+2r-2l)! \, \Gamma(\frac{n-k+1}{2}-l)}{l!(n+r-l)!(n-k-2l)!\Gamma(\frac{k+n}{2}+\lambda-l+1)} C_k^{(\lambda)}(x)
$$

$$
= \frac{\Gamma(\lambda)}{\sqrt{\pi}(2r-1)!!2^{n+r}} \sum_{j=0}^{[\frac{n}{2}]} \frac{(n-2j+\lambda)}{2^{n-2j}}
$$

$$
\times \sum_{l=0}^{j} \frac{(-1)^l(2n+2r-2l)! \, \Gamma(j-l+\frac{1}{2})}{l! \, (n+r-l)! \, (2j-2l)! \, \Gamma(n+\lambda-j-l+1)} C_{n-2j}^{(\lambda)}(x)
$$

$$
= \frac{\Gamma(\lambda)}{(2r-1)!! \, 2^{2n+r}} \sum_{j=0}^{[\frac{n}{2}]} \frac{(n-2j+\lambda)}{\Gamma(n+\lambda-j+1)}
$$

$$
\times \sum_{l=0}^{j} \frac{(-4)^l(2n+2r-2l)! \, (n+\lambda-j)_l}{l! \, (n+r-l)! \, (j-l)!} C_{n-2j}^{(\lambda)}(x)
$$

$$
= \frac{2^r \, \Gamma(\lambda)(n+r-\frac{1}{2})_{n+r}}{(2r-1)!!} \sum_{j=0}^{[\frac{n}{2}]} \frac{(n-2j+\lambda)}{\Gamma(n+\lambda-j+1) \, j!}
$$

$$
\times \sum_{l=0}^{j} \frac{\langle -j \rangle_l \, \langle j-n-r \rangle_l}{l! \, \langle \frac{1}{2}-n-r \rangle_l} C_{n-2j}^{(\lambda)}(x)
$$

$$
= \frac{2^r \, \Gamma(\lambda)(n+r-\frac{1}{2})_{n+r}}{(2r-1)!!}
$$

$$
\times \sum_{j=0}^{[\frac{n}{2}]} \frac{(n-2j+\lambda) \, {}_2F_1(-j,j-n-r;\frac{1}{2}-n-r;1)}{\Gamma(n+\lambda-j+1) \, j!} C_{n-2j}^{(\lambda)}(x).
$$

(60)

This completes the proof for (38) in Theorem 1.

4. Proof of Theorem 2

The proofs for (42), (43) and (45) are left to the reader as an exercise and we will show only (44) and (46) in Theorem 2.

The following lemma is important for our discussion in this section and can be derived by differentiating (11).

Lemma 2. *Let n, r be nonnegative integers. Then we have the following identity.*

$$
\sum_{i_1+i_2+\cdots+i_{r+1}=n} L_{i_1}\left(\frac{x}{r+1}\right) L_{i_2}\left(\frac{x}{r+1}\right) \cdots L_{i_{r+1}}\left(\frac{x}{r+1}\right) = (-1)^r L_{n+r}^{(r)}(x),
$$

(61)

where the sum runs over all nonnegative integers $i_1, i_2, \ldots, i_{r+1}$, with $i_1 + i_2 + \cdots + i_{r+1} = n$.

From (17), it is immediate to see that the rth derivative of $L_n(x)$ is given by

$$
L_n^{(r)}(x) = \sum_{l=0}^{n-r} (-1)^{n-l} \binom{n}{l} \frac{1}{(n-l-r)!} x^{n-l-r}.
$$

(62)

In particular, we have

$$L_{n+r}^{(r+k)}(x) = \sum_{l=0}^{n-k} (-1)^{n+r-l} \binom{n+r}{l} \frac{1}{(n-k-l)!} x^{n-k-l}. \tag{63}$$

With $\varepsilon_{n,r}(x)$ as in (33), let us set

$$\varepsilon_{n,r}(x) = \sum_{k=0}^{n} C_{k,3} P_k(x). \tag{64}$$

Then, from (c) of Proposition 1, (61), (63), by integration by parts k times and using (b) of Proposition 2, we get

$$C_{k,3} = \frac{(2k+1)(-1)^{r+k}}{2^{k+1}k!} \int_{-1}^{1} L_{n+r}^{(r+k)}(x)(x^2-1)^k dx$$

$$= \frac{(-1)^{n+k}(2k+1)(n+r)!}{2^{k+1}k!} \sum_{l=0}^{n-k} \frac{(-1)^l}{l!(n+r-l)!(n-k-l)!}$$

$$\times \begin{cases} 0, & \text{if } l \not\equiv n-k \ (mod\ 2) \\ \frac{2^{2k+2}k!(n-k-l)!(\frac{n+k-l}{2}+1)!}{(\frac{n-k-l}{2})!(n+k-l+2)!}, & \text{if } l \equiv n-k \ (mod\ 2) \end{cases}$$

$$= (-1)^{n+k}(2k+1)2^{k+1}(n+r)!$$

$$\times \sum_{\substack{0 \le l \le n-k \\ l \equiv n-k \ (mod\ 2)}} \frac{(-1)^l (\frac{n+k-l}{2}+1)!}{l!(n+r-l)!(\frac{n-k-l}{2})!(n+k-l+2)!}$$

$$= (n+r)!(2k+1)2^{k+1}$$

$$\times \sum_{j=0}^{[\frac{n-k}{2}]} \frac{(k+j+1)!}{j!(n-k-2j)!(r+k+2j)!(2k+2j+2)!}.$$

By combining (64) and (65) we get the following result.

$$\varepsilon_{n,r}(x) = (n+r)! \sum_{k=0}^{n} (2k+1) 2^{k+1}$$

$$\times \sum_{j=0}^{[\frac{n-k}{2}]} \frac{(k+j+1)!}{j!(n-k-2j)!(r+k+2j)!(2k+2j+2)!} P_k(x). \tag{66}$$

This completes the proof for (44).
Finally, we put

$$\varepsilon_{n,r}(x) = \sum_{k=0}^{n} C_{k,5} P_k^{(\alpha,\beta)}(x). \tag{67}$$

Then, from (e) of Proposition 1, (61), (63), integration by parts k times and exploiting (d) of Proposition 2, we have

$$
\begin{aligned}
C_{k,5} &= \frac{(-1)^r \, (2k+\alpha+\beta+1)\, \Gamma(k+\alpha+\beta+1)}{2^{\alpha+\beta+k+1}\Gamma(\alpha+k+1)\Gamma(\beta+k+1)} \\
&\quad \times \int_{-1}^{1} L_{n+r}^{(r+k)}(x)(1-x)^{k+\alpha}(1+x)^{k+\beta}dx \\
&= \frac{(-1)^r (2k+\alpha+\beta+1)\Gamma(k+\alpha+\beta+1)}{2^{\alpha+\beta+k+1}\Gamma(\alpha+k+1)\Gamma(\beta+k+1)} \\
&\quad \times \sum_{l=0}^{n-k}(-1)^{n+r-l}\binom{n+r}{l}\frac{1}{(n-k-l)!} \\
&\quad \times \int_{-1}^{1} x^{n-k-l}(1-x)^{k+\alpha}(1+x)^{k+\beta}dx \\
&= \frac{(n+r)!\,(-2)^k(2k+\alpha+\beta+1)\Gamma(k+\alpha+\beta+1)}{\Gamma(\beta+k+1)} \\
&\quad \times \sum_{l=0}^{n-k}\frac{1}{l!\,(n+r-l)!}\sum_{s=0}^{n-k-l}\frac{(-2)^s\,\Gamma(k+\beta+s+1)}{s!\,(n-k-l-s)!\,\Gamma(2k+\alpha+\beta+s+2)} \\
&= \frac{(n+r)!\,(-2)^k\Gamma(k+\alpha+\beta+1)}{\Gamma(2k+\alpha+\beta+1)} \\
&\quad \times \sum_{l=0}^{n-k}\frac{1}{l!\,(n+r-l)!\,(n-k-l)!}\sum_{s=0}^{n-k-l}\frac{2^s\,\langle k+l-n\rangle_s\,\langle k+\beta+1\rangle_s}{s!\,\langle 2k+\alpha+\beta+2\rangle_s} \\
&= \frac{(n+r)!\,(-2)^k\Gamma(k+\alpha+\beta+1)}{\Gamma(2k+\alpha+\beta+1)} \\
&\quad \times \sum_{l=0}^{n-k}\frac{{}_2F_1(k+l-n,k+\beta+1;2k+\alpha+\beta+2;2)}{l!\,(n+r-l)!\,(n-k-l)!}
\end{aligned}
\tag{68}
$$

We now obtain

$$
\begin{aligned}
\varepsilon_{n,r}(x) &= (n+r)!\sum_{k=0}^{n}\frac{(-2)^k\,\Gamma(k+\alpha+\beta+1)}{\Gamma(2k+\alpha+\beta+1)} \\
&\quad \times \sum_{l=0}^{n-k}\frac{{}_2F_1(k+l-n,k+\beta+1;2k+\alpha+\beta+2;2)}{l!\,(n+r-l)!\,(n-k-l)!}P_k^{(\alpha,\beta)}(x).
\end{aligned}
\tag{69}
$$

This verifies (46) in Theorem 2.

5. Conclusions

Let $\gamma_{m,r}(x)$, $\varepsilon_{m,r}(x)$, and $\alpha_m(x)$ denote the following sums of finite products given by

$$
\gamma_{n,r}(x) = \sum_{i_1+\cdots+i_{2r+1}=n} P_{i_1}(x)P_{i_2}(x)\ldots P_{i_{2r+1}}(x),
$$

$$
\varepsilon_{n,r}(x) = \sum_{i_1+\cdots+i_{r+1}=n} L_{i_1}\left(\frac{x}{r+1}\right)L_{i_2}\left(\frac{x}{r+1}\right)\ldots L_{i_{r+1}}\left(\frac{x}{r+1}\right),
$$

$$
\alpha_m(x) = \sum_{k=1}^{m-1}\frac{1}{k(m-k)}B_k(x)B_{m-k}(x),\ (m\geq 2),
$$

where $P_n(x)$, $L_n(x)$, $B_n(x)$, $(n\geq 0)$ are respectively Legendre, Laguerre and Bernoulli polynomials. In this paper, we studied sums of finite products of Legendre polynomials $\gamma_{m,r}(x)$ and those of Laguerre polynomials $\varepsilon_{m,r}(x)$, and expressed them as linear combinations of the orthogonal polynomials $H_n(x), L_n^\alpha(x), P_n(x), C_n^{(\lambda)}(x)$, and $P_n^{(\alpha,\beta)}(x)$. These have been done by carrying out explicit computations. In recent years, we have obtained similar results for many other special polynomials.

For example, we considered sums of finite products of Bernoulli, Euler and Genocchi polynomials and represented them in terms of Bernoulli polynomials. In addition, as for Chebyshev polynomials of the second, third, and fourth kinds, and Fibonacci, Legendre and Laguerre polynomials, we expressed them not only in terms of Bernoulli polynomials but also of Chebyshev polynomials of all kinds and Hermite, generalized Laguerre, Legendre, Gegenbauer and Jacobi polynomials.

We gave twofold justification for studying such sums of finite products of special polynomials. Firstly, it can be viewed as a generalization of the classical connection problem in which one wants to determine the connection coefficients in the expansion of a product of two polynomials in terms of any given sequence of polynomials. Secondly, from the representation of $\alpha_m(x)$ in terms of Bernoulli polynomials we can derive the famous Faber-Pandharipande-Zagier identity and a slightly different variant of the Miki's identity. We emphasized that these identities had been obtained by several different methods which are quite involved and not elementary, while our previous method used only elementary Fourier series expansions.

Along the same line of the present paper, we would like to continue to work on representing sums of finite products of some special polynomials in terms of various kinds of special polynomials and to find interesting applications of them in mathematics, science and engineering areas.

Author Contributions: All authors contributed equally to the manuscript and typed, read, and approved the final manuscript.

References

1. Andrews, G.E.; Askey, R.; Roy, R. *Special Functions, Encyclopedia of Mathematics and Its Applications*; Cambridge University Press: Cambridge, UK, 1999.
2. Beals, R.; Wong, R. Special functions and orthogonal polynomials. In *Cambridge Studies in Advanced Mathematics*; Cambridge University Press: Cambridge, UK, 2016.
3. Kim, D.S.; Kim, T.; Dolgy, D.V. Some identities on Laguerre polynomials in connection with Bernoulli and Euler numbers. *Discret. Dyn. Nat. Soc.* **2012**, *2012*, 619197. [CrossRef]
4. Kim, D.S.; Kim, T.; Rim, S. Some identities involving Gegenbauer polynomials. *Adv. Differ. Equ.* **2012**, *2012*, 219. [CrossRef]
5. Kim, D.S.; Kim, T.; Rim, S.; Lee, S.H. Hermite polynomials and their applications associated with Bernoulli and Euler numbers. *Discret. Dyn. Nat. Soc.* **2012**, *2012*, 974632. [CrossRef]
6. Kim, D.S.; Rim, S.; Kim, T. Some identities on Bernoulli and Euler polynomials arising from orthogonality of Legendre polynomials. *J. Inequal. Appl.* **2012**, *2012*, 227. [CrossRef]
7. Kim, T.; Kim, D.S. Extended Laguerre polynomials associated with Hermite, Bernoulli, and Euler numbers and polynomials. *Abstr. Appl. Anal.* **2012**, *2012*, 957350. [CrossRef]
8. Kim, T.; Kim, D.S.; Dolgy, D.V. Some identities on Bernoulli and Hermite polynomials associated with Jacobi polynomials. *Discret. Dyn. Nat. Soc.* **2012**, *2012*, 584643. [CrossRef]
9. Kim, T.; Kim, D.S.; Dolgy, D.V.; Park, J.-W. Sums of finite products of Chebyshev polynomials of the second kind and of Fibonacci polynomials. *J. Inequal. Appl.* **2018**, *2018*, 148. [CrossRef] [PubMed]
10. Kim, T.; Kim, D.S.; Dolgy, D.V.; Park, J.-W. Sums of finite products of Legendre and Laguerre polynomials. *Adv. Differ. Equ.* **2018**, *2018*, 277. [CrossRef]
11. Kim, T.; Kim, D.S.; Dolgy, D.V.; Ryoo, C.S. Representing sums of finite products of Chebyshev polynomials of third and fourth kinds by Chebyshev polynomials. *Symmetry* **2018**, *10*, 258. [CrossRef]
12. Kim, T.; Kim, D.S.; Jang, G.-W.; Kwon, J. Sums of finite products of Legendre and Laguerre polynomials by Chebyshev polynomials. *Adv. Stud. Contemp. Math.* **2018**, *28*, 551–565.
13. Kim, T.; Kim, D.S.; Kwon, J.; Dolgy, D.V. Expressing sums of finite products of Chebyshev polynomials of the second kind and of Fibonacci polynomials by several orthogonal polynomials. *Mathematics* **2018**, *6*, 210. [CrossRef]

14. Cesarano, C.; Fornaro, C. A note on two-variable Chebyshev polynomials. *Georgian Math. J.* **2017**, *24*, 339–349. [CrossRef]

15. Cesarano, C.; Fornaro, C.; Vazquez, L. A note on a special class of Hermite polynomials. *Int. J. Pure Appl. Math.* **2015**, *98*, 261–273. [CrossRef]

16. Kim, T.; Kim, D.S.; Jang, L.C.; Jang, G. Sums of finite products of Genocchi functions. *Adv. Differ. Equ.* **2017**; *2017*, 268. [CrossRef]

17. Agarwal, R.P.; Kim, D.S.; Kim, T.; Kwon, J. Sums of finite products of Bernoulli functions. *Adv. Differ. Equ.* **2017**, *2017*, 237. [CrossRef]

18. Kim, T.; Kim, D.S.; Jang, G.-W.; Kwon, J. Sums of finite products of Euler functions. In *Advances in Real and Complex Analysis with Applications*; Trends in Mathematics; Springer: Berlin, Germany, 2017, pp. 243–260.

19. Kim, T.; Dolgy, D.V.; Kim, D.S. Representing sums of finite products of Chebyshev polynomials of second kind and Fibonacci polynomials in terms of Chebyshev polynomials. *Adv. Stud. Contemp. Math. (Kyungshang)* **2018**, *28*, 321–335.

20. Araci, S.; Acikgoz, M.; Bagdasaryan, A.; Sen, E. The Legendre polynomials associated with Bernoulli, Euler, Hermite and Bernstein polynomials. *Turk. J. Anal. Number Theory* **2013**, *1*, 1–3. [CrossRef]

21. Khan, W.A.; Araci, S.; Acikgoz, M. A new class of Laguerre-based Apostol type polynomials. *Cogent Math.* **2016**, *3*, 1243839. [CrossRef]

17

A New Identity Involving Balancing Polynomials and Balancing Numbers

Yuanyuan Meng

School of Mathematics, Northwest University, Xi'an 710127, Shaanxi, China; yymeng@stumail.nwu.edu.cn

Abstract: In this paper, a second-order nonlinear recursive sequence $M(h, i)$ is studied. By using this sequence, the properties of the power series, and the combinatorial methods, some interesting symmetry identities of the structural properties of balancing numbers and balancing polynomials are deduced.

Keywords: balancing numbers; balancing polynomials; combinatorial methods; symmetry sums

1. Introduction

For any positive integer $n \geq 2$, we denote the balancing number by B_n and the balancer corresponding to it by $r(n)$ if

$$1 + 2 + \cdots + (B_n - 1) = (B_n + 1) + (B_n + 2) + \cdots + (B_n + r(n))$$

holds for some positive integer $r(n)$ and B_n. It is clear that $r(n) = \frac{B_n - B_{n-1} - 1}{2}$, for example, $r(2) = 2$, $r(3) = 14$, $r(4) = 84$, $r(5) = 492\ldots$

It is found that the balancing numbers satisfy the second order linear recursive sequence $B_{n+1} = 6B_n - B_{n-1}$ ($n \geq 1$), providing $B_0 = 0$ and $B_1 = 1$ [1].

The balancing polynomials $B_n(x)$ are defined by $B_0(x) = 1$, $B_1(x) = 6x$, $B_2(x) = 36x^2 - 1$, $B_3(x) = 216x^3 - 12x$, $B_4(x) = 1296x^4 - 108x^2 + 1$, and the second-order linear difference equation:

$$B_{n+1}(x) = 6xB_n(x) - B_{n-1}(x), n \geq 1,$$

where x is any real number. While $n \geq 1$, we get $B_{n+1} = 6B_n - B_{n-1}$ with $B_n(1) = B_{n+1}$. Such balancing numbers have been widely studied in recent years. G. K. Panda and T. Komatsu [2] studied the reciprocal sums of the balancing numbers and proved the following inequation holds for any positive integer n:

$$\frac{1}{B_n - B_{n-1}} < \sum_{k=n}^{\infty} \frac{1}{B_k} < \frac{1}{B_n - B_{n-1} - 1}.$$

G. K. Panda [3] studied some fascinating properties of balancing numbers and gave the following result for any natural numbers $m > n$:

$$(B_m + B_n)(B_m - B_n) = B_{m+n} \cdot B_{m-n}.$$

Other achievements related to balancing numbers can be found in [4–7].

It is found that the balancing polynomials $B_n(x)$ can be generally expressed as

$$B_n(x) = \frac{1}{2\sqrt{9x^2 - 1}} \left[\left(3x + \sqrt{9x^2 - 1}\right)^{n+1} - \left(3x - \sqrt{9x^2 - 1}\right)^{n+1} \right],$$

and the generating function of the balancing polynomials $B_n(x)$ is given by

$$\frac{1}{1 - 6xt + t^2} = \sum_{n=0}^{\infty} B_n(x) \cdot t^n. \tag{1}$$

Recently, our attention was drawn to the sums of polynomials calculating problem [8–11], which is important in mathematical application. We are going to study the computational problem of the symmetry summation:

$$\sum_{a_1+a_2+\cdots+a_{h+1}=n} B_{a_1}(x) B_{a_2}(x) \cdots B_{a_{h+1}}(x),$$

where h is any positive integer. We shall prove the following theorem holds.

Theorem 1. *For any specific positive integer h and any integer $n \geq 0$, the following identity stands:*

$$\sum_{a_1+a_2+\cdots+a_{h+1}=n} B_{a_1}(x) B_{a_2}(x) \cdots B_{a_{h+1}}(x)$$

$$= \frac{1}{2^h \cdot h!} \cdot \sum_{j=1}^{h} \frac{M(h,j)}{(3x)^{2h-j}} \sum_{i=0}^{n} \frac{(n-i+j)!}{(n-i)!} \cdot \frac{B_{n-i+j}(x)}{(3x)^i} \cdot \binom{2h+i-j-1}{i},$$

where $M(h,i)$ is defined by $M(h,0) = 0$, $M(h,i) = \frac{(2h-i-1)!}{2^{h-i} \cdot (h-i)! \cdot (i-1)!}$ for all positive integers $1 \leq i \leq h$.

In particular, for $n = 0$, the following corollary can be deduced.

Corollary 1. *For any positive integer $h \geq 1$, the following formula holds:*

$$\sum_{j=1}^{h} M(h,j) \cdot j! \cdot (3x)^j \cdot B_j(x) = 2^h \cdot h! \cdot (3x)^{2h}.$$

The formula in Corollary 1 shows the close relationship among the balancing polynomials. For $h = 2$, the following corollary can be inferred by Theorem 1.

Corollary 2. *For any integer $n \geq 0$, we obtain*

$$\sum_{a+b+c=n} B_a(x) \cdot B_b(x) \cdot B_c(x) = \frac{1}{216x^3} \sum_{i=0}^{n} (n-i+1)(i+1)(i+2) \cdot \frac{B_{n-i+2}}{(3x)^i}$$

$$+ \frac{1}{72x^2} \sum_{i=0}^{n} (n-i+1)(n-i+2)(i+1) \cdot \frac{B_{n-i+3}}{(3x)^i}.$$

For $x = 1$, $h = 2$ *and* 3, according to Theorem 1 we can also infer the following corollaries:

Corollary 3. *For any integer $n \geq 0$, we obtain*

$$\sum_{a+b+c=n} B_{a+1} \cdot B_{b+1} \cdot B_{c+1} = \frac{1}{216} \sum_{i=0}^{n} (n-i+1)(i+1)(i+2) \cdot \frac{B_{n-i+2}}{3^i}$$

$$+ \frac{1}{72} \sum_{i=0}^{n} (n-i+1)(n-i+2)(i+1) \cdot \frac{B_{n-i+3}}{3^i}.$$

Corollary 4. *For any integer $n \geq 0$, we obtain:*

$$\sum_{a+b+c+d=n} B_{a+1} \cdot B_{b+1} \cdot B_{c+1} \cdot B_{d+1}$$

$$= \frac{1}{3888} \sum_{i=0}^{n} (n-i+1)(i+1)(i+2)(i+3)(i+4) \cdot \frac{B_{n-i+2}}{3^i}$$

$$+ \frac{1}{1296} \sum_{i=0}^{n} (n-i+1)(n-i+2)(i+1)(i+2)(i+3) \cdot \frac{B_{n-i+3}}{3^i}$$

$$+ \frac{1}{1296} \sum_{i=0}^{n} (n-i+1)(n-i+2)(n-i+3)(i+1)(i+2) \cdot \frac{B_{n-i+4}}{3^i}.$$

Corollary 5. *For any odd prime p, we have the congruence $M(p,i) \equiv 0 (mod p), 0 \leq i \leq p-1$.*

Corollary 6. *The balancing polynomials are essentially Chebyshev polynomials of the second kind, specifically $B_n(x) = U_n(3x)$. Taking $x = \frac{1}{3}x$ in Theorem 1, we can get the following:*

$$\sum_{a_1+a_2+\cdots+a_{h+1}=n} U_{a_1}(x)U_{a_2}(x)\cdots U_{a_{h+1}}(x)$$

$$= \frac{1}{2^h \cdot h!} \cdot \sum_{j=1}^{h} \frac{(2h-j-1)!}{2^{h-j} \cdot (h-j)! \cdot (j-1)! \cdot x^{2h-j}} \sum_{i=0}^{n} \frac{(n-i+j)!}{(n-i)!} \cdot \frac{U_{n-i+j}(x)}{x^i} \cdot \binom{2h+i-j-1}{i}.$$

Compared with [8], we give a more precise result for $\sum_{a_1+a_2+\cdots+a_{h+1}=n} U_{a_1}(x)U_{a_2}(x)\cdots U_{a_{h+1}}(x)$ with the specific expressions of $M(h,i)$. This shows our novelty.

Here, we list the first several terms of $M(h,i)$ in Table 1 in order to demonstrate the properties of the sequence $M(h,i)$ clearly.

Table 1. Values of $M(h,i)$.

$M(h,i)$	$i=1$	$i=2$	$i=3$	$i=4$	$i=5$	$i=6$	$i=7$	$i=8$
$h=1$	1							
$h=2$	1	1						
$h=3$	3	3	1					
$h=4$	15	15	6	1				
$h=5$	105	105	45	10	1			
$h=6$	945	945	420	105	15	1		
$h=7$	10,395	10,395	4725	1260	210	21	1	
$h=8$	135,135	135,135	62,370	17,325	3150	378	28	1

2. Several Lemmas

For the sake of clarity, several lemmas that are necessary for proving our theorem will be given in this section.

Lemma 1. *For the sequence $M(n,i)$, the following identity holds for all $1 \leq i \leq n$:*

$$M(n,i) = \frac{(2n-i-1)!}{2^{n-i} \cdot (n-i)! \cdot (i-1)!}.$$

Proof. We present a straightforward proof of this lemma by using mathematical introduction. It is obvious that

$$M(1,1) = \frac{0!}{1 \cdot 0! \cdot 0!} = 1.$$

This means Lemma 1 is valid for $n = 1$. Without loss of generality, we assume that Lemma 1 holds for $1 \le n = h$ and all $1 \le i \le h$. Then, we have

$$M(h, i) = \frac{(2h - i - 1)!}{2^{h-i} \cdot (h - i)! \cdot (i - 1)!},$$

$$M(h, i + 1) = \frac{(2h - i - 2)!}{2^{h-i-1} \cdot (h - i - 1)! \cdot i!}.$$

According to the definitions of $M(n, i)$, it is easy to find that

$$
\begin{aligned}
M(h + 1, i + 1) &= (2h - 1 - i) \cdot M(h, i + 1) + M(h, i) \\
&= (2h - 1 - i) \cdot \frac{2(h - i)}{(2h - i - 1)i} \cdot M(h, i) + M(h, i) \\
&= \frac{2h - i}{i} M(h, i) = \frac{(2h - i)!}{2^{h-i} \cdot (h - i)! \cdot i!} \\
&= \frac{(2(h + 1) - (i + 1) - 1)!}{2^{h-i} \cdot (h - i)! \cdot i!}.
\end{aligned}
$$

Thus, Lemma 1 is also valid for $n = h + 1$. From now on, Lemma 1 has been proved. $\square$

Lemma 2. *If we have a function* $f(t) = \frac{1}{1-6xt+t^2}$, *then for any positive integer n, real numbers x and t with* $|t| < |3x|$, *the following identity holds:*

$$2^n \cdot n! \cdot f^{n+1}(t) = \sum_{i=1}^{n} M(n, i) \cdot \frac{f^{(i)}(t)}{(3x - t)^{2n-i}},$$

where $f^{(i)}(t)$ *denotes the i-th order derivative of* $f(t)$, *with respect to variable t and* $M(n, i)$, *which is defined in the theorem.*

Proof. Similarly, Lemma 2 will be proved by mathematical induction. We start by showing that Lemma 2 is valid for $n = 1$. Using the properties of the derivative, we have:

$$f'(t) = (6x - 2t) \cdot f^2(t),$$

or

$$2f^2(t) = \frac{f'(t)}{3x - t} = M(1, 1) \cdot \frac{f'(t)}{3x - t}.$$

This is in fact true and provides the main idea to show the following steps. Without loss of generality, we assume that Lemma 2 holds for $1 \le n = h$. Then, we have

$$2^h \cdot h! \cdot f^{h+1}(t) = \sum_{i=1}^{h} M(h, i) \cdot \frac{f^{(i)}(t)}{(3x - t)^{2h-i}}. \tag{2}$$

As an immediate consequence, we can tell by (2), the properties of $M(n, i)$, and the derivative, we get

$$
\begin{aligned}
2^h \cdot (h + 1)! \cdot f^h(t) \cdot f'(t) &= 2^{h+1} \cdot (h + 1)! \cdot (3x - t) \cdot f^{h+2}(t) \\
&= \sum_{i=1}^{h} \frac{M(h, i)}{(3x - t)^{2h-i}} \cdot f^{(i+1)}(t) + \sum_{i=1}^{h} \frac{(2h - i)M(h, i)}{(3x - t)^{2h-i+1}} \cdot f^{(i)}(t)
\end{aligned}
$$

$$\begin{aligned}
= \ & \frac{M(h,h)}{(3x-t)^h} \cdot f^{(h+1)}(t) + \sum_{i=1}^{h-1} \frac{M(h,i)}{(3x-t)^{2h-i}} \cdot f^{(i+1)}(t) + \frac{(2h-1)M(h,1)}{(3x-t)^{2h}} \cdot f'(t) \\
& + \sum_{i=1}^{h-1} \frac{(2h-i-1)M(h,i+1)}{(3x-t)^{2h-i}} \cdot f^{(i+1)}(t) \\
= \ & \frac{M(h+1,h+1)}{(3x-t)^h} \cdot f^{(h+1)}(t) + \frac{M(h+1,1)}{(3x-t)^{2h}} \cdot f'(t) + \sum_{i=1}^{h-1} \frac{M(h+1,i+1)}{(3x-t)^{2h-i}} \cdot f^{(i+1)}(t) \\
= \ & \frac{M(h+1,h+1)}{(3x-t)^h} \cdot f^{(h+1)}(t) + \frac{M(h+1,1)}{(3x-t)^{2h}} \cdot f'(t) + \sum_{i=2}^{h} \frac{M(h+1,i)}{(3x-t)^{2h+1-i}} \cdot f^{(i)}(t) \\
= \ & \sum_{i=1}^{h+1} M(h+1,i) \cdot \frac{f^{(i)}(t)}{(3x-t)^{2h+1-i}}.
\end{aligned}$$

(3)

Then, it is deduced that

$$2^{h+1} \cdot (h+1)! \cdot (3x-t) \cdot f^{h+2}(t) = \sum_{i=1}^{h+1} M(h+1,i) \cdot \frac{f^{(i)}(t)}{(3x-t)^{2h+1-i}},$$

or

$$2^{h+1} \cdot (h+1)! \cdot f^{h+2}(t) = \sum_{i=1}^{h+1} M(h+1,i) \cdot \frac{f^{(i)}(t)}{(3x-t)^{2h+2-i}}.$$

Thus, Lemma 2 is also valid for $n = h+1$. From now on, Lemma 2 has been proved. $\square$

Lemma 3. *The following power series expansion holds for arbitrary positive integers h and k:*

$$\frac{f^{(h)}(t)}{(3x-t)^k} = \frac{1}{(3x)^k} \sum_{n=0}^{\infty} \left(\sum_{i=0}^{n} \frac{(n-i+h)!}{(n-i)!} \cdot \frac{B_{n-i+h}(x)}{(3x)^i} \cdot \binom{i+k-1}{i} \right) t^n,$$

where t and x are any real numbers with $|t| < |3x|$.

Proof. According to the definition of the balancing polynomials $B_n(x)$, we have:

$$f(t) = \frac{1}{1 - 6xt + t^2} = \sum_{n=0}^{\infty} B_n(x) \cdot t^n.$$

For any positive integer h, from the properties of the power series, we can obtain

$$f^{(h)}(t) = \sum_{n=0}^{\infty} (n+h)(n+h-1)\cdots(n+1) \cdot B_{n+h}(x) \cdot t^n$$

$$= \sum_{n=0}^{\infty} \frac{(n+h)!}{n!} \cdot B_{n+h}(x) \cdot t^n.$$

(4)

For all real t and x with $|t| < |3x|$, we have the following power series expansion:

$$\frac{1}{3x-t} = \frac{1}{3x} \cdot \sum_{n=0}^{\infty} \frac{t^n}{(3x)^n},$$

and

$$\frac{1}{(3x-t)^k} = \frac{1}{(3x)^k} \cdot \sum_{n=0}^{\infty} \binom{n+k-1}{n} \cdot \frac{t^n}{(3x)^n},$$

(5)

with any positive integer k. Then, it is found that

$$
\frac{f^{(h)}(t)}{(3x-t)^k}
$$

$$
= \frac{1}{(3x)^k} \cdot \left(\sum_{n=0}^{\infty} \frac{(n+h)!}{n!} \cdot B_{n+h}(x) \cdot t^n \right) \left(\sum_{n=0}^{\infty} \binom{n+k-1}{n} \cdot \frac{t^n}{(3x)^n} \right)
$$

$$
= \frac{1}{(3x)^k} \sum_{n=0}^{\infty} \left(\sum_{i+j=n} \frac{(j+h)!}{j!} \cdot B_{j+h}(x) \cdot \binom{i+k-1}{i} \cdot \frac{1}{(3x)^i} \right) t^n
$$

$$
= \frac{1}{(3x)^k} \sum_{n=0}^{\infty} \left(\sum_{i=0}^{n} \frac{(n-i+h)!}{(n-i)!} \cdot B_{n-i+h}(x) \cdot \binom{i+k-1}{i} \cdot \frac{1}{(3x)^i} \right) t^n,
$$

where we have used the multiplicative of the power series. Lemma 3 has been proved. $\square$

3. Proof of Theorem

Based on the lemmas in the above section, it is easy to deduce the proof of Theorem 1. For any positive integer h, we can derive

$$
2^h \cdot h! \cdot f^{h+1}(t) = 2^h \cdot h! \cdot \left(\sum_{n=0}^{\infty} B_n(x) \cdot t^n \right)^{h+1}
$$

$$
= 2^h \cdot h! \cdot \sum_{n=0}^{\infty} \left(\sum_{a_1+a_2+\cdots+a_{h+1}=n} B_{a_1}(x) B_{a_2}(x) \cdots B_{a_{h+1}}(x) \right) \cdot t^n. \tag{6}
$$

On the other hand, by the observation made in Lemma 3, it is deduced that

$$
2^h \cdot h! \cdot f^{h+1}(t) = \sum_{j=1}^{h} M(h,j) \cdot \frac{f^{(j)}(t)}{(3x-t)^{2h-j}}
$$

$$
= \sum_{j=1}^{h} \frac{M(h,j)}{(3x)^{2h-j}} \cdot \left(\sum_{n=0}^{\infty} \left(\sum_{i=0}^{n} \frac{(n-i+j)!}{(n-i)!} \cdot B_{n-i+j}(x) \cdot \binom{2h+i-j-1}{i} \cdot \frac{1}{(3x)^i} \right) t^n \right)
$$

$$
= \sum_{n=0}^{\infty} \left(\sum_{j=1}^{h} \frac{M(h,j)}{(3x)^{2h-j}} \sum_{i=0}^{n} \frac{(n-i+j)!}{(n-i)!} \cdot \frac{B_{n-i+j}(x)}{(3x)^i} \cdot \binom{2h+i-j-1}{i} \right) \cdot t^n. \tag{7}
$$

Altogether, we obtain the identity:

$$
2^h \cdot h! \sum_{a_1+a_2+\cdots+a_{h+1}=n} B_{a_1}(x) B_{a_2}(x) \cdots B_{a_{h+1}}(x)
$$

$$
= \sum_{j=1}^{h} \frac{M(h,j)}{(3x)^{2h-j}} \sum_{i=0}^{n} \frac{(n-i+j)!}{(n-i)!} \cdot \frac{B_{n-i+j}(x)}{(3x)^i} \cdot \binom{2h+i-j-1}{i}.
$$

This proves Theorem 1.

4. Conclusions

In this paper, a representation of a linear combination of balancing polynomials $B_i(x)$ (see Theorem 1) is obtained. Moreover, the specific expressions of $M(h,i)$ is given by using mathematical induction (see Lemma 1).

Theorem 1 can be reduced to various studies for the specific values of x, n, and h in the literature. For example, if $n=0$, our results reduce to Corollary 1. Taking $h=2$, our results reduce to Corollary 2. Taking $x=1$, $h=2,3$, our results reduce to Corollary 3 and Corollary 4, respectively.

Acknowledgments: The author would like to thank the Editor and the referees for their very helpful and detailed comments, which have significantly improved the presentation of this paper.

References

1. Behera, A.; Panda, G.K. On the square roots of triangular numbers. *Fibonacci Quart.* **1999**, *37*, 98–105.
2. Panda, G.K.; Komatsu, T.; Davala, R.K. Reciprocal sums of sequences involving balancing and lucas-balancing numbers. *Math. Rep.* **2018**, *20*, 201–214.
3. Panda, G.K. Some fascinating properties of balancing numbers. *Congr. Numer.* **2009**, *194*, 185–189.
4. Patel, B.K.; Ray, P.K. The period, rank and order of the sequence of balancing numbers modulo m. *Math. Rep.* **2016**, *18*, 395–401.
5. Komatsu, T.; Szalay, L. Balancing with binomial coefficients. *Int. J. Number. Theory* **2014**, *10*, 1729–1742. [CrossRef]
6. Finkelstein, R. The house problem. *Am. Math. Mon.* **1965**, *72*, 1082–1088. [CrossRef]
7. Ray, P.K. Balancing and Cobalancing Numbers. Ph.D. Thesis, National Institute of Technology, Rourkela, India, 2009.
8. Zhang, Y.X.; Chen, Z.Y. A new identity involving the Chebyshev polynomials. *Mathematics* **2018**, *6*, 244. [CrossRef]
9. Zhao, J.H.; Chen, Z.Y. Some symmetric identities involving Fubini polynomials and Euler numbers. *Symmetry* **2018**, *10*, 303.
10. Ma, Y.K.; Zhang, W.P. Some identities involving Fibonacci polynomials and Fibonacci numbers. *Mathematics* **2018**, *6*, 334. [CrossRef]
11. Kim, D.S.; Kim, T. A generalization of power and alternating power sums to any Appell polynomials. *Filomat* **2017**, *1*, 141–157. [CrossRef]

18

A Note on the Sequence Related to Catalan Numbers

Jin Zhang [1] **and Zhuoyu Chen** [2,*]

[1] School of Information Engineering, Xi'an University, Xi'an 710127, China; zhangjin0921@xawl.edu.cn
[2] School of Mathematics, Northwest University, Xi'an 710127, China
* Correspondence: chenzymath@stumail.nwu.edu.cn

Abstract: The main purpose of this paper is to find explicit expressions for two sequences and to solve two related conjectures arising from the recent study of sums of finite products of Catalan numbers by Zhang and Chen.

Keywords: new sequence; Catalan numbers; elementary and combinatorial methods; congruence; conjecture

MSC: 11B83; 11B75

1. Introduction

Let n be any non-negative integer. Then, $C_n = \frac{1}{n+1} \cdot \binom{2n}{n}$ ($n = 0, 1, 2, 3, \cdots$) are defined as the Catalan numbers. For example, the first several values of the Catalan numbers are $C_0 = 1$, $C_1 = 1$, $C_2 = 2$, $C_3 = 5$, $C_4 = 14$, $C_5 = 42$, $C_6 = 132$, $C_7 = 429$, $C_8 = 1430$, $\cdots$. The generating function of the sequence $\{C_n\}$ is:

$$\frac{2}{1 + \sqrt{1 - 4x}} = \sum_{n=0}^{\infty} \frac{\binom{2n}{n}}{n+1} \cdot x^n = \sum_{n=0}^{\infty} C_n \cdot x^n. \tag{1}$$

This sequence occupies a pivotal position in combinatorial mathematics, so lots of counting problems are closely related to it. A great number of examples can be found in a study by Stanley [1]. Because of these, plenty of scholars have researched the properties of Catalan numbers and obtained a large number of vital and meaningful results. Interested readers can refer to the relevant references [2–26], which is not an exhaustive list. Very recently, Zhang and Chen [27] researched the calculation problem of the following convolution sums:

$$\sum_{a_1 + a_2 + \cdots + a_h = n} C_{a_1} \cdot C_{a_2} \cdot C_{a_3} \cdots C_{a_h}, \tag{2}$$

where the summation has taken over all h-dimension non-negative integer coordinates $(a_1, a_2, \cdots, a_h)$, such that the equation $a_1 + a_2 + \cdots + a_h = n$.

They first introduced two new recursive sequences, $C(h, i)$ and $D(h, i)$, and after the elementary and combinatorial methods, they proved the following two significant conclusions:

Theorem 1. *For any positive integer h, one gets the identity:*

$$\sum_{a_1 + a_2 + \cdots + a_{2h+1} = n} C_{a_1} \cdot C_{a_2} \cdot C_{a_3} \cdots C_{a_{2h+1}}$$
$$= \frac{1}{(2h)!} \sum_{i=0}^{h} C(h, i) \sum_{j=0}^{\min(n,i)} \frac{(n-j+h+i)! \cdot C_{n-j+h+i}}{(n-j)!} \cdot \binom{i}{j} \cdot (-4)^j,$$

where the sequence $C(h, i)$ is defined as $C(1,0) = -2$, $C(h,h) = 1$, $C(h+1,h) = C(h,h-1) - (8h + 2) \cdot C(h,h)$, $C(h+1,0) = 8 \cdot C(h,1) - 2 \cdot C(h,0)$, and for all integers $1 \le i \le h-1$, we acquire the recursive formula:

$$C(h+1,i) = C(h,i-1) - (8i+2) \cdot C(h,i) + (4i+4)(4i+2) \cdot C(h,i+1).$$

Theorem 2. For any positive integer h and non-negative n, one can obtain:

$$\sum_{a_1+a_2+\cdots+a_{2h}=n} C_{a_1} \cdot C_{a_2} \cdot C_{a_3} \cdots C_{a_{2h}}$$
$$= \frac{1}{(2h-1)!} \sum_{i=0}^{h-1} \sum_{j=0}^{n} D(h, i+1) \cdot \binom{i+\frac{1}{2}}{j} \cdot (-4)^j \cdot \frac{(n-j+h+i)! \cdot C_{n-j+h+i}}{(n-j)!},$$

where $\binom{n+\frac{1}{2}}{i} = \left(n+\frac{1}{2}\right) \cdot \left(n-1+\frac{1}{2}\right) \cdots \left(n-i+1+\frac{1}{2}\right) / i!$, the sequence $D(k, i)$ are defined as $D(k, 0) = 0$, $D(k, k) = 1$, $D(k+1, k) = D(k, k-1) - (8k-2)$, $D(k+1, 1) = 24D(k, 2) - 6D(k, 1)$, and for all integers $1 \le i \le k-1$,

$$D(k+1, i) = D(k, i-1) - (8i-2) \cdot D(k, i) + 4i(4i+2) \cdot D(k, i+1).$$

Meanwhile, through numerical observation, Zhang and Chen [27] also proposed the following two conjectures:

Conjecture 1. Let p be a prime. Then, for any integer $0 \le i < \frac{p+1}{2}$, we obtain the congruence:

$$C\left(\frac{p+1}{2}, i\right) \equiv 0 \bmod p(p+1).$$

Conjecture 2. Let p be a prime. Then, for any integer $0 \le i < \frac{p+1}{2}$, we obtain the congruence:

$$D\left(\frac{p+1}{2}, i\right) \equiv 0 \bmod p(p-1).$$

For easy comparison, here we list some of the values of $C(h, i)$ and $D(h, i)$ with $1 \le h \le 6$ and $0 \le i \le h$ in the following Tables 1 and 2.

Table 1. Values of $C(k, i)$.

$C(k, i)$	$i=0$	$i=1$	$i=2$	$i=3$	$i=4$	$i=5$	$i=6$
$k=1$	-2	1					
$k=2$	12	-12	1				
$k=3$	-120	180	-30	1			
$k=4$	1680	-3360	840	-56	1		
$k=5$	$-30,240$	75,600	$-25,200$	2520	-90	1	
$k=6$	665,280	$-1,995,840$	831,600	$-110,880$	5940	-132	1

Table 2. Values of $D(k, i)$.

$D(k, i)$	$i=0$	$i=1$	$i=2$	$i=3$	$i=4$	$i=5$	$i=6$
$k=1$	0	1					
$k=2$	0	-6	1				
$k=3$	0	60	-20	1			
$k=4$	0	-840	420	-42	1		
$k=5$	0	15,120	$-10,080$	1512	-72	1	
$k=6$	0	-332640	277,200	$-55,440$	3960	-110	1

Based on these two tables and a large number of numerical calculations, we found that these conjectures are not only correct, but also have generalized conclusions. Actually, they provide a simpler and clearer representation.

In this paper, by using some notes from Zhang and Chen's work [27] as well as some basic and combinatorial methods, we are going to prove the following:

Theorem 3. *Let h be a positive integer. Then, for any integer i with $0 \leq i \leq h$, we acquire the identity:*

$$C(h, i) = (-1)^{h-i} \cdot \frac{(2h)!}{(h-i)! \cdot (2i)!}.$$

Theorem 4. *Let h be a positive integer. Then, for any integer i with $1 \leq i \leq h$, we acquire the identity:*

$$D(h, i) = (-1)^{h-i} \cdot \frac{(2h-1)!}{(h-i)! \cdot (2i-1)!}.$$

Based on the above two theorems, we may instantly deduce the following two corollaries:

Corollary 1. *Let h be any positive integer. Then, for any integer $0 \leq i \leq h - 1$, we gain the congruence:*

$$C\,(h, i) \equiv 0 \bmod 2h(2h - 1).$$

Corollary 2. *Let h be any positive integer. Then, for any integer $0 \leq i \leq h - 1$, we gain the congruence:*

$$D\,(h, i) \equiv 0 \bmod (2h - 1)(2h - 2).$$

Suppose that we consider p an odd prime, and that when $h = \frac{p+1}{2}$ in Corollary 1 and Corollary 2, combined with the identities $2h(2h - 1) = p(p + 1)$ and $(2h - 1)(2h - 2) = p(p - 1)$, our Corollary 1 and Corollary 2 proves Conjecture 1 and Conjecture 2, respectively. Practically, they prove two more general conclusions.

Taking $n = 0$ in Theorem 1 and Theorem 2 and applying our theorems, we may instantly deduce the following two identities:

Corollary 3. *Let h be any positive integer. Then, we get the identity:*

$$\sum_{i=0}^{h} (-1)^{h-i} \binom{h+i}{2i} \cdot C_{h+i} = 1.$$

Corollary 4. *Let h be any positive integer. Then, we get the identity:*

$$\sum_{i=1}^{h} (-1)^{h-i} \binom{h+i-1}{2i-1} \cdot C_{h+i-1} = 1.$$

Some notes: If we replace $C(h, i)$ ($D(h, i)$) in Theorem 1 (Theorem 2) with the formula for $C(h, i)$ ($D(h, i)$) in our Theorem 3 (Theorem 4), then we can get a more accurate representation for convolution sums (2).

The proof of the results in this paper is uncomplicated, but guessing their specific forms is not easy.

2. Proofs of the Theorems

Actually, the recursive form of the sequence $C(h, i)$ or $D(h, i)$ is more complex, but as long as we are able to guess its accurate representation, it is not difficult to prove. First of all, combining the mathematical induction method, we are going to prove:

$$C(h, i) = (-1)^{h-i} \cdot \frac{(2h)!}{(h-i)! \cdot (2i)!}. \tag{3}$$

According to Table 1, we know that $C(1,0) = -2$, $C(1,1) = 1$, $C(2,0) = -12$, $C(2,1) = 12$, $C(2,2) = 1$, $C(3,0) = -120$, $C(3,1) = 180$, $C(3,2) = -30$, $C(3,3) = 1$. This means that (3) is correct for $h = 1$, 2, 3, and $0 \le i \le h$.

Assume that (3) is correct for integer $h = k$ and all $0 \le i \le k$. That is,

$$C(k, i) = (-1)^{k-i} \cdot \frac{(2k)!}{(k-i)! \cdot (2i)!}, \quad 0 \le i \le k. \tag{4}$$

Then, for $h = k + 1$, if $i = h + 1$, applying the definition of $C(h, i)$, we acquire $C(k+1, k+1) = 1$. If $i = 0$, combining the inductive hypothesis (4) and noting that $C(k+1, 0) = 8C(k, 1) - 2C(k, 0)$, we obtain:

$$C(k+1, 0) = 8 \cdot (-1)^{k-1} \cdot \frac{(2k)!}{(k-1)! \cdot 2!} - (-1)^k \cdot 2 \cdot \frac{(2k)!}{k!} = (-1)^{k+1} \frac{(2k+2)!}{(k+1)!}. \tag{5}$$

Suppose that $1 \le i \le k$. From (4) and the recursive properties of $C(h, i)$, we gain:

$$\begin{aligned}
C(k+1, i) &= C(k, i-1) - (8i+2) \cdot C(k, i) + (4i+4)(4i+2) \cdot C(k, i+1) \\
&= (-1)^{k-i+1} \frac{(2k)!}{(k-i+1)!(2i-2)!} - (-1)^{k-i}(8i+2) \frac{(2k)!}{(k-i)!(2i)!} \\
&\quad + (-1)^{k-i-1}(4i+4)(4i+2) \cdot \frac{(2k)!}{(k-i-1)!(2i+2)!} \\
&= (-1)^{k+1-i} \cdot \frac{(2k+2)!}{(k+1-i!) \cdot (2i)!}.
\end{aligned} \tag{6}$$

According to (5) and (6), we know that the Formula (3) is correct for $h = k + 1$ and all integers $0 \le i \le k + 1$. Theorem 3 can then be proved by mathematical induction.

In a similar way, we can also prove Theorem 4 by mathematical induction. Since the proof process is the same as the proof of Theorem 3, it is omitted.

3. Conclusions

The main purpose of this paper was to give two specific expressions for the sequences $C(h, i)$ and $D(h, i)$. As for some applications of our results, we proved two conjectures proposed by Zhang and Chen in [27].

As a matter of fact, our results are more general and not subject to prime conditions. Meanwhile, using our formulae for $C(h, i)$ and $D(h, i)$ in the theorems, we can simplify the variety of results that appear in Reference [27].

This paper not only enriches the research content of the Catalan numbers, but can also be regarded as a supplement and further improvement to Zhang and Chen's work in [27].

Author Contributions: All authors have equally contributed to this work. All authors read and approved the final manuscript.

Acknowledgments: The author would like to thank the referees for their very helpful and detailed comments, which have significantly improved the presentation of this paper.

References

1. Stanley, R.P. *Enumerative Combinatorics: Volume 2 (Cambridge Studieds in Advanced Mathematics)*; Cambridge University Press: Cambridge, UK, 1997; p. 49.
2. Chu, W.C. Further identities on Catalan numbers. *Discr. Math.* **2018**, *341*, 3159–3164. [CrossRef]

3.   Anthony, J.; Polyxeni, L. A new interpretation of the Catalan numbers arising in the theory of crystals. *J. Algebra* **2018**, *504*, 85–128.

4.   Liu, J.C. Congruences on sums of super Catalan numbers. *Results Math.* **2018**, *73*, 73–140. [CrossRef]

5.   Kim, D.S.; Kim, T. A new approach to Catalan numbers using differential equations. *Russ. J. Math. Phys.* **2017**, *24*, 465–475. [CrossRef]

6.   Kim, T.; Kim, D.S. Some identities of Catalan-Daehee polynomials arising from umbral calculus. *Appl. Comput. Math.* **2017**, *16*, 177–189.

7.   Kim, T.; Kim, D.S. Differential equations associated with Catalan-Daehee numbers and their applications. *Revista de la Real Academia de Ciencias Exactas, Físicas y Naturales. Serie A. Matemáticas* **2017**, *111*, 1071–1081. [CrossRef]

8.   Kim, D.S.; Kim, T. Triple symmetric identities for w-Catalan polynomials. *J. Korean Math. Soc.* **2017**, *54*, 1243–1264.

9.   Kim, T.; Kwon, H.-I. Revisit symmetric identities for the λ-Catalan polynomials under the symmetry group of degree n. *Proc. Jangjeon Math. Soc.* **2016**, *19*, 711–716.

10. Kim, T. A note on Catalan numbers associated with p-adic integral on Zp. *Proc. Jangjeon Math. Soc.* **2016**, *19*, 493–501.

11. Basic, B. On quotients of values of Euler's function on the Catalan numbers. *J. Number Theory* **2016**, *169*, 160–173. [CrossRef]

12. Qi, F.; Shi, X.T.; Liu, F.F. An integral representation, complete monotonicity, and inequalities of the Catalan numbers. *Filomat* **2018**, *32*, 575–587. [CrossRef]

13. Qi, F.; Shi, X.T.; Mahmoud, M. The Catalan numbers: A generalization, an exponential representation, and some properties. *J. Comput. Anal. Appl.* **2017**, *23*, 937–944.

14. Qi, F.; Guo, B.N. Integral representations of the Catalan numbers and their applications. *Mathematics* **2017**, *5*, 40. [CrossRef]

15. Aker, K.; Gursoy, A.E. A new combinatorial identity for Catalan numbers. *Ars Comb.* **2017**, *135*, 391–398.

16. Dolgy, D.V.; Jang, G.-W.; Kim, D.S.; Kim, T. Explicit expressions for Catalan-Daehee numbers. *Proc. Jangjeon Math. Soc.* **2017**, *20*, 1–9.

17. Ma, Y.K.; Zhang, W.P. Some identities involving Fibonacci polynomials and Fibonacci numbers. *Mathematics* **2018**, *6*, 334. [CrossRef]

18. Zhang, W.P.; Lin, X. A new sequence and its some congruence properties. *Symmetry* **2018**, *10*, 359. [CrossRef]

19. Zhang, Y.X.; Chen, Z.Y. A new identity involving the Chebyshev polynomials. *Mathematics* **2018**, *6*, 244. [CrossRef]

20. Zhao, J.H.; Chen, Z.Y. Some symmetric identities involving Fubini polynomials and Euler numbers. *Symmetry* **2018**, *10*, 303.

21. Zhang, W.P. Some identities involving the Fibonacci numbers and Lucas numbers. *Fibonacci Q.* **2004**, *42*, 149–154.

22. Guariglia, E. Entropy and Fractal Antennas. *Entropy* **2016**, *18*, 84. [CrossRef]

23. Guido, R.C. Practical and useful tips on discrete wavelet transforms. *IEEE Signal Process. Mag.* **2015**, *32*, 162–166. [CrossRef]

24. Berry, M.V.; Lewis, Z.V. On the Weierstrass-Mandelbrot fractal function. *Proc. R. Soc. Lond. Ser. A* **1980**, *370*, 459–484. [CrossRef]

25. Guido, R.C.; Addison, P.; Walker, J. Introducing wavelets and time-frequency analysis. *IEEE Eng. Biol. Med. Mag.* **2009**, *28*, 13. [CrossRef] [PubMed]

26. Guariglia, E. Harmonic Sierpinski Gasket and Applications. *Entropy* **2018**, *20*, 714. [CrossRef]

27. Zhang, W.P.; Chen, L. On the Catalan numbers and some of their identities. *Symmetry* **2019**, *11*, 62. [CrossRef]

19

Symmetric Identities for Fubini Polynomials

Taekyun Kim [1,2], Dae San Kim [3], Gwan-Woo Jang [2] and Jongkyum Kwon [4,*

[1] Department of Mathematics, College of Science, Tianjin Polytechnic University, Tianjin 300160, China; tkkim@kw.ac.kr or kwangwoonmath@hanmail.net
[2] Department of Mathematics, Kwangwoon University, Seoul 139-701, Korea; gwjang@kw.ac.kr
[3] Department of Mathematics, Sogang University, Seoul 121-742, Korea; dskim@sogang.ac.kr
[4] Department of Mathematics Education and ERI, Gyeongsang National University, Jinju, Gyeongsangnamdo 52828, Korea
* Correspondence: mathkjk26@gnu.ac.kr

Abstract: We represent the generating function of w-torsion Fubini polynomials by means of a fermionic p-adic integral on $\mathbb{Z}_p$. Then we investigate a quotient of such p-adic integrals on $\mathbb{Z}_p$, representing generating functions of three w-torsion Fubini polynomials and derive some new symmetric identities for the w-torsion Fubini and two variable w-torsion Fubini polynomials.

Keywords: Fubini polynomials; w-torsion Fubini polynomials; fermionic p-adic integrals; symmetric identities

1. Introduction and Preliminaries

In recent years, various p-adic integrals on $\mathbb{Z}_p$ have been used in order to find many interesting symmetric identities related to some special polynomials and numbers. The relevant p-adic integrals are the Volkenborn, fermionic, q-Volkenborn, and q-fermionic integrals of which the last three were discovered by the first author T. Kim (see [1–3]). They have been used by a good number of researchers in various contexts and especially in unfolding new interesting symmetric identities. This verifies the usefulness of such p-adic integrals. Moreover, we can expect that people will find some further applications of these p-adic integrals in the years to come. The present paper is an effort in this direction. Assume that p is any fixed odd prime number. Throughout our discussion, we will use the standard notations $\mathbb{Z}_p$, $\mathbb{Q}_p$, and $\mathbb{C}_p$ to denote the ring of p-adic integers, the field of p-adic rational numbers and the completion of the algebraic closure of $\mathbb{Q}_p$, respectively. The p-adic norm $|\cdot|_p$ is normalized as $|p|_p = \frac{1}{p}$. Assume that $f(x)$ is a continuous function on $\mathbb{Z}_p$. Then the fermionic p-adic integral of $f(x)$ on $\mathbb{Z}_p$ was introduced by Kim (see [2]) as

$$\int_{\mathbb{Z}_p} f(x)d\mu_{-1}(x) = \lim_{N\to\infty} \sum_{x=0}^{p^N-1} f(x)(-1)^x, \tag{1}$$

where $\mu_{-1}(x + p^N\mathbb{Z}_p) = (-1)^x$.

We can easily deduce from (1) that (see [2,3])

$$\int_{\mathbb{Z}_p} f(x+1)d\mu_{-1}(x) + \int_{\mathbb{Z}_p} f(x)d\mu_{-1}(x) = 2f(0). \tag{2}$$

By invoking (2), we easily get (see [2,4])

$$\int_{\mathbb{Z}_p} e^{(x+y)t}d\mu_{-1}(y) = \frac{2}{e^t+1}e^{xt} = \sum_{n=0}^{\infty} E_n(x)\frac{t^n}{n!}, \tag{3}$$

where $E_n(x)$ are the usual Euler polynomials.

As is known, the two variable Fubini polynomials are defined by means of the following (see [5,6])

$$\sum_{n=0}^{\infty} F_n(x,y)\frac{t^n}{n!} = \frac{1}{1-y(e^t-1)}e^{xt}. \tag{4}$$

When $x = 0$, $F_n(y) = F_n(0,y)$, $(n \geq 0)$, are called Fubini polynomials. Further, if $y = 1$, then $Ob_n = F_n(0,1)$ are the ordered Bell numbers (also called Frobenius numbers). They first appeared in Cayley's work on a combinatorial counting problem in 1859 and have many different combinatorial interpretations. For example, the ordered Bell numbers count the possible outcomes of a multi-candidate election. From (3) and (4), we note that $F_n(x,-1/2) = E_n(x)$, $(n \geq 0)$. By (4), we easily get (see [6]),

$$F_n(y) = \sum_{k=0}^{n} S_2(n,k)k!y^k, \quad (n \geq 0), \tag{5}$$

where $S_2(n,k)$ are the Stirling numbers of the second kind.

For $w \in \mathbb{N}$, we define the two variable w-torsion Fubini polynomials given by

$$\frac{1}{1-y^w(e^t-1)^w}e^{xt} = \sum_{n=0}^{\infty} F_{n,w}(x,y)\frac{t^n}{n!}. \tag{6}$$

In particular, for $x = 0$, $F_{n,w}(y) = F_{n,w}(0,y)$ are called the w-torsion Fubini polynomials. It is obvious that $F_{n,1}(x,y) = F_n(x,y)$.

We represent the generating function of w-torsion Fubini polynomials by means of a fermionic p-adic integral on $\mathbb{Z}_p$. Then we investigate a quotient of such p-adic integrals on $\mathbb{Z}_p$, representing generating functions of three w-torsion Fubini polynomials and derive some new symmetric identities for the w-torsion Fubini and two variable w-torsion Fubini polynomials. Recently, a number of researchers have studied symmetric identities for some special polynomials. The reader may refer to [7–11] as an introduction to this active area of research. Some symmetric identities for q-special polynomials and numbers were treated in [12–15], including q-Bernoulli, q-Euler, and q-Genocchi numbers and polynomials. While some identities of symmetry for degenerate special polynomials were discussed in the more recent papers [6,16,17]. Finally, interested readers may want to have a glance at [18,19] as general references on polynomials.

2. Symmetric Identities for w-torsion Fubini and Two Variable w-torsion Fubini Polynomials

From (2), we note that

$$\int_{\mathbb{Z}_p}(-1)^x(y(e^t-1))^x d\mu_{-1}(x) = \frac{2}{1-y(e^t-1)} = 2\sum_{n=0}^{\infty} F_n(y)\frac{t^n}{n!}, \tag{7}$$

and

$$e^{xt}\int_{\mathbb{Z}_p}(-1)^z(y(e^t-1))^z d\mu_{-1}(z) = \frac{2}{1-y(e^t-1)}e^{xt} = 2\sum_{n=0}^{\infty} F_n(x,y)\frac{t^n}{n!}. \tag{8}$$

From (7) and (8), we note that

$$\left(\sum_{l=0}^{\infty} x^l \frac{t^l}{l!}\right) \left(\sum_{m=0}^{\infty} 2F_m(y) \frac{t^m}{m!}\right) = e^{xt} \int_{\mathbb{Z}_p} (-1)^z (y(e^t - 1))^z d\mu_{-1}(z)$$

$$= \sum_{n=0}^{\infty} 2F_n(x, y) \frac{t^n}{n!}. \tag{9}$$

Thus, by (9), we easily get

$$\sum_{l=0}^{n} \binom{n}{l} x^l F_{n-l}(y) = F_n(x, y), \quad (n \geq 0). \tag{10}$$

Now, we observe that

$$\frac{1 - y^k(e^t - 1)^k}{1 - y(e^t - 1)} = \sum_{i=0}^{k-1} y^i(e^t - 1)^i = \sum_{i=0}^{k-1} \sum_{l=0}^{i} \binom{i}{l} (-1)^{i-l} y^i e^{lt}$$

$$= \sum_{n=0}^{\infty} \left(\sum_{i=0}^{k-1} \sum_{l=0}^{i} \binom{i}{l} (-1)^{i-l} y^i l^n\right) \frac{t^n}{n!} \tag{11}$$

$$= \sum_{n=0}^{\infty} \left(\sum_{i=0}^{k-1} y^i \Delta^i 0^n\right) \frac{t^n}{n!},$$

where $\Delta f(x) = f(x + 1) - f(x)$.

For $w \in \mathbb{N}$, the w-torsion Fubini polynomials are represented by means of the following fermionic p-adic integral on $\mathbb{Z}_p$:

$$\int_{\mathbb{Z}_p} (-y^w(e^t - 1)^w)^x d\mu_{-1}(x) = \frac{2}{1 - y^w(e^t - 1)^w} = \sum_{n=0}^{\infty} 2F_{n,w}(y) \frac{t^n}{n!}. \tag{12}$$

From (7) and (12), we have

$$\frac{\int_{\mathbb{Z}_p} (-y(e^t - 1))^x d\mu_{-1}(x)}{\int_{\mathbb{Z}_p} (-y^{w_1}(e^t - 1)^{w_1})^x d\mu_{-1}(x)} = \frac{1 - y^{w_1}(e^t - 1)^{w_1}}{1 - y(e^t - 1)} = \sum_{i=0}^{w_1-1} y^i(e^t - 1)^i$$

$$= \sum_{n=0}^{\infty} \left(\sum_{i=0}^{w_1-1} y^i \Delta^i 0^n\right) \frac{t^n}{n!}, \quad (w_1 \in \mathbb{N}). \tag{13}$$

For $w_1, w_2 \in \mathbb{N}$, we let

$$I = \frac{\int_{\mathbb{Z}_p} (-y^{w_1}(e^t - 1)^{w_1})^{x_1} d\mu_{-1}(x_1) \int_{\mathbb{Z}_p} (-y^{w_2}(e^t - 1)^{w_2})^{x_2} d\mu_{-1}(x_2)}{\int_{\mathbb{Z}_p} (-y^{w_1 w_2}(e^t - 1)^{w_1 w_2})^x d\mu_{-1}(x)}. \tag{14}$$

Here it is important to observe that (14) has the built-in symmetry. Namely, it is invariant under the interchange of w_1 and w_2.

Then, by (14), we get

$$I = \left(\int_{\mathbb{Z}_p} (-y^{w_1}(e^t - 1)^{w_1})^x d\mu_{-1}(x)\right) \times \left(\frac{\int_{\mathbb{Z}_p} (-y^{w_2}(e^t - 1)^{w_2})^x d\mu_{-1}(x)}{\int_{\mathbb{Z}_p} (-y^{w_1 w_2}(e^t - 1)^{w_1 w_2})^x d\mu_{-1}(x)}\right). \tag{15}$$

First, we observe that

$$\frac{\int_{\mathbb{Z}_p}(-y^{w_2}(e^t-1)^{w_2})^x d\mu_{-1}(x)}{\int_{\mathbb{Z}_p}(-y^{w_1w_2}(e^t-1)^{w_1w_2})^x d\mu_{-1}(x)} = \frac{1-y^{w_1w_2}(e^t-1)^{w_1w_2}}{1-y^{w_2}(e^t-1)^{w_2}} = \sum_{i=0}^{w_1-1} y^{w_2i}(e^t-1)^{w_2i}$$

$$= \sum_{i=0}^{w_1-1} y^{w_2i} \sum_{l=0}^{w_2i}\binom{w_2i}{l}(-1)^{w_2i-l}e^{lt} \qquad (16)$$

$$= \sum_{n=0}^{\infty}\left(\sum_{i=0}^{w_1-1} y^{w_2i}\Delta^{w_2i}0^n\right)\frac{t^n}{n!}.$$

From (15) and (16), we can derive the following equation.

$$I = \left(\int_{\mathbb{Z}_p}(-y^{w_1}(e^t-1)^{w_1})^x d\mu_{-1}(x)\right) \times \left(\frac{\int_{\mathbb{Z}_p}(-y^{w_2}(e^t-1)^{w_2})^x d\mu_{-1}(x)}{\int_{\mathbb{Z}_p}(-y^{w_1w_2}(e^t-1)^{w_1w_2})^x d\mu_{-1}(x)}\right)$$

$$= \left(\sum_{m=0}^{\infty} 2F_{m,w_1}(y)\frac{t^m}{m!}\right) \times \left(\sum_{k=0}^{\infty}\left(\sum_{i=0}^{w_1-1} y^{w_2i}\Delta^{w_2i}0^k\right)\frac{t^k}{k!}\right) \qquad (17)$$

$$= \sum_{n=0}^{\infty}\left(2\sum_{k=0}^{n}\sum_{i=0}^{w_1-1} y^{w_2i}\Delta^{w_2i}0^k F_{n-k,w_1}(y)\binom{n}{k}\right)\frac{t^n}{n!}.$$

Interchanging the roles of w_1 and w_2, by (14), we get

$$I = \left(\int_{\mathbb{Z}_p}(-y^{w_2}(e^t-1)^{w_2})^x d\mu_{-1}(x)\right) \times \left(\frac{\int_{\mathbb{Z}_p}(-y^{w_1}(e^t-1)^{w_1})^x d\mu_{-1}(x)}{\int_{\mathbb{Z}_p}(-y^{w_1w_2}(e^t-1)^{w_1w_2})^x d\mu_{-1}(x)}\right). \qquad (18)$$

We note that

$$\frac{\int_{\mathbb{Z}_p}(-y^{w_1}(e^t-1)^{w_1})^x d\mu_{-1}(x)}{\int_{\mathbb{Z}_p}(-y^{w_1w_2}(e^t-1)^{w_1w_2})^x d\mu_{-1}(x)} = \frac{1-y^{w_1w_2}(e^t-1)^{w_1w_2}}{1-y^{w_1}(e^t-1)^{w_1}} = \sum_{i=0}^{w_2-1} y^{w_1i}(e^t-1)^{w_1i}$$

$$= \sum_{n=0}^{\infty}\left(\sum_{i=0}^{w_2-1} y^{w_1i}\Delta^{w_1i}0^n\right)\frac{t^n}{n!}. \qquad (19)$$

Thus, by (18) and (19), we get

$$I = \left(\int_{\mathbb{Z}_p}(-y^{w_2}(e^t-1)^{w_2})^x d\mu_{-1}(x)\right) \times \left(\frac{\int_{\mathbb{Z}_p}(-y^{w_1}(e^t-1)^{w_1})^x d\mu_{-1}(x)}{\int_{\mathbb{Z}_p}(-y^{w_1w_2}(e^t-1)^{w_1w_2})^x d\mu_{-1}(x)}\right)$$

$$= \left(\sum_{m=0}^{\infty} 2F_{m,w_2}(y)\frac{t^m}{m!}\right) \times \left(\sum_{k=0}^{\infty}\left(\sum_{i=0}^{w_2-1} y^{w_1i}\Delta^{w_1i}0^k\right)\frac{t^k}{k!}\right) \qquad (20)$$

$$= \sum_{n=0}^{\infty}\left(2\sum_{k=0}^{n}\sum_{i=0}^{w_2-1} y^{w_1i}\Delta^{w_1i}0^k F_{n-k,w_2}(y)\binom{n}{k}\right)\frac{t^n}{n!}.$$

The following theorem is now obtained by Equations (17) and (20).

Theorem 1. *For $w_1, w_2 \in \mathbb{N}$ with $w_1 \equiv 1$ (mod 2), $w_2 \equiv 1$ (mod 2), $n \geq 0$, we have*

$$\sum_{k=0}^{n}\sum_{i=0}^{w_1-1}\binom{n}{k}F_{n-k,w_1}(y)y^{w_2i}\Delta^{w_2i}0^k = \sum_{k=0}^{n}\sum_{i=0}^{w_2-1}\binom{n}{k}F_{n-k,w_2}(y)y^{w_1i}\Delta^{w_1i}0^k. \qquad (21)$$

Remark 1. *In particular, for $w_1 = 1$, we have*

$$F_n(y) = \sum_{k=0}^{n} \sum_{i=0}^{w_2-1} \binom{n}{k} F_{n-k,w_2}(y) y^i \Delta^i 0^k. \tag{22}$$

By expressing I in a different way, we have

$$
\begin{aligned}
I &= \left(\int_{\mathbb{Z}_p} (-y^{w_1}(e^t - 1)^{w_1})^x d\mu_{-1}(x) \right) \times \left(\frac{\int_{\mathbb{Z}_p} (-y^{w_2}(e^t - 1)^{w_2})^x d\mu_{-1}(x)}{\int_{\mathbb{Z}_p} (-y^{w_1 w_2}(e^t - 1)^{w_1 w_2})^x d\mu_{-1}(x)} \right) \\
&= \left(\int_{\mathbb{Z}_p} (-y^{w_1}(e^t - 1)^{w_1})^x d\mu_{-1}(x) \right) \times \left(\frac{1 - y^{w_1 w_2}(e^t - 1)^{w_1 w_2}}{1 - y^{w_2}(e^t - 1)^{w_2}} \right) \\
&= \left(\sum_{i=0}^{w_1-1} y^{w_2 i}(e^t - 1)^{w_2 i} \right) \times \left(\frac{2}{1 - y^{w_1}(e^t - 1)^{w_1}} \right) \\
&= \sum_{i=0}^{w_1-1} \sum_{l=0}^{w_2 i} \binom{w_2 i}{l} y^{w_2 i}(-1)^l \frac{2}{1 - y^{w_1}(e^t - 1)^{w_1}} e^{(w_2 i - l)t} \\
&= 2 \sum_{n=0}^{\infty} \left(\sum_{i=0}^{w_1-1} \sum_{l=0}^{w_2 i} \binom{w_2 i}{l} y^{w_2 i}(-1)^l F_{n,w_1}(w_2 i - l, y) \right) \frac{t^n}{n!}.
\end{aligned} \tag{23}
$$

Interchanging the roles of w_1 and w_2, by (14), we get

$$
\begin{aligned}
I &= \left(\int_{\mathbb{Z}_p} (-y^{w_2}(e^t - 1)^{w_2})^x d\mu_{-1}(x) \right) \times \left(\frac{\int_{\mathbb{Z}_p} (-y^{w_1}(e^t - 1)^{w_1})^x d\mu_{-1}(x)}{\int_{\mathbb{Z}_p} (-y^{w_1 w_2}(e^t - 1)^{w_1 w_2})^x d\mu_{-1}(x)} \right) \\
&= \left(\int_{\mathbb{Z}_p} (-y^{w_2}(e^t - 1)^{w_2})^x d\mu_{-1}(x) \right) \times \left(\frac{1 - y^{w_1 w_2}(e^t - 1)^{w_1 w_2}}{1 - y^{w_1}(e^t - 1)^{w_1}} \right) \\
&= \left(\sum_{i=0}^{w_2-1} y^{w_1 i}(e^t - 1)^{w_1 i} \right) \times \left(\frac{2}{1 - y^{w_2}(e^t - 1)^{w_2}} \right) \\
&= \sum_{i=0}^{w_2-1} \sum_{l=0}^{w_1 i} y^{w_1 i} \binom{w_1 i}{l} (-1)^l \frac{2}{1 - y^{w_2}(e^t - 1)^{w_2}} e^{(w_1 i - l)t} \\
&= 2 \sum_{n=0}^{\infty} \left(\sum_{i=0}^{w_2-1} \sum_{l=0}^{w_1 i} y^{w_1 i} \binom{w_1 i}{l} (-1)^l F_{n,w_2}(w_1 i - l, y) \right) \frac{t^n}{n!}.
\end{aligned} \tag{24}
$$

Hence, by Equations (23) and (24), we obtain the following theorem.

Theorem 2. *For $w_1, w_2 \in \mathbb{N}$ with $w_1 \equiv 1 \pmod 2$, $w_2 \equiv 1 \pmod 2$, $n \geq 0$, we have*

$$\sum_{i=0}^{w_1-1} \sum_{l=0}^{w_2 i} y^{w_2 i} \binom{w_2 i}{l} (-1)^l F_{n,w_1}(w_2 i - l, y) = \sum_{i=0}^{w_2-1} \sum_{l=0}^{w_1 i} y^{w_1 i} \binom{w_1 i}{l} (-1)^l F_{n,w_2}(w_1 i - l, y). \tag{25}$$

Remark 2. *Especially, if we take $w_1 = 1$, then by Theorem 2, we get*

$$F_n(y) = \sum_{i=0}^{w_2-1} \sum_{l=0}^{i} \binom{i}{l} y^i (-1)^l F_{n,w_2}(i - l, y). \tag{26}$$

3. Conclusions

In this paper, we introduced w-torsion Fubini polynomials as a generalization of Fubini polynomials and expressed the generating function of w-torsion Fubini polynomials by means of a fermionic p-adic integral on $\mathbb{Z}_p$. Then we derived some new symmetric identities for the w-torsion Fubini and two variable w-torsion Fubini polynomials by investigating a quotient of such p-adic integrals on $\mathbb{Z}_p$, representing generating functions of three w-torsion Fubini polynomials. It seems that they are the first double symmetric identities on Fubini polynomials. As was done, for example in [4,20,21], we expect that this result can be extended to the case of triple symmetric identities. That is one of our next projects.

Author Contributions: T.K. and D.S.K. conceived the framework and structured the whole paper; T.K. wrote the paper; G.-W.J. and J.K. checked the results of the paper; D.S.K. and J.K. completed the revision of the article.

References

1. Kim, T. q-Volkenborn integration. *Russ. J. Math. Phys.* **2002**, *9*, 288–299.
2. Kim, T. Symmetry of power sum polynomials and multivariate fermionic p-adic invariant integral on $\mathbb{Z}_p$. *Russ. J. Math. Phys.* **2009**, *16*, 93–96. [CrossRef]
3. Kim, T. A study on the q-Euler numbers and the fermionic q-integral of the product of several type q-Bernstein polynomials on $\mathbb{Z}_p$. *Adv. Stud. Contemp. Math.* **2013**, *23*, 5–11.
4. Kim, D.S.; Park, K.H. Identities of symmetry for Bernoulli polynomials arising from quotients of Volkenborn integrals invariant under S_3. *Appl. Math. Comput.* **2013**, *219*, 5096–5104. [CrossRef]
5. Kilar, N.; Simsek, Y. A new family of Fubini type numbers and polynomials associated with Apostol-Bernoulli numbers and polynomials. *J. Korean Math. Soc.* **2017**, *54*, 1605–1621.
6. Kim, T.; Kim, D.S.; Jang, G.-W. A note on degenerate Fubini polynomials. *Proc. Jangjeon Math. Soc.* **2017**, *20*, 521–531.
7. Kim, Y.-H.; Hwang, K.-H. Symmery of power sum and twisted Bernoulli polynomials. *Adv. Stud. Contemp. Math. (Kyungshang)* **2009**, *18*, 127–133.
8. Lee, J.G.; Kwon, J.; Jang, G.-W.; Jang, L.-C. Some identities of λ-Daehee polynomials. *J. Nonlinear Sci. Appl.* **2017**, *10*, 4137–4142. [CrossRef]
9. Rim, S.-H.; Jeong, J.-H.; Lee, S.-J.; Moon, E.-J.; Jin, J.-H. On the symmetric properties for the generalized twisted Genocchi polynomials. *ARS Comb.* **2012**, *105*, 267–272.
10. Rim, S.-H.; Moon, E.-J.; Jin, J.-H.; Lee, S.-J. On the symmetric properties for the generalized Genocchi polynomials. *J. Comput. Anal. Appl.* **2011**, *13*, 1240–1245.
11. Seo, J.J.; Kim, T. Some identities of symmetry for Daehee polynomials arising from p-adic invariant integral on $\mathbb{Z}_p$. *Proc. Jangjeon Math. Soc.* **2016**, *19*, 285–292.
12. Ağyüz, E.; Acikgoz, M.; Araci, S. A symmetric identity on the q-Genocchi polynomials of higher-order under third dihedral group D_3. *Proc. Jangjeon Math. Soc.* **2015**, *18*, 177–187.
13. He, Y. Symmetric identities for Calitz's q-Bernoulli numbers and polynomials. *Adv. Differ. Equ.* **2013**, *2013*, 246. [CrossRef]
14. Moon, E.-J.; Rim, S.-H.; Jin, J.-H.; Lee, S.-J. On the symmetric properties of higher-order twisted q-Euler numbers and polynoamials. *Adv. Differ. Equ.* **2010**, *2010*, 765259. [CrossRef]
15. Ryoo, C.S. An identity of the symmetry for the second kind q-Euler polynomials. *J. Comput. Anal. Appl.* **2013**, *15*, 294–299.
16. Kim, T.; Kim, D.S. An identity of symmetry for the degenerate Frobenius-Euler polynomials. *Math. Slovaca* **2018**, *68*, 239–243. [CrossRef]
17. Kim, T.; Kim, D.S. Identities of symmetry for degenerate Euler polynomials and alternating generalized falling factorial sums. *Iran J. Sci. Technol. Trans. A Sci.* **2017**, *41*, 939–949. [CrossRef]
18. Carlitz, L. Eulerian numbers and polynomials. *Math. Mag.* **1959**, *32*, 247–260. [CrossRef]
19. Milovanović, G.V.; Mitrinović, D.S.; Rassias, T.M. *Topics in Polynomials: Extremal Problems, Inequalities, Zeros*; World Scientific Publishing Co., Inc.: River Edge, NJ, USA, 1994.

20. Kim, D.S.; Kim, T. Triple symmetric identities for w-Catalan polynomials. *J. Korean Math. Soc.* **2017**, *54*, 1243–1264.
21. Kim, D.S.; Lee, N.; Na, J.; Park, K.H. Identities of symmetry for higher-order Euler polynomials in the three varibles (I). *Adv. Stud. Contemp. Math. (Kyungshang)* **2012**, *22*, 51–74.

20

A Modified PML Acoustic Wave Equation

Dojin Kim

Department of Mathematics, Pusan National University, Busan 46241, Korea; kimdojin@pusan.ac.kr

Abstract: In this paper, we consider a two-dimensional acoustic wave equation in an unbounded domain and introduce a modified model of the classical un-split perfectly matched layer (PML). We apply a regularization technique to a lower order regularity term employed in the auxiliary variable in the classical PML model. In addition, we propose a staggered finite difference method for discretizing the regularized system. The regularized system and numerical solution are analyzed in terms of the well-posedness and stability with the standard Galerkin method and *von Neumann* stability analysis, respectively. In particular, the existence and uniqueness of the solution for the regularized system are proved and the Courant-Friedrichs-Lewy (CFL) condition of the staggered finite difference method is determined. To support the theoretical results, we demonstrate a non-reflection property of acoustic waves in the layers.

Keywords: well-posedness; stability; acoustic wave equation; perfectly matched layer

1. Introduction

It is quite important to effectively truncate an unbounded domain in wave propagation simulations in open space, where the perfectly matched layer (PML) methods that surround the domain of interest with thin artificial absorbing layers are popularly used in easy and effective ways. After the method was introduced by J. P. Bérenger [1], which involves splitting a field into two nonphysical electromagnetic fields, many studies were conducted regarding the PML method and its modified reformulations in many different wave-type equations. These include Maxwell's equations [2,3], elastodynamics [4,5], linearized Euler equations [6–9], Helmholtz equations [10], and other types of wave equations [10–12]. Most PML models by the splitting technique, named a split PML method, yield a hyperbolic system of first order partial differential equations [1,6,13–15]. It is known that the split PML models demonstrate excellent overall performance from the viewpoint of applications. However, it was pointed out in [7,16,17] that Bérenger's split, as well as other split models, transform Maxwell's equations from being strongly hyperbolic into weakly hyperbolic. These transforms imply a transition from strong to weak well-posedness in the Cauchy problem and may lead to ill-posedness under certain low-order damping functions in PML layers [18]. The authors of [6,19] mention that the use of artificial dissipation is necessary to stabilize the numerical scheme of such formulations for long-time simulations.

The resulting concerns about the well-posedness and stability of the split PML models have prompted the development of other PMLs. Some examples of such developments, without splitting the fields, include un-split PML models using convolution integrals [20,21] and auxiliary variables [17,22,23]. In contrast to the split PML models, it is known that the un-split PML wave equations are more effective at time discretization [22] and does not make the use of additional memory for the nonphysical field variables. However, it has also been found that the un-split PML models are susceptible to developing gradual instabilities in long-time simulations [10,19]. To overcome this instability issue, various studies are reported: a low-pass filter inside the absorbing layer [6], selective damping coefficients [24], a new layer by regularizing the damping terms [8], a change of variable [25], etc. These issues are the motivation for the mathematical study of the well-posedness and stability for the un-split PML acoustic wave model in various sound speed. A time-domain analysis of PML acoustic wave equation with a constant sound speed is presented with a time-dependent point source in two dimensions using the Cagniard-de-Hoop method [25,26], which includes the time-stability and error estimates. However, it is not easy to extend the analysis to general initial value problems in variable sound speed, because those include not only straight propagating but also evanescent waves [27]. There is another approach to demonstrating the well-posedness and stability by investigating the eigenvalues of the Cauchy hyperbolic problems for the PML wave equations [4,7,12,16–18,28]. This approach gives a restricted result when the original formulation of the PML wave equation is considered in a bounded domain, in which the solutions should be affected by boundary conditions.

Alternatively, energy techniques are used to analyze the issue of stability for the PML wave equations by presenting the energy behavior for the solution in each model [12,16,29]. In general, the restriction of the PML equations to the computational domain coincides with the original problem [12], so that damping terms are required to vanish identically in the computational region. As the constant damping function can be considered as the Heaviside function, the equation $(\partial_t + \sigma_x)\partial_x = \partial_x(\partial_t + \sigma_x)$ used in [12,16,29] is not valid at the interface between the domain of interest and the layers for the constant damping case from a discontinuity. However, all these approaches only provide its well-posedness, the stability has not been clearly proved in finite PML acoustic wave equations with variable sound speed.

The main contribution of this manuscript is not only to introduce a regularized system of the second order PML acoustic wave equation that exhibits well-posedness without losing the non-reflection property of PMLs, but also to demonstrate its numerical stability. To construct the system, we adopt a regularization technique for the term $\nabla \cdot \vec{q}$ that has a lower regularity, which is introduced in [8], to regularize the PML model for the Maxwell equation, where $\vec{q}$ is the auxiliary variable (see (2)). The standard Galerkin approximation and energy estimation of the solution are used to show the well-posedness of the regularized system. A concrete energy estimate yields the boundedness of the solution (see Theorem 1) together with the existence and uniqueness of the solution under the regularity assumption of the damping terms $\sigma_x, \sigma_y \in L^\infty(\Omega)$ (see Theorem 2). As a numerical scheme for the regularized system, a family of finite difference schemes using half-step staggered grids in space and time is used. All spatial and temporal derivatives are discretized with central finite differences that maintain the second order approximation in both space and time, respectively. A concrete *von Neumann* stability analysis for the numerical scheme indicates that the scheme is stable under the Courant-Friedrichs-Lewy (CFL) condition between the temporal and spatial grids (see Theorem 3). The novel features of this study include the good performance of the solution that present not only the well-posedness and stability but also the non-reflection property of the wave propagation compared to the classical PML model; even the regularized system does not possess PMLs in the original wave equation. This novelty is numerically illustrated in Section 4.

The remainder of the manuscript is organized as follows. Section 2 describes a regularized system for the un-split PML model of the acoustic wave equation and also contains the well-posedness of its solution based on the energy estimation. In Section 3, we develop a staggered finite difference scheme for the regularized system and determine the CFL condition for the numerical stability. In Section 4, several numerical results are presented to support our theoretical analysis and demonstrate the efficiency of the regularized system. Finally, some discussions are given in Section 5.

2. Regularized System

The aim of this section is to introduce a modified PML system using a regularization technique in a classical PML model for the acoustic wave equation. For the sake of argument, we let $H^1(\Omega) := \{\varphi : \varphi, \partial_x\varphi, \partial_y\varphi \in L^2(\Omega)\}$ and $H^{-1}(\Omega)$ be the Sobolev space and dual space of $H_0^1(\Omega)$, respectively.

The target problem we consider with here is a general second order acoustic wave equation with a variable sound speed $c(\mathbf{x}) > 0$ described by

$$u_{tt}(\mathbf{x},t) - c^2(\mathbf{x})\Delta u(\mathbf{x},t) = 0 \quad \forall (\mathbf{x},t) \in \mathbb{R}^2 \times (0,T]$$

with initial conditions $u(\cdot,0) = f$ and $u_t(\cdot,0) = 0$, where $supp(f) \subset \Omega_0$ with a domain $\Omega_0 \subset\subset [-a,a] \times [-b,b] \subset \mathbb{R}^2$. Here, $T > 0$ and the sound speed $c(\mathbf{x})$ is bounded by

$$0 < c_* \leq c(\mathbf{x}) \leq c^* < \infty. \tag{1}$$

Let the domain $\Omega := [-a - L_x, a + L_x] \times [-b - L_y, b + L_y]$ consist of the computational domain $[-a,a] \times [-b,b]$ surrounded by PML layers, where $a, b, L_x, L_y > 0$. Using a complex coordinate stretch, we consider the following system of the PML wave equation which is introduced in [28]: find $(u, \vec{q})$ satisfying

$$\begin{cases} \dfrac{1}{c^2} u_{tt}(\mathbf{x},t) + \alpha(\mathbf{x})u_t(\mathbf{x},t) + \beta(\mathbf{x})u(\mathbf{x},t) - \nabla \cdot \vec{q}(\mathbf{x},t) - \Delta u(\mathbf{x},t) = 0 & \forall (\mathbf{x},t) \in \Omega \times (0,T], \\ \vec{q}_t(\mathbf{x},t) + A(\mathbf{x})\vec{q}(\mathbf{x},t) + B(\mathbf{x})\nabla u(\mathbf{x},t) = 0 & \forall (\mathbf{x},t) \in \Omega \times (0,T], \end{cases} \tag{2}$$

with the initial conditions

$$u(\cdot,0) := u_0 = f, \quad u_t(\cdot,0) := u_1 = 0, \quad \vec{q}(\cdot,0) := \vec{q}_0 = \vec{0},$$

and the zero *Dirichlet* boundary condition $u(\mathbf{x},\cdot)|_{\partial\Omega} = 0$, where

$$\alpha(\mathbf{x}) := \frac{\sigma_x + \sigma_y}{c^2}, \quad \beta(\mathbf{x}) := \frac{\sigma_x \sigma_y}{c^2}, \quad A(\mathbf{x}) =: \begin{bmatrix} \sigma_x & 0 \\ 0 & \sigma_y \end{bmatrix}, \quad B(\mathbf{x}) := \begin{bmatrix} \sigma_x - \sigma_y & 0 \\ 0 & \sigma_y - \sigma_x \end{bmatrix}.$$

Here, the damping terms $\sigma_x := \sigma_x(x)$ and $\sigma_y := \sigma_y(y)$ are assumed to be nonnegative C^0 functions which vanish in the computational domain in the sense of the analytical continuation of the PML.

Please note that a weak solution $(u, \vec{q})$ of (2) is in $H_0^1(\Omega) \times \mathbb{L}^2(\Omega)$, i.e., $\nabla \cdot \vec{q} \in H^{-1}(\Omega)$, which regularity is not enough to show the existence. In order to provide regularity on the term by an operator, we introduce a mollifier ρ_ϵ. Let $\rho \in C^\infty(\mathbb{R}^2)$ with $supp(\rho) \subseteq B_1(0)$ satisfying $\int_{\mathbb{R}^2} \rho(\mathbf{x})d\mathbf{x} = 1$. Then, for $\epsilon > 0$, one can define a mollifier $\rho_\epsilon(\mathbf{x})$ on $\mathbb{R}^2$ by

$$\rho_\epsilon(\mathbf{x}) = \epsilon^{-2}\rho\left(\frac{|\mathbf{x}|}{\epsilon}\right) \text{ and satisfies } \int_{\mathbb{R}^2} \rho_\epsilon(\mathbf{x})d\mathbf{x} = 1 \text{ with } supp(\rho_\epsilon) \subseteq \overline{B_\epsilon(0)}.$$

Remark 1. *Let $\mathcal{R} := -\Delta + I$ be the Riesz map from $H_0^1(\Omega) \to H^{-1}(\Omega)$. Then, we consider the operator $\delta_\epsilon : H^{-1}(\Omega) \to L^2(\Omega)$ given by*

$$\delta_\epsilon(\varphi) = \mathcal{R} \circ \delta_\epsilon^* \circ \mathcal{R}^{-1}(\varphi) \text{ for all } \varphi \in H^{-1}(\Omega), \tag{3}$$

where $\delta_\epsilon^* : H_0^1(\Omega) \to H_0^1(\Omega) \cap H^2(\Omega)$ is a linear bounded operator such that $\delta_\epsilon^* \to \mathbf{1}$, the identity operator in $H_0^1(\Omega)$, as $\epsilon \to 0$ in the strong operator topology (see, for detail, Theorem 3 on page 7 in [30]). Then, we obtain

$$\delta_\epsilon \to \mathbf{1} \text{ as } \epsilon \to 0 \text{ in the strong operator topology}$$

and $\|\delta_\epsilon(\varphi)\|_{L^2(\Omega)} \leq C_{\delta_\epsilon} \|\varphi\|_{H^{-1}(\Omega)}$ for some $C_{\delta_\epsilon} > 0$. Furthermore, by the isometry of $\mathcal{R}$,

$$\|\delta_\epsilon(\varphi) - \varphi\|_{H^{-1}(\Omega)} = \|\delta_\epsilon^* u - u\|_{H_0^1(\Omega)} \to 0 \text{ as } \epsilon \to 0$$

for $u \in H_0^1(\Omega)$ such that $\mathcal{R}(u) = \varphi$. Please note that δ_ε is a linear and bounded operator from $H^{-1}(\Omega)$ to $L^2(\Omega)$.

Now, following [8,31], we introduce a regularized system of the classical PML model (2) by using δ_ε in the term $\nabla \cdot \vec{q}$, which is given by

$$\begin{cases} \dfrac{1}{c^2} u_{tt}(\mathbf{x}, t) + \alpha(\mathbf{x}) u_t(\mathbf{x}, t) + \beta(\mathbf{x}) u(\mathbf{x}, t) - \delta_\varepsilon \nabla \cdot \vec{q}(\mathbf{x}, t) - \Delta u(\mathbf{x}, t) = 0, \\ \vec{q}_t(\mathbf{x}, t) + A(\mathbf{x})\vec{q}(\mathbf{x}, t) + B(\mathbf{x})\nabla u(\mathbf{x}, t) = 0, \end{cases} \tag{4}$$

with initial and boundary conditions

$$u(\cdot, 0) := u_0 = f, \quad u_t(\cdot, 0) := u_1 = 0, \quad \vec{q}(\cdot, 0) := \vec{q}_0 = \vec{0}, \quad u(\mathbf{x}, \cdot)|_{\partial\Omega} = 0.$$

The remainder of this section details the analysis of the well-posedness of the solution to the regularized system (4) based on the energy estimation under the assumption that the dampings σ_x and σ_y are in $L^\infty(\Omega)$.

2.1. Energy Estimate of Weak Solution

We assume that the damping functions σ_x, σ_y satisfy $\sigma_x, \sigma_y \in L^\infty(\Omega)$, which implies that

$$\begin{aligned} \|\alpha\|_\infty &= \|\sigma_x + \sigma_y\|_\infty < \infty, & \|\beta\|_\infty &\leq \|\sigma_x \sigma_y\|_\infty < \infty, \\ \|A\|_2 &:= \max\{\|\sigma_x\|_\infty, \|\sigma_y\|_\infty\} < \infty, & \|B\|_2 &\leq \sqrt{2}(\|\sigma_x\|_\infty + \|\sigma_y\|_\infty) < \infty \end{aligned} \tag{5}$$

under the condition of $c(\mathbf{x}) = 1$ in the layers of the PML model (2), where $\|\cdot\|_\infty$ denotes the L^∞-norm. Under these assumptions, the aim of this subsection is to provide an energy estimation of the weak solution of (4) in the sense that

$$u \in L^2(0, T; H_0^1(\Omega)), \quad \vec{q} \in L^2(0, T; \mathbb{L}^2(\Omega)) \tag{6}$$

with

$$u_t \in L^2(0, T; L^2(\Omega)), \quad u_{tt} \in L^2(0, T; H^{-1}(\Omega)), \quad \vec{q}_t \in L^2(0, T; \mathbb{L}^2(\Omega)),$$

which satisfies

$$\begin{cases} \left\langle \dfrac{1}{c^2} u_{tt}, w \right\rangle + (\alpha u_t, w) + (\beta u, w) - (\delta_\varepsilon \nabla \cdot \vec{q}, w) + (\nabla u, \nabla w) = 0, \\ (\vec{q}_t, \vec{v}) + (A\vec{q}, \vec{v}) + (B\nabla u, \vec{v}) = 0 \end{cases} \tag{7}$$

for each $w \in H_0^1(\Omega), \vec{v} \in \mathbb{L}^2(\Omega)$, and almost everywhere $0 \leq t \leq T$ and the initial data satisfy

$$(u(0), w) = (u_0, w), \quad <u_t(0), w> = (u_1, w), \quad \text{and } (\vec{q}(0), \vec{v}) = (\vec{q}_0, \vec{v}) \tag{8}$$

for each $w \in H_0^1(\Omega), \vec{v} \in \mathbb{L}^2(\Omega)$. Here, $< \cdot, \cdot >$ denotes the duality pairing between $H^{-1}(\Omega)$ and $H_0^1(\Omega)$, and $(\cdot, \cdot)$ is the inner product in $L^2(\Omega)$. In addition, the time derivatives are understood in a distributional sense.

Remark 2. We note that $u \in C([0, T]; L^2(\Omega))$, $u_t \in C([0, T]; H^{-1}(\Omega))$, and $\vec{q} \in C([0, T]; \mathbb{L}^2(\Omega))$. (see Theorem 2, Chapter 5.9.2 [32] for detail). Consequently, the equalities in (7), (8) make sense.

To investigate the weak solution of (4) that satisfies (7) and (8), we use the standard Galerkin approximation and estimate the energy of the solution, which will be used to show the well-posedness of the regularized system (4) in the subsequent subsection. Let $\{w_j | j \in \mathbb{N}\}$ be an c^{-2}-weighted orthonormal basis in $L^2(\Omega)$, i.e., $(c^{-2}w_j, w_k) = \delta_{jk}$, where the Kronecker delta is given by $\delta_{jk} = \begin{cases} 0 & \text{if } j \neq k, \\ 1 & \text{if } j = k \end{cases}$ of the eigenfunctions of the eigenvalue problem

$$\begin{cases} c^2 \Delta w = \lambda w & \text{in } \Omega, \qquad \lambda \in \mathbb{C}, \\ w = 0 & \text{on } \partial\Omega. \end{cases}$$

Let $\mathcal{U}_k$ be the subspace generated by the orthonormal system $\{w_1, w_2, \cdots, w_k\}$ of $L^2(\Omega)$. Then, one can see that $\mathcal{U}_k$ also becomes the c^{-2}-weighted orthogonal basis of $H_0^1(\Omega)$ in the sense that

$$\left(c^{-2}w_j, w_k\right) + (\nabla w_j, \nabla w_k) = 0 \quad \text{if } j \neq k.$$

Let us also denote $\mathcal{Q}_k$, which is the space generated by the smooth functions $\{\vec{v}_1, \vec{v}_2, \cdots, \vec{v}_k\}$ such that $\{\vec{v}_k : k \in \mathbb{N}\}$ is an orthonormal basis of $\mathbb{L}^2(\Omega)$. We now construct approximate solutions $\left(u^k, \vec{q}^k\right), k = 1, 2, 3, \cdots$, in the form

$$u^k(t) = \sum_{j=1}^{k} g_j^k(t) w_j, \qquad \vec{q}^k(t) = \sum_{j=1}^{k} h_j^k(t) \vec{v}_j, \tag{9}$$

whose coefficients $g_j^k(t)$, $h_j^k(t)$, $j = 1, 2, \cdots, k$, are chosen so that

$$g_j^k(0) = (u_0, w_j), \quad (g_j^k)_t(0) = (u_1, w_j), \quad h_j^k(0) = (\vec{q}_0, \vec{v}_j)$$

and

$$\begin{cases} \left(\dfrac{1}{c^2}u_{tt}^k, w_j\right) + \left(\alpha u_t^k + \beta u^k - \delta_\varepsilon \nabla \cdot \vec{q}^k, w_j\right) + \left(\nabla u^k, \nabla w_j\right) = 0, \\ \left(\vec{q}_t^k, \vec{v}_j\right) + \left(A\vec{q}^k, \vec{v}_j\right) + \left(B\nabla u^k, \vec{v}_j\right) = 0 \end{cases} \tag{10}$$

are satisfied for all $w_j \in \mathcal{U}_k$, $\vec{v}_j \in \mathcal{Q}_k$, $j = 1, \cdots, k$. For each integer $k = 1, 2, \cdots$, the standard theory of ordinary differential equations guarantees the existence of the approximation $\left(u^k(t), \vec{q}^k(t)\right)$ satisfying (9) and (10).

The following theorem gives a uniform bound of energy of the approximate solutions (9), which allows us to send $k \to \infty$.

Theorem 1. *There exists a constant $C_T > 0$ that depends only on $\sigma_x, \sigma_y, \Omega$, and T such that for $k \geq 1$*

$$\max_{0 \leq t \leq T} E_k(t) + \left\|u_{tt}^k\right\|_{L^2(0,T;H^{-1}(\Omega))} + \left\|\vec{q}_t^k\right\|_{L^2(0,T;\mathbb{L}^2(\Omega))} \leq C_T \left(\|u_0\|_{H_0^1(\Omega)}^2 + \|u_1\|_{L^2(\Omega)}^2 + \|\vec{q}_0\|_{\mathbb{L}^2(\Omega)}^2\right),$$

where the energy $E_k(t)$ is defined by

$$E_k(t) = \|\tfrac{1}{c}u_t^k(t)\|_{L^2(\Omega)}^2 + \|\nabla u^k(t)\|_{\mathbb{L}^2(\Omega)}^2 + \|\vec{q}^k(t)\|_{\mathbb{L}^2(\Omega)}^2.$$

Proof. Please note that $u_t^k \in \mathcal{U}^k$ and $\vec{q}^k \in \mathcal{Q}^k$. Hence, we apply u_t^k and $\vec{q}^k$ in the first and second equations of (10), respectively, to obtain

$$\begin{cases} \left(\dfrac{1}{c^2}u_{tt}^k, u_t^k\right) + \left(\alpha u_t^k + \beta u^k - \delta_\varepsilon \nabla \cdot \vec{q}^k, u_t^k\right) + \left(\nabla u^k, \nabla u_t^k\right) = 0, \\ \left(\vec{q}_t^k, \vec{q}^k\right) + \left(A\vec{q}^k, \vec{q}^k\right) + \left(B\nabla u^k, \vec{q}^k\right) = 0 \end{cases}$$

for almost everywhere $0 \leq t \leq T$. Combining the two equations with the equality $\left(\dfrac{1}{c^2}u_{tt}^k, u_t^k\right) = \dfrac{d}{dt}\left(\dfrac{1}{2}\|\dfrac{1}{c}u_t^k\|_{L^2(\Omega)}^2\right)$, we obtain

$$\frac{1}{2}\frac{d}{dt}E_k + F_k^1 + F_k^2 = 0,$$

where

$$F_k^1 = \left(\alpha u_t^k, u_t^k\right) + \left(\beta u^k, u_t^k\right) - \left(\delta_\varepsilon \nabla \cdot \vec{q}^k, u_t^k\right), \quad F_k^2 = \left(A\vec{q}^k, \vec{q}^k\right) + \left(B\nabla u^k, \vec{q}^k\right).$$

Based on the linear bounded operator $\varphi \longmapsto \delta_\varepsilon(\varphi)$, Hölder's inequality, assumptions for $\sigma_x, \sigma_y,$ and Poincaré inequality, it can be noted that $E_k(t)$ satisfies the inequality

$$\frac{dE_k}{dt} \leq C_k^\varepsilon E_k \quad \text{for a suitable constant } C_k^\varepsilon > 0.$$

Furthermore, by applying Gronwall's inequality, Poincaré inequality, and (1) in the above equation, one can obtain

$$\max_{0 \leq t \leq T} \left(\|u^k(t)\|_{H_0^1(\Omega)}^2 + \|u_t^k(t)\|_{L^2(\Omega)}^2 + \|\vec{q}^k\|_{\mathbb{L}^2(\Omega)}^2\right) \leq C \left(\|u_0\|_{H_0^1(\Omega)}^2 + \|u_1\|_{L^2(\Omega)}^2 + \|\vec{q}_0\|_{\mathbb{L}^2(\Omega)}^2\right) \quad (11)$$

for some $C > 0$.

Fix any $w \in H_0^1(\Omega)$ with $\|w\|_{H_0^1(\Omega)} \leq 1$ and $\vec{v} \in \mathbb{L}^2(\Omega)$ with $\|\vec{v}\|_{\mathbb{L}^2(\Omega)} \leq 1$, and write $w = w^1 + w^2$ and $\vec{v} = \vec{v}^1 + \vec{v}^2$, where

$$w^1 \in span\{w_j\}_{j=1}^k, \quad \left(\frac{1}{c^2}w^2, w_j\right) = 0 \text{ for } j = 1, \cdots, k$$

and

$$\vec{v}^1 \in span\{\vec{v}_j\}_{j=1}^k, \quad \left(\vec{v}^2, \vec{v}_j\right) = 0 \text{ for } j = 1, \cdots, k.$$

From (9) and (10), we have

$$\left\langle \frac{1}{c^2}u_{tt}^k, w\right\rangle = \left(\frac{1}{c^2}u_{tt}^k, w\right) = \left(\frac{1}{c^2}u_{tt}^k, w^1\right)$$
$$= -(\alpha u_t^k + \beta u^k, w^1) - (\delta_\varepsilon \nabla \cdot \vec{q}^k, w^1) + (\nabla u^k, \nabla w^1),$$
$$\left(\vec{q}_t^k, \vec{v}\right) = \left(\vec{q}_t^k, \vec{v}^1\right) = -\left(A\vec{q}^k, \vec{v}^1\right) - \left(B\nabla u^k, \vec{v}^1\right).$$

Thus, we have

$$\left|\left\langle u_{tt}^k, w\right\rangle\right| + \left|\left(\vec{q}_t^k, \vec{v}\right)\right| \leq C \left(\|u^k\|_{H_0^1(\Omega)} + \|u_t^k\|_{L^2(\Omega)} + \|\vec{q}^k\|_{\mathbb{L}^2(\Omega)}\right).$$

Consequently, we obtain

$$\int_0^T \left(\|u_{tt}^k\|_{H^{-1}(\Omega)} + \|\vec{q}_t\|_{\mathbb{L}^2(\Omega)}\right) dt \leq C \int_0^T \left(\|u^k\|_{H_0^1(\Omega)}^2 + \|u_t^k\|_{L^2(\Omega)}^2 + \|\vec{q}^k\|_{\mathbb{L}^2(\Omega)}^2\right) dt$$
$$\leq C_T \left(\|u_0\|_{H_0^1(\Omega)}^2 + \|u_1\|_{L^2(\Omega)}^2 + \|\vec{q}_0\|_{\mathbb{L}^2(\Omega)}^2\right). \quad (12)$$

The proof is carried out by combining (11) and (12). □

2.2. Existence and Uniqueness

In this subsection, we will discuss the well-posedness of the regularized system by demonstrating the existence and uniqueness of the solution (6) based on the result of Theorem 1.

Theorem 2. *(Existence and Uniqueness) Assume that the initial data $(u_0, u_1, \vec{q}_0)$ are in $H_0^1(\Omega) \times L^2(\Omega) \times \mathbb{L}^2(\Omega)$. Then, the system (4) has a unique weak solution provided by $\sigma_x, \sigma_y \in L^\infty(\Omega)$.*

Proof. The energy estimates of Theorem 1 and the standard Galerkin method enable the existence of a weak solution using the fact that $\nabla\cdot : \mathbb{L}^2(\Omega) \to H^{-1}(\Omega)$ and $\delta_\varepsilon : H^{-1}(\Omega) \to L^2(\Omega)$ are continuous almost everywhere $t \in [0,T]$ (see [31] for detail proof of uniqueness). $\square$

Remark 3. *The most important concern in the proof is the estimation of the term $\delta_\epsilon \nabla \cdot \vec{q}$ in the regularized system, which has roles of a convolution, improving the stability of the system from the regularization of the term from $H^{-1}(\Omega)$ to $\mathbb{L}^2(\Omega)$.*

3. Numerical Scheme

The aim of this section is to introduce a staggered finite difference method for discretizing the regularized system and to find a stability condition for the numerical scheme. For the staggered finite difference method, we use a family of finite difference schemes [33] with half-step staggered grids in space and time. All spatial derivatives are discretized with the centered finite differences over two or three cells, which guarantees a second order approximation in space. For the time discretization, we also use the centered finite differences for the first and second order time derivatives on a uniform mesh, which is also of the second order approximation in time. Based on the standard *von Neumann* stability analysis technique, we analyze the stability of the numerical scheme and obtain its CFL condition.

3.1. Staggered Finite Differences

Let $\triangle t > 0$ denote the time step size and $\triangle x > 0$ and $\triangle y > 0$ denote the spatial mesh sizes in the x and y directions, respectively. In addition, we also introduce the time step $t_n = n\triangle t$ and the spatial nodes $x_i = i\triangle x$ and $y_j = j\triangle y$ for $n \in \mathbb{N} \cup \{0\}$ and $i,j \in \mathbb{Z}$. We also define staggered nodes in the time direction and the x and y directions, respectively, as $t_{n\pm\frac{1}{2}} = t_n \pm \frac{1}{2}\triangle t$, $x_{i\pm\frac{1}{2}} = x_i \pm \frac{1}{2}\triangle x$, and $y_{j\pm\frac{1}{2}} = y_j \pm \frac{1}{2}\triangle y$ for $n,i,j \in \mathbb{N}$. To simplify the notation, we denote $u_{i,j}^n := u(t_n, x_i, y_j)$ and $q_{\alpha_{i+\frac{1}{2},j+\frac{1}{2}}}^{n+\frac{1}{2}} := q_\alpha(t_{n+\frac{1}{2}}, x_{i+\frac{1}{2}}, y_{j+\frac{1}{2}})$ for $\vec{q} = (q_x, q_y)$, $\alpha = x,y$. For the discretization of the regularization defined in Remark 1 for the regularized system, the smooth function $\rho_\epsilon(x,y)$ chosen in the following examples is constant on a rectangle centered at zero,

$$\rho_\epsilon(x,y) = \rho_{\epsilon 1}(x)\rho_{\epsilon 2}(y), \tag{13}$$

where

$$\rho_{\epsilon_k}(\xi) = \begin{cases} \frac{1}{\epsilon_k} & \text{if } \xi \in [-\frac{\epsilon_k}{2}, \frac{\epsilon_k}{2}], \quad k=1,2, \\ 0 & \text{elsewhere.} \end{cases}$$

For a given two-dimensional finite difference grid with spatial sizes $\triangle x$ and $\triangle y$, a possible choice of ϵ_k is $\epsilon_1 = n_x\triangle x$ and $\epsilon_2 = n_y\triangle y$ with $n_x, n_y \in \mathbb{N}$. For instance, with $n_x = n_y = 1$ and the usual integration formula (see Chapter 3 in [34]), we discretize the regularized term $\delta_\varepsilon(v)_{i,j} := (\rho_\epsilon * v)_{i,j}$, using the 9-point central difference formula, as follows:

$$(\rho_\epsilon * v)_{i,j} = \frac{1}{16}\left(4v_{i,j} + 2v_{i\pm1,j} + 2v_{i,j\pm1} + v_{i\pm1,j+1} + v_{i\pm1,j-1}\right).$$

Let us now introduce new notations

$$A_{i+\frac{1}{2}}^{x\pm} := 1 \pm \frac{\triangle t}{2}\sigma_{x_{i+\frac{1}{2}}}, \quad A_{j+\frac{1}{2}}^{y\pm} := 1 \pm \frac{\triangle t}{2}\sigma_{y_{j+\frac{1}{2}}},$$

and for $k = i,j$, $\alpha = x,y$,

$$A_{i,j}^{xy\pm} := 1 \pm \frac{\triangle t}{2}(\sigma_{x_i} + \sigma_{y_j}), \quad \sigma_{\alpha_k} := \sigma_\alpha(\alpha_k), \quad \sigma_{\alpha_{k+\frac{1}{2}}} := \sigma_\alpha(\alpha_{k+\frac{1}{2}}).$$

Based on these notations, the staggered finite difference scheme for discretizing the regularized system is defined in the following steps.

Step 1. Compute $\left(q x_{i+\frac{1}{2},j+\frac{1}{2}}^{n+\frac{1}{2}}, q y_{i+\frac{1}{2},j+\frac{1}{2}}^{n+\frac{1}{2}} \right)$,

$$A_{i+\frac{1}{2}}^{x+} q x_{i+\frac{1}{2},j+\frac{1}{2}}^{n+\frac{1}{2}} = A_{i+\frac{1}{2}}^{x-} q x_{i+\frac{1}{2},j+\frac{1}{2}}^{n-\frac{1}{2}} - \triangle t (\sigma_{x_{i+\frac{1}{2}}} - \sigma_{y_{j+\frac{1}{2}}}) \tilde{\partial}_x u_{i+\frac{1}{2},j+\frac{1}{2}}^{n},$$

$$A_{j+\frac{1}{2}}^{y+} q y_{i+\frac{1}{2},j+\frac{1}{2}}^{n+\frac{1}{2}} = A_{j+\frac{1}{2}}^{y-} q y_{i+\frac{1}{2},j+\frac{1}{2}}^{n-\frac{1}{2}} - \triangle t (\sigma_{y_{j+\frac{1}{2}}} - \sigma_{x_{i+\frac{1}{2}}}) \tilde{\partial}_y u_{i+\frac{1}{2},j+\frac{1}{2}}^{n},$$

where the cell averages $\tilde{\partial}_x u_{i+\frac{1}{2},j+\frac{1}{2}}^n$ and $\tilde{\partial}_y u_{i+\frac{1}{2},j+\frac{1}{2}}^n$ are defined as

$$\tilde{\partial}_x u_{i+\frac{1}{2},j+\frac{1}{2}}^n = \frac{u_{i+1,j+1}^n - u_{i,j+1}^n + u_{i+1,j}^n - u_{i,j}^n}{2\triangle x}, \quad \tilde{\partial}_y u_{i+\frac{1}{2},j+\frac{1}{2}}^n = \frac{u_{i+1,j+1}^n - u_{i+1,j}^n + u_{i,j+1}^n - u_{i,j}^n}{2\triangle y}.$$

The definition of the cell averages allows us to compute the regularized term in (3)

$$(\delta_\varepsilon \partial_x q x)_{i,j}^n := (\rho_\varepsilon * \partial_x q x)_{i,j}^n, \quad (\delta_\varepsilon \partial_y q y)_{i,j}^n := (\rho_\varepsilon * \partial_x q y)_{i,j}^n$$

for $\partial_x q_{x_{i,j}}^n = \frac{1}{2} \left(\tilde{\partial}_x q_{x_{i,j}}^{n+\frac{1}{2}} + \tilde{\partial}_x q_{x_{i,j}}^{n-\frac{1}{2}} \right)$ and $\partial_y q_{y_{i,j}}^n = \frac{1}{2} \left(\tilde{\partial}_y q_{y_{i,j}}^{n+\frac{1}{2}} + \tilde{\partial}_y q_{y_{i,j}}^{n-\frac{1}{2}} \right)$, where the cell averages of the derivatives of the function $(q_{x_{i,j}}^{n\pm\frac{1}{2}}, q_{y_{i,j}}^{n\pm\frac{1}{2}})$ are defined as

$$\tilde{\partial}_x q_{x_{i,j}}^{n\pm\frac{1}{2}} = \frac{1}{2\triangle x} \left(q x_{i+\frac{1}{2},j+\frac{1}{2}}^{n\pm\frac{1}{2}} - q x_{i-\frac{1}{2},j+\frac{1}{2}}^{n\pm\frac{1}{2}} + q x_{i+\frac{1}{2},j-\frac{1}{2}}^{n\pm\frac{1}{2}} - q x_{i-\frac{1}{2},j-\frac{1}{2}}^{n\pm\frac{1}{2}} \right),$$

$$\tilde{\partial}_y q_{y_{i,j}}^{n\pm\frac{1}{2}} = \frac{1}{2\triangle y} \left(q y_{i+\frac{1}{2},j+\frac{1}{2}}^{n\pm\frac{1}{2}} - q y_{i+\frac{1}{2},j-\frac{1}{2}}^{n\pm\frac{1}{2}} + q y_{i-\frac{1}{2},j+\frac{1}{2}}^{n\pm\frac{1}{2}} - q y_{i-\frac{1}{2},j-\frac{1}{2}}^{n\pm\frac{1}{2}} \right).$$

Step 2. Compute $u_{i,j}^{n+1}$,

$$A_{i,j}^{xy+} u_{i,j}^{n+1} = 2u_{i,j}^n - A_{i,j}^{xy-} u_{i,j}^{n-1} + \triangle t^2 \left(-\sigma_{i,j}^{xy} u_{i,j}^n + c_{i,j}^2 \left((\delta_\varepsilon \partial_x q x)_{i,j}^n + (\delta_\varepsilon \partial_y q y)_{i,j}^n \right) + c_{i,j}^2 \triangle_n u_{i,j}^n \right), \quad (14)$$

where

$$\sigma_{i,j}^{xy} = \sigma_{x_i} \sigma_{y_j}, \quad c_{i,j} = c(x_i, y_j), \quad \triangle_n u_{i,j}^n = \frac{u_{i+1,j}^n - 2u_{i,j}^n + u_{i-1,j}^n}{\triangle x^2} + \frac{u_{i,j+1}^n - 2u_{i,j}^n + u_{i,j-1}^n}{\triangle y^2}.$$

3.2. Stability Analysis

To obtain the stability condition of the staggered finite difference scheme defined above, we restrict our concern to the constant damping case with $\sigma_x = \sigma_y = \sigma_0 \geq 0$ for simplicity in our analysis. The stability condition for the scheme in the computational domain is as follows.

Remark 4. *The CFL condition of scheme* (13)–(14) *in the computational area (i.e., $\sigma_x = \sigma_y = 0$) is*

$$c\frac{\triangle t}{h} \leq \frac{1}{\sqrt{2}}$$

for $\triangle x = \triangle y = h$ from the standard von Neumann *stability analysis technique.*

Generally the stability condition for the staggered finite difference scheme developed in Section 3.1 can be obtained as follows.

Theorem 3. *Assume that $\sigma_x = \sigma_y = \sigma_0 > 0$ and the sound speed c are constants. Then, the discrete scheme (13)–(14) is stable if the CFL condition*

$$c \triangle t \leq \frac{h}{\sqrt{2}} \frac{1}{(1 + \frac{\sigma_0^2 h^2}{8c^2})^{1/2}} \tag{15}$$

is satisfied for $\triangle x = \triangle y = h$.

To prove Theorem 3 and use the technique of the standard *von Neumann* stability analysis, we recall the definition of the simple *von Neumann* polynomial and some of its properties as follows.

Definition 1. *A polynomial is a simple* von Neumann *polynomial if all its roots, r, lie on the unit disk $(|B(0,r)| < 1)$ and its roots on the unit circle are simple roots.*

The following theorem demonstrates that a simple *von Neumann* polynomial can be a sufficient stability condition.

Theorem 4. *A sufficient stability condition is that ϕ be a simple* von Neumann *polynomial, where ϕ is the characteristic polynomial (see [35] for the proof).*

With Theorem 4, the stability condition for a polynomial is presented in the following.

Theorem 5. *Let ϕ be a polynomial of degree p written as*

$$\phi(z) = c_0 + c_1 z + \cdots + c_p z^p,$$

where $c_0, c_1, \cdots, c_p \in \mathbb{C}$ and $c_p \neq 0$. The polynomial ϕ is a simple von Neumann *polynomial if and only if ϕ^0 is a simple* von Neumann *polynomial and $|\phi(0)| \leq |\bar{\phi}(0)|$, where ϕ^0 is defined as*

$$\phi^0(z) = \frac{\bar{\phi}(0)\phi(z) - \phi(0)\bar{\phi}(z)}{z},$$

and the conjugate polynomial $\bar{\phi}$ is defined as

$$\bar{\phi}(z) = \bar{c}_p + \bar{c}_{p-1} z + \cdots + \bar{c}_0 z^p,$$

where $\bar{c}$ is the complex conjugate of c. The main ingredient in the proof of the theorem is Rouché's theorem; the proof is detailed in [36].

Now, we can computationally verify the stability condition (15) in Theorem 3 using Theorems 4 and 5.

Proof of Theorem 3 . Assume that $\sigma_x = \sigma_y = \sigma_0$ in scheme (13)–(14) and we rewrite the scheme as the second order central difference scheme of the variables u and $\vec{q}$.

$$\frac{1}{c_{i,j}^2} \left[\frac{u_{i,j}^{n+1} - 2u_{i,j}^n + u_{i,j}^{n-1}}{\triangle t^2} + 2\sigma_0 \frac{u_{i,j}^{n+1} - u_{i,j}^{n-1}}{2\triangle t} + \sigma_0^2 u_{i,j}^n \right] \tag{16}$$

$$= \frac{u_{i+1,j}^n - 2u_{i,j}^n + u_{i-1,j}^n}{\triangle x^2} + \frac{u_{i,j+1}^n - 2u_{i,j}^n + u_{i,j-1}^n}{\triangle y^2} + (\rho_\epsilon * \partial_x q_x)_{i,j}^n + (\rho_\epsilon * \partial_y q_y)_{i,j}^n,$$

$$\frac{\vec{q}_{i+\frac{1}{2},j+\frac{1}{2}}^{n+\frac{1}{2}} - \vec{q}_{i+\frac{1}{2},j+\frac{1}{2}}^{n-\frac{1}{2}}}{\triangle t} + \sigma_0 \frac{\vec{q}_{i+\frac{1}{2},j+\frac{1}{2}}^{n+\frac{1}{2}} + \vec{q}_{i+\frac{1}{2},j+\frac{1}{2}}^{n-\frac{1}{2}}}{2} = \vec{0}. \tag{17}$$

By *von Neumann* analysis, we can assume a spatial dependence of the following form in the field quantities:

$$u_{i,j}^{n+1} = \hat{u}^{n+1}(k_x, k_y)e^{ik_x x_i + ik_y y_j}, \quad u_{i,j}^n = \hat{u}^n(k_x, k_y)e^{ik_x x_i + ik_y y_j},$$

$$\vec{q}_{i+\frac{1}{2}, j+\frac{1}{2}}^{n+\frac{1}{2}} = \hat{\vec{q}}_{i+\frac{1}{2}, j+\frac{1}{2}}^{n+\frac{1}{2}}(k_x, k_y)e^{ik_x x_{i+\frac{1}{2}} + ik_y y_{j+\frac{1}{2}}},$$

where k_x, k_y, is the component of the wave vector $\vec{k}$, i.e., $\vec{k} = (k_x, k_y)^T$, and the wave number is $k = \sqrt{k_x^2 + k_y^2}$. Then, we have the system $\left[\hat{u}^{n+1}, \hat{u}^n, \hat{q}_x^{n+\frac{1}{2}}, \hat{q}_y^{n+\frac{1}{2}}\right]^T = G\left[\hat{u}^n, \hat{u}^{n-1}, \hat{q}_x^{n-\frac{1}{2}}, \hat{q}_y^{n-\frac{1}{2}}\right]^T$, where the amplification matrix G of scheme (16), (17) is given by

$$G = \begin{bmatrix} -\frac{c_1}{c_2} & -\frac{c_0}{c_2} & C_{\hat{q}_x} & C_{\hat{q}_y} \\ 1 & 0 & 0 & 0 \\ 0 & 0 & \eta & 0 \\ 0 & 0 & 0 & \eta \end{bmatrix},$$

where $C_{\hat{q}_x}$ and $C_{\hat{q}_y}$ satisfy $c_2\hat{u}^{n+1} + c_1\hat{u}^n + c_0\hat{u}^{n-1} = C_{\hat{q}_x}\hat{q}_x^{n-\frac{1}{2}} + C_{\hat{q}_y}\hat{q}_y^{n-\frac{1}{2}}$ with $c_0 = \frac{1}{\triangle t^2} - \frac{\sigma_0}{\triangle t}$, $c_1 = -\frac{2}{\triangle t^2} - 2c^2\frac{\cos(k_x \triangle x) - 1}{\triangle x^2} - 2c^2\frac{\cos(k_y \triangle y) - 1}{\triangle y^2} + \sigma_0^2$, $c_2 = \frac{1}{\triangle t^2} + \frac{\sigma_0}{\triangle t}$, and $\eta = \frac{1 - \frac{\triangle t}{2}\sigma_0}{1 + \frac{\triangle t}{2}\sigma_0}$. Then, it is noted that the characteristic function of G is given by

$$\phi(G) = \left(G^2 + \frac{c_1}{c_2}G + \frac{c_0}{c_2}\right)(G - \eta)^2.$$

Please note that $|\eta| < 1$ by the assumption. It can be observed from Theorem 5 that $\phi(G)$ is a simple *von Neumann* polynomial if and only if $|c_1| \leq |c_0 + c_2|$, i.e.,

$$\left|\frac{2}{\triangle t^2} + 2c^2\frac{\cos(k_x h) + \cos(k_y h) - 2}{h^2} - \sigma_0^2\right| \leq \frac{2}{\triangle t^2}, \quad \text{for } h = \triangle x = \triangle y.$$

This inequality gives the CFL condition (15), which completes the proof. $\square$

Remark 5. *From the proof of Theorem 3, we notice that the characteristic function ϕ of the amplification matrix G does not depend on any quantity related to the regularized term. That is, the staggered finite difference scheme corresponding to the classical PML model (2) with a constant damping in the layers is stable under the CFL condition (15).*

4. Numerical Result

The aim of this section is to provide numerical evidence of the well-posedness of the regularized system and the non-reflection properties of the acoustic wave in the layers of the classical PML model. For the discussion of the non-reflection properties, we demonstrate the behavior of the maximum error at t_n defined as the maximum of the differences between the numerical solution and a reference solution in the computational domain $\Omega_0 := [0,1] \times [0,1]$. Here, the reference solution is taken in the same computational domain instead of the layers with an additional large domain, for example, 15 times wider in the x and y directions in our experiment, causing the wave in the computational domain to be unaffected by the wave propagating from outside in the chosen long-time step. Furthermore, we use the energy method introduced in [37] and numerically examine the well-posedness or stability of the model (4) by observing the long-time behavior of the acoustic wave energy defined by

$$\mathcal{E}(t) = \frac{1}{2}\int_{\Omega_0}\left(\frac{1}{c^2}u_t(t)^2 + \nabla u(t) \cdot \nabla u(t)\right)dx. \tag{18}$$

For the numerical simulation, we use the same initial condition defined by (4) and, in the absorbing layer, the damping function of the form given by

$$
\sigma_{x_k}(x_k) = \begin{cases} 0, & |x_k| < a_k = 1, \\ \sigma_0 \left(\dfrac{|x_k - 1|}{L} \right)^\beta, & 1 \le |x_k| \le 1 + L, \end{cases} \tag{19}
$$

where $\beta = 0, 1, 2$, σ_0 is a given constant and L denotes the thickness of the layers.

For the comparisons of non-reflection property, we first demonstrate the maximum error for both Formulas (2) and (4) with two sets of thickness and damping as $(L, \sigma_0) = (0.25, 30)$ and $(L, \sigma_0) = (0.1875, 30)$. The numerical results are displayed in Figure 1. The classical PML has slightly smaller errors than the modified one in both cases, as shown in Figure 1, but it can be observed that these errors of the modified one can be reduced by simply increasing small amounts of thickness or damping such as $L = 0.27$ or $\sigma_0 = 35$.

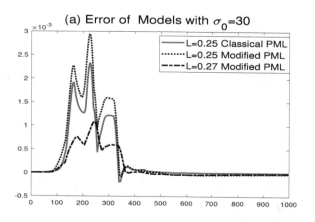 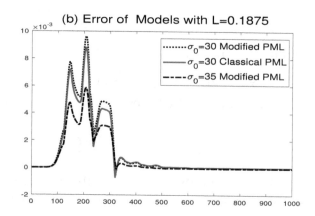

Figure 1. Comparison of errors: **(a)** a fixed damping $\sigma_0 = 30$, **(b)** a thickness $L = 0.1875$ ($\beta = 2$)

To see the influence of absorbing property by incidence angle, we demonstrate both formulas with different positions of source function. The resulted differences between reference and computed values of the solution during simulation at one point within the computational domain are plotted in Figure 2. The errors of the classical PML have relatively smaller than the modified one and both formulas have slightly better absorbing property when the angle of incidence to the interface between the computational domain and the layers is bigger.

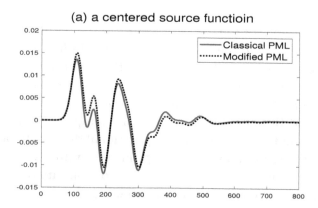 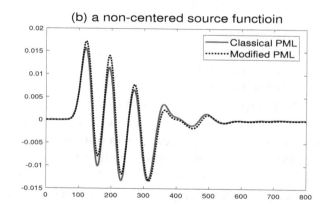

Figure 2. Comparison of the difference at a point from different positions of source function with $\sigma_0 = 35$ and $L = 0.1$

Next, to investigate the energy $\mathcal{E}(t)$ behavior, we choose a time step size $\triangle t$ of $\triangle x/3$, which satisfies the CFL condition (15) to guarantee the stability of the staggered finite difference scheme (see Remark 4). Here, the first order backward and second order central finite differences in time and space, respectively, are used to discretize the energy $\mathcal{E}(t_n)$ of (18) at each time step t_n. We investigate the behavior of the energy for a long-time simulation at time $t_n = 10{,}000$ according to the thickness of the layers and magnitude of the damping. The numerical results are displayed in Figure 3: (a) the energy with various dampings $\sigma_0 = 40, 50, 50, 60, 70$ for a fixed thickness $L = 0.0625$ and (b) the energy with various thicknesses $L = 0.0625, 0.1, 0.125, 0.15$ for a fixed damping $\sigma_0 = 50$. The results indicate that the numerical stability of the modified formula is consistently stable in the long-time simulation regardless of the magnitudes of damping and thickness of the layer. This provides proof of the well-posedness of the developed system and numerical stability for the finite difference method.

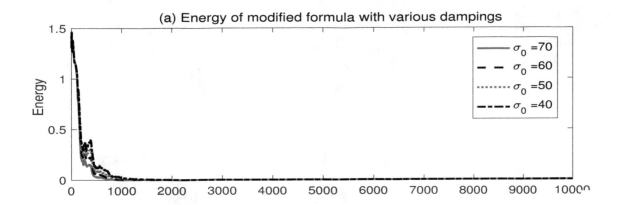

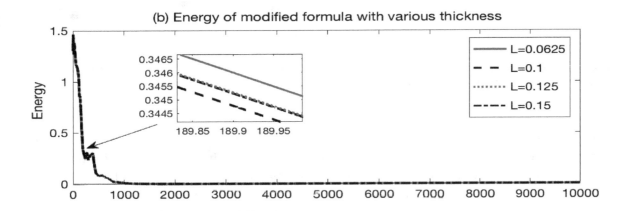

Figure 3. $\mathcal{E}(t)$ with (a) various damping values $\sigma_0 = 40, 50, 60, 70$ for a fixed thickness $L = 0.0625$ ($\beta = 0$), (b) various thickness $L = 0.0625, 0.1, 0.125, 0.15$ for a fixed damping $\sigma_0 = 50$ ($\beta = 0$).

Lastly, in order to illustrate this visual investigation, we consider the damping $\beta = 2$ and display the snap shots of the wave propagation at times $t_n = 1, 30, 60, 100, 130, 150, 200, 300, 500$ with $\sigma_0 = 35, L = 0.25$ in Figure 4. One can see that the regularized system displays a good property of non-reflection in the layers, which is the purpose of building the layers . It is remarkable that from a mathematical point of view, the analytical well-posedness without losing the non-reflection property in the layers of that the classical PML model.

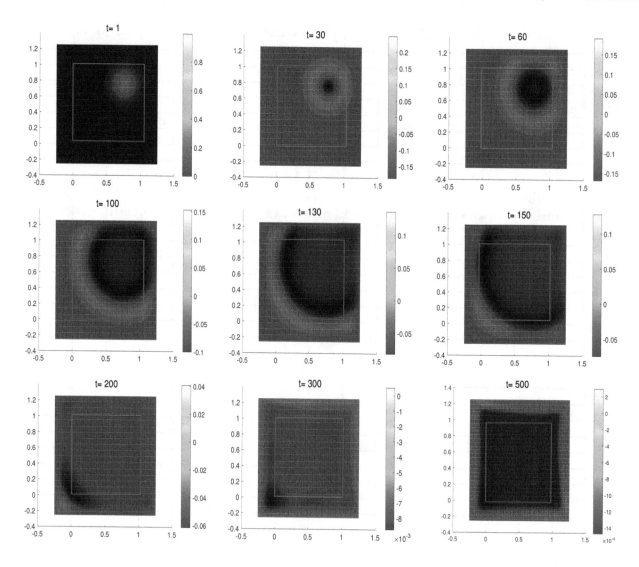

Figure 4. Snap shots of the regularized system at time $t_n = 1, 30, 60, 100, 130, 150, 200, 300, 500$ with $\sigma_0 = 35, \beta = 2, L = 0.25$ (Red rectangular box represents the computational domain.)

5. Discussion

We have introduced a new and efficient formulation related to the acoustic wave equation based on the regularization of the un-split PML wave equation. By regularizing the lower order regularity term in the original equation and the standard *von Neumann* stability analysis, we have achieved well-posedness as well as numerical stability of the solution in the new formulation. We summarize the main novelty and results of this study as follows: (1) We have proved the analytical well-posedness of our formulation without any restriction of damping terms; (2) a staggered finite difference scheme for the formulation is introduced and numerical stability is also analyzed; (3) several numerical tests are exhibited to show the numerical stability and a non-reflection property.

Abbreviations

The following abbreviations are used in this manuscript:

PML   Perfectly Matched Layers
CFL   Courant-Friedrichs-Lewy

References

1. Bérenger, J.P. A perfectly matched layer for the absorption of electromagnetic waves. *J. Comput. Phys.* **1994**, *114*, 185–200. [CrossRef]
2. Chew, W.C.; Weedon, W.H. A 3D Perfectly matched medium from modified Maxwell's equations with stretched coordinates. *Microw. Opt. Technol. Lett.* **1994**, *7*, 599–604. [CrossRef]
3. Sjögreen, B.; Petersson, N.A. Perfectly matched layers for Maxwell's equations in second order formulation. *J. Comput. Phys.* **2005**, *209*, 19–46. [CrossRef]
4. Collino, F.; Tsogka, C. Application of the PML absorbing layer model to the linear elastodynamic problem in anisotropic heterogeneous medias. *Geophysics* **2001**, *88*, 43–73.
5. Chew, W.C.; Liu, Q.H. Perfectly matched layers for elastodynamics: A new absorbing boundary condition. *J. Comput. Acoust.* **1996**, *4*, 341–359. [CrossRef]
6. Hu, F.Q. On absorbing boundary conditions for linearized Euler equations by a perfectly matched layer. *J. Comput. Phys.* **1996**, *129*, 201–219. [CrossRef]
7. Hesthaven, J.S. On the analysis and construction of perfectly matched layers for the linearized Euler equations. *J. Comput. Phys.* **1998**, *142*, 129–147. [CrossRef]
8. Lions, J.-L.; Métral, J.; Vacus, O. Well-posed absorbing layer for hyperbolic problems. *Numer. Math.* **2002**, *92*, 535–562. [CrossRef]
9. Nataf, F. A new approach to perfectly matched layers for the linearized Euler system. *J. Comput. Phys.* **2006**, *214*, 757–772. [CrossRef]
10. Turkei, E.; Yefet, A. Absorbing PML boundary layers for wave-like equations. *Appl. Num. Math.* **1998**, *27*, 533–557. [CrossRef]
11. Barucq, H.; Diaz, J.; Tlemcani, M. New absorbing layers conditions for short water waves. *J. Comput. Phys.* **2010**, *229*, 58–72. [CrossRef]
12. Appelö, D.; Hagstrom, T.; Kress, G. Perfectly matched layer for hyperbolic systems: General formulation, well-posedness, and stability. *J. Appl. Math.* **2006**, *67*, 1–23. [CrossRef]
13. Hu, F.Q. A stable perfectly matched layer for linearized Euler equations in unsplit physical variables. *J. Comput. Phys.* **2001**, *173*, 455–480. [CrossRef]
14. Cohen, G.C. *Higher-Order Numerical Methods for Transient Wave Equations*; Springer: Berlin, Germany, 2002.
15. Zhao, L.; Cangellaris, A.C. A general approach for the development of unsplit-field time-domain implementations of perfectly matched layers for FDTD grid truncation. *IEEE Microw. Guided Lett.* **1996**, *6*, 209–211. [CrossRef]
16. Bécache, E.; Joly, P. On the analysis of Bérenger's Perfectly Matched Layers for Maxwell's equations. *Math. Model. Numer. Anal.* **2002**, *36*, 87–120. [CrossRef]
17. Abarbanel, S.; Gottlieb, D. A mathematical analysis of the PML method. *J. Comput. Phys.* **1997**, *134*, 357–363. [CrossRef]
18. Halpern, L.; Petit-Bergez, S.; Rauch, J. The analysis of matched layers. *Conflu. Math.* **2011**, *3*, 159–236. [CrossRef]
19. Abarbanel, S.; Qasimov, H.; Tsynkov, S. Long-time performance of unsplit PMLs with explicit second order schemes. *J. Sci. Comput.* **2009**, *41*, 1–12. [CrossRef]
20. Roden, J.A.; Gedney, S.D. Convolution PML (CPML): An efficient FDTD implementation of the CFS–PML for arbitrary media. *Microw. Opt. Technol. Lett.* **2000**, *27*, 334–339. [CrossRef]
21. Rylander, T.; Jin, J. Perfectly matched layer for the time domain finite element method. *J. Comput. Phys.* **2004**, *200*, 238–250. [CrossRef]
22. Komatitsch, D.; Tromp, J. A perfectly matched layer absorbing boundary condition for the second-order seismic wave equation. *Geophys. J. Int.* **2003**, *154*, 146–153. [CrossRef]
23. Appelö, D.; Kress, G. Application of a perfectly matched layer to the nonlinear wave equation. *Wave Motion* **2007**, *44*, 531–548. [CrossRef]
24. Tam, C.K.W.; Auriault, L.; Cambuli, F. Perfectly matched layer as absorbing condition for the linearized Euler equations in open and ducted domains. *J. Comput. Phys.* **1998**, *114*, 213–234. [CrossRef]
25. Diaz, J.; Joly, P. A time domain analysis of PML models in acoustics. *Comput. Methods Appl. Mech. Eng.* **2006**, *195*, 3820–3853. [CrossRef]

26. Johnson, S. *Notes on Perfectly Matched Layers*; Technical Report; Massachusetts Institute of Technology: Cambridge, MA, USA, 2010.

27. Hagstrom, T.; Hariharan, S.I. A formulation of asymptotic and exact boundary conditions using local operators. *Appl. Numer. Math.* **1998**, *27*, 403–416. [CrossRef]

28. Grote, M.J.; Sim, I. Efficient PML for the wave equation. *arXiv* **2010**, arXiv:1001.0319v1.

29. Bécache, E.; Petropoulos, P.G.; Gedney, S.D. On the long-time behavior of unsplit perfectly matched layers. *IEEE Trans. Antennas Propag.* **2004**, *52*, 1335–1342. [CrossRef]

30. Petersen, B.E. *Introduction to the Fourier Transform and Pseudo-Differential Operators*; Series: Monographs and studies in mathematics; Pitman Advanced Pub. Program: Boston, MA, USA, 1983.

31. Kim, D. The Variable Speed Wave Equation and Perfectly Matched Layers. Ph.D. Thesis, Oregon State University, Corvallis, OR, USA, 2015.

32. Evans, L.C. *Partial Differential Equations*, 2nd ed.; Graduate Series in Mathematics; Springer: Berlin, Germany, 2010.

33. LeVeque, R.J. *Finite Difference Methods for Ordinary and Partial Differential Equations*; Society for Industrial and Applied Mathematics: Philadelphia, PA, USA, 2007.

34. Trucco, E.; Verri, A. *Introductory Techniques for 3-D Computer Vision*; Prentice Hall: Upper Saddle River, NJ, USA, 1998.

35. Bidégaray-Fesquet, B. Stability of FD-TD schemes for Maxwell-Debye and Maxwell-Lorentz equations. *SIAM J. Numer. Anal.* **2008**, *46*, 2551–2566. [CrossRef]

36. Miller, J.J.H. On the location of zeros of certain classes of polynomials with applications to numerical analysis. *J. Inst. Math. Appl.* **1971**, *8*, 397–406. [CrossRef]

37. Kaltenbacher, B.; Kaltenbacher, M.; Sim, I. A modified and stable version of a Perfectly Matched Layer technique for the 3-d second order wave equation in time domain with an application to aeroacoustics. *J. Comput. Phys.* **2013**, *35*, 407–422. [CrossRef]

Permissions

List of Contributors

Luis E. Garza
Facultad de Ciencias, Universidad de Colima, Colima 28045, Mexico

Noé Martínez and Gerardo Romero
Unidad Académica Multidisciplinaria Reynosa Rodhe, Universidad Autónoma de Tamaulipas, Reynosa 88779, Mexico

Jorge Arvesú and Andys M. Ramírez-Aberasturis
Department of Mathematics, Universidad Carlos III de Madrid, Avda. de la Universidad, 30, 28911 Leganés, Madrid, Spain

Jinjiang Li, Chao Liu and Zhuo Zhang
Department of Mathematics, China University of Mining and Technology, Beijing 100083, China

Min Zhang
School of Applied Science, Beijing Information Science and Technology University, Beijing 100192, China

Joohee Jeong and Seog-Hoon Rim
Department of Mathematics Education, Kyungpook National University, Daegu 41566, Korea

Dong-Jin Kang
Department of Computer Engineering, Information Technology Services, Kyungpook National University, Daegu 41566, Korea

Li Chen
School of Mathematics, Northwest University, Xi'an 710127, China

Jeong Gon Lee
Division of Applied Mathematics, Nanoscale Science and Technology Institute, Wonkwang University, Iksan 54538, Korea

Wonjoo Kim
Department of Applied Mathematics, Kyunghee University, Yongin 17104, Korea

Lee-Chae Jang
Graduate School of Education, Konkuk University, Seoul 05029, Korea

Yunyun Qu
School of Mathematical Sciences, Xiamen University, Xiamen 361005, China
School of Mathematical Sciences, Guizhou Normal University, Guiyang 550001, China

Jiwen Zeng
School of Mathematical Sciences, Xiamen University, Xiamen 361005, China

Yongfeng Cao
School of Big Data and Computer Science, Guizhou Normal University, Guiyang 550001, China

Dmitry V. Dolgy
Kwangwoon Glocal Education Center, Kwangwoon University, Seoul 139-701, Korea

Serkan Araci
Department of Economics, Faculty of Economics, Administrative and Social Sciences, Hasan Kalyoncu University, Gaziantep TR-27410, Turkey

Mumtaz Riyasat, Shahid Ahmad Wani and Subuhi Khan
Department of Mathematics, Faculty of Science, Aligarh Muslim University, Aligarh 202 002, India

Dug Hun Hong
Department of Mathematics, Myongji University, Yongin Kyunggido 449-728, Korea

Zhao Jianhong
Department of Teachers Education, Lijiang Teachers College, Lijiang 674199, China

Chen Zhuoyu
School of Mathematics, Northwest University, Xi'an 710127, China

Waseem A. Khan
Department of Mathematics and Natural Sciences, Prince Mohammad Bin Fahd University, Al Khobar 31952, Kingdom of Saudi Arabia

Sunil Kumar Sharma
College of Computer and Information Sciences, Majmaah University, Majmaah 11952, Saudi Arabia

Mohd Ghayasuddin
Department of Mathematics, Integral University Campus, Shahjahanpur 242001, India

Can Kızılateş
Faculty of Art and Science, Department of Mathematics, Zonguldak Bülent Ecevit University, Zonguldak 67100, Turkey

Bayram Çekim and Naim Tuğlu
Faculty of Science, Department of Mathematics, Gazi University, Teknikokullar, Ankara 06500, Turkey

Wenpeng Zhang and Xin Lin
School of Mathematics, Northwest University, Xi'an 710127, China

Kyung-Won Hwang
Department of Mathematics, Dong-A University, Busan 49315, Korea

Dae San Kim
Department of Mathematics, Sogang University, Seoul 04107, Korea

Yuanyuan Meng
School of Mathematics, Northwest University, Xi'an 710127, Shaanxi, China

Jin Zhang
School of Information Engineering, Xi'an University, Xi'an 710127, China

Zhuoyu Chen
School of Mathematics, Northwest University, Xi'an 710127, China

Taekyun Kim
Department of Mathematics, College of Science, Tianjin Polytechnic University, Tianjin 300160, China
Department of Mathematics, Kwangwoon University, Seoul 139-701, Korea

Gwan-Woo Jang
Department of Mathematics, Kwangwoon University, Seoul 139-701, Korea

Jongkyum Kwon
Department of Mathematics Education and ERI, Gyeongsang National University, Jinju, Gyeongsangnamdo 52828, Korea

Dojin Kim
Department of Mathematics, Pusan National University, Busan 46241, Korea

Index

Printed in the USA
CPSIA information can be obtained
at www.ICGtesting.com
JSHW051404091023
49903JS00006B/272

9 781647 285296